Understanding
Solid State
Physics

Understanding
Solid State
Physics

Sharon Ann Holgate

CRC Press
Taylor & Francis Group
Boca Raton   London   New York

CRC Press is an imprint of the
Taylor & Francis Group, an **informa** business

A TAYLOR & FRANCIS BOOK

This book contains activities that may be dangerous if you do not follow suitable safety procedures. These activities should be carried out under the supervision of a qualified instructor and using suitable safety precautions and equipment. The authors and publishers exclude all liability associated with this book to the extent permitted by law.

Taylor & Francis
6000 Broken Sound Parkway NW, Suite 300
Boca Raton, FL 33487-2742

© 2010 by Sharon Ann Holgate
Taylor & Francis is an Informa business

No claim to original U.S. Government works

Printed in the United States of America on acid-free paper
10 9 8 7 6 5 4 3 2 1

International Standard Book Number: 978-0-7503-0972-1 (Hardback)

Library of Congress Cataloging-in-Publication Data

Holgate, Sharon Ann, Dr.
    Understanding solid state physics / Sharon Ann Holgate.
        p. cm.
    Includes bibliographical references and index.
    ISBN 978-0-7503-0972-1 (hardcover : alk. paper)
    1. Solid state physics. I. Title.

QC176.H626 2010
530.4'1--dc22   2009025307

Visit the Taylor & Francis Web site at
http://www.taylorandfrancis.com

and the CRC Press Web site at
http://www.crcpress.com

Dedication

*In memory of my grandmother Louisa Edmondson,
and my friend John Wilson.*

Contents

Preface

I first came up with the idea for this book when I was a postgraduate student, and was working part-time as a teaching assistant for my physics department. I could see a lot of undergraduates struggling to understand their courses simply because they were unable to grasp the basic principles behind the various topics. It seemed to me that this was partly because many textbooks assumed a prior knowledge and mathematical aptitude that not all of their readers had, and which was obscuring the main points very early on in the explanation process. From that moment of realisation onwards, I wanted to write a truly accessible solid state physics textbook for introductory courses that concentrated on explaining the basics, and gave students a firm grounding in the subject. It was also very important to me to relate the theories and concepts to the real world, so that anyone reading it could see the point of learning the physics and how it was likely to be used once they had left university.

To help achieve these objectives, I have endeavoured to highlight the technological applications of the physics being discussed and to point out the multidisciplinary nature of much scientific research. I have also tried to keep the number of equations to a minimum wherever possible, in the hope that this will allow the book to provide a useful introduction to solid state physics for any physics and materials science undergraduates who feel daunted by a highly mathematical approach. The mathematics that does appear is presented in small logical steps, and problems are tackled as worked examples, in the hope that if an ex-maths struggler like myself can understand it like that, everyone else who is baffled will understand it too! The small number of more complicated derivations that were impossible to avoid are also presented in a step-by-step fashion in an appendix.

I hope this approach will also aid students or researchers in other scientific disciplines who may cross over into the field at some stage. With readers like the geologists and geographers I used to work alongside in a thermoluminescence laboratory in mind, in addition to the maths appendix, I have included some short appendices on related physics topics.

After spending the last decade working as a freelance journalist and broadcaster, I felt it was important to include some magazine-style boxes on interesting research. This allowed me to cover topics that might otherwise not feature in a textbook of this size, and also to give some idea of what it is like researching either in industry or academia. Cowriting my first book (with the late Robin Kerrod), a picture-based popular science book for children, has also influenced this project in the respect that I was keen to include images that I hope will inspire, and in some cases amuse.

The questions with answers overleaf are intended to help readers test their knowledge as they make their way through the book, and should prove equally useful for revision. Further questions can be found on the accompanying website, and there is a solutions manual available for qualifying instructors. The website also houses a light-hearted video quiz for readers, and downloadable supplementary material including further references for reading and web links. Finally the book includes a glossary of widely used terms, and I have used underlining to highlight the first instance of each use throughout the main text.

Writing this book has reminded me of a story about a magic pudding that I enjoyed reading as a child. This was a pudding with attitude. He walked about on skinny legs with his bowl on his head, and waved his fists at his enemies when they tried to eat him. But he had no need to be so aggressive because his magical powers meant that when anyone ate a slice of him, he re-formed into a complete round pudding again. In a similar way, just when I thought I had finished writing a section of this book, the remainder seemed to re-form into a whole book waiting to be written, as there were so many more things that had come into my mind that I wanted to include. However, despite the frustrations that the sheer size of the project produced, I have now, I hope, achieved my original aims.

I am indebted in no small way to Alan Piercy, who having taught solid state physics for over 30 years probably thought he would escape the subject in his retirement. Instead, he has spent the last few years as my academic advisor, steering me in the right direction, answering my myriad of questions, and putting up with my occasional rants when things went awry. I would also like to thank the friends and colleagues—including Jim Al-Khalili, John Barrow, Bill Buckley, Sue Bullock, Sue Crossfield, David Culpeck, Nicki Dennis, Colin Humphries, Steve Keevil, Peter Main, David Mowbray, Derek Palmer, Manoj Patairiya, John and Alan Robbins, Tom Spicer, Dianne Stilwell, and Tracey and Alice de Whalley—who have helped in various ways to make this book possible.

Thanks are also due to the many press officers at institutions and companies, including Amanda Bowie at the U.S. Naval Research Laboratory, Kathryn Klein from Lakeland Limited, Jane Koropsak at Brookhaven National Laboratory, Keith Lumley at Network Rail, and Leigh Rees at Oxford Diffraction, who have aided my quest for interesting photographs and have kindly provided additional information. I extend similar thanks to all the researchers around the world who have kindly given permission for me to write about their work, and feature their results.

My third editor, John Navas, has provided an immeasurable amount of support and advice during the last two years of this project, and my heartfelt thanks go to him. In addition, my mother Joan and friends—including Dawson Chance, Larry Crockett, Andrew Fisher, Amanda Kernot, David King, Julian Mayers, Darren Naylor, Ian Rennison, and Emma Winder—have provided a welcome distraction at evenings and weekends, and prevented me from becoming even more obsessed with solid state physics than I was before I began writing this.

Sharon Ann Holgate, Sussex, U.K., 2009

Corrections:
Whilst great care and much time has been taken in the creation of this book, mistakes may have slipped through the net, and any corrections or suggestions for improvement can be sent to: John.Navas@informa.com

Author

Sharon Ann Holgate has a DPhil in physics from the University of Sussex, where she is a Visiting Fellow in physics and astronomy. She has worked for over a decade as a science writer and broadcaster, with 50 broadcast appearances including presenting on the BBC World Service and BBC Radio 4, and competing in a 'Boffins Special' of The Weakest Link. Her numerous articles have appeared in *New Scientist*, *The Times Higher Education Supplement*, *E&T*, *Flipside*, *Focus*, *Physics World*, *Interactions*, *Modern Astronomer*, and *Astronomy Now*, while her first book, *The Way Science Works* (coauthored with Robin Kerrod) was shortlisted for the Royal Society Junior Books Prize. She has also written and developed brochures, national careers material, and press releases for various scientific institutions, and given talks at venues including the Science Museum in London. Dr. Holgate was the Institute of Physics Young Professional Physicist of the Year for 2006, awarded for her "passionate and talented promotion of physics and the public perception of physics through her books, articles, talks and broadcast work".

Further Acknowledgements

With special thanks to Alan Piercy and David Culpeck for their assistance with the line diagrams.

Hand and foot modelling (images 1.3, 4.7, and part (d) of Example question 4.4): the author.

1 Introduction

It is impossible to escape solid state physics. Solids are all around us, and their properties affect our everyday lives in many ways from the mundane to the sophisticated. For example, we can only pick up a hot metal saucepan if it has a handle made of an insulator such as a plastic, or if we wrap our hands in a thick cloth, because metals conduct heat so well. By contrast, it is the piezoelectric properties of certain crystals that allow them to be used as sensors on bridges to warn engineers of impending structural failure.

Solid state physics tells us why some solids can conduct heat and electricity well and why other solids cannot. It explains magnetism, the ways in which light and other types of radiation interact with solids, and reveals the processes that enable electronic components and devices to work. It also tells us how the atoms are arranged within different types of solids, and how the tiny forces holding the atoms in these arrangements affect much larger scale properties of solids such as melting point and hardness.

In some ways research in the field of solid state physics can move forward relatively slowly, and it certainly lacks the glamour that, say, the discovery of a new subatomic particle or a new M-class planet brings. But when solid state physics does produce an important result, the seemingly small step can have a major influence on all our lives. It is unlikely, for instance, that the group of people witnessing a demonstration of the transistor on a December day in 1947 could have predicted the size and influence of the modern electronics industry that would result from this invention.

Of course in electronics as in many other fields, there has been considerable progress in the last 50 years. The continuing decrease in the sizes of transistors and other circuit components—brought about not only by experimental work, but also by theoretical studies revealing information such as the influence of impurities on the semiconductor materials circuit components are made from—has allowed computer chips to become faster and faster. Smaller components also mean more information can be stored in a given space. These improvements in processing speed and data storage have allowed new products including mobile phones and digital cameras to be developed, as well as helping enable computers to shrink from the size of a room to something we can balance on our laps (see Figure 1.1). And if nanotechnology—the building of materials, structures, and devices on the nanometre scale by manipulating individual atoms and molecules—lives up to its promises, laptop computers will soon seem as large and cumbersome to us as those early mainframe computers.

Progress in solid state physics can actually be charted quite well by improvements in computer technology. An understanding of the optical properties of liquid crystals, and an ability to manufacture them on an industrial scale, made flat-screen displays for laptops, electronic organisers, and calculators possible. In addition, the discovery of powerfully magnetic rare earth materials has created a decrease in the

(a)

(b)

FIGURE 1.1 One of the world's first electronic computers, the UNIAC (a) (U.S. Army Photo). By the mid 1990s, 5 million transistors could be fitted onto a silicon chip that you could balance on your fingertip, and there were electronic organisers on the market no larger than early calculators. Chips are now many times smaller than an ant (b) (© Philips.)

Volume of Magnet for a Given Application

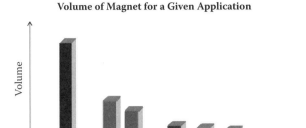

1930   1940   1950   1960   1970   1980   1990   2000

FIGURE 1.2 A decrease in the size of permanent magnets, made possible by the discovery of more and more powerful magnetic materials over the last 50 years, has enabled small electric motors to be developed. These have applications ranging from car windscreen wipers and electric toothbrushes to magnetic storage devices such as computer hard disks and video recorders.

size of permanent magnets (see Figure 1.2) and so allowed much smaller electric motors to be designed. This enabled devices including hard disks, floppy disks, CDs, and DVDs to be developed—which use tiny motors to spin the disk when reading and writing—and replace cumbersome magnetic tape for computer data storage. However, CDs and DVDs would not have been possible without another breakthrough in solid state physics: the invention of the diode laser.

Nowadays we think nothing of using a laser pointer when giving a presentation (see Figure 1.3), and while magnetic tape is still used for both audio and video recording, compact discs have become standard for audio recording, while DVDs provide high-quality video recording. These devices use tiny solid-state diode lasers made from semiconductor materials to write and read data, but just a few decades ago—before enough was understood about semiconductors and the way they interact with light—the only lasers that existed were huge gas or crystal lasers confined to

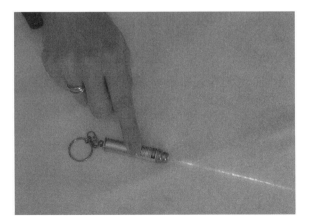

FIGURE 1.3 A laser pointer is now an everyday article, but years ago the only lasers were laboratory-based devices.

laboratories. The next generation of so-called "quantum dot" lasers should be even smaller than solid-state lasers, which could lead to a whole new range of optoelectronic devices and applications. It is also likely that we will see a range of new applications in the future for an existing optoelectronic device—the light-emitting diode (LED). Improvements in the brightness and cost of LEDs are leading us to the point where they are becoming viable as a replacement for conventional tungsten light bulbs for domestic lighting, and can be used for traffic lights.

In order to make new types of devices, new manufacturing methods have to be developed. Epitaxial growth techniques, which allow electronic components to be built up layer by layer, have made a huge impact by helping make the continued development of smaller electronic components possible. Meanwhile, new ways of growing crystals, and of making other materials such as composites and polymers, have enabled a range of modern materials to replace more traditional choices in many applications; for example, plastic bottles are now more widely used than glass for soft drinks. There have also been improvements in manufacturing more traditional materials. In fact it was the ability to produce high-quality glass that made optical fibres a practical proposition for the telecommunications industry.

Important as all these developments have been, there are many other areas in which solid state physics has already made a significant impact on our lives, and in which huge improvements may only be a short time away. The ongoing quest for higher temperature superconductors is a good example, as it could eventually lead to domestic power cables with almost no resistance, while further development of solar cells may also help reduce the amount of nonrenewable energy we use.

It is hard to predict what breakthroughs solid state physics will produce in the next few decades, but one thing seems certain. As we all become increasingly dependent on technology, it is likely that this fascinating area of physics will play an important part in providing the sort of future our societies will demand.

2 Crystal Clear
Bonding and Crystal Structures

CONTENTS

2.1   BONDING IN SOLIDS

If we were faced with an unknown substance, but had no experimental apparatus to investigate it with, we should at least be able to figure out which of the three most common states of matter—solid, liquid, or gas—that substance was in. We know that we can pick up solids with no fear of them pouring through our fingers like a liquid, and if we take a solid out of a container it will not expand to fill the room like a gas. Having said that, we may get caught out if we used these criteria for our investigation, as some materials which appear solid at room temperature, like pitch (a sticky black substance obtained from tar), can in fact flow like liquids over extremely long time periods (see Figure 2.1).

Under normal conditions and shorter timescales, however, materials with a definite shape that are firm enough for us to hold on to can be regarded as solids. Unlike gases (in which the molecules are essentially independent of one another) and liquids (whose atoms or molecules are only very loosely bound to each other), solids are solid because their component atoms are held closely together by atomic bonds.

There are several different ways in which atoms can bond together, but it is worth bearing in mind that descriptions of each of these types of bonding are only models of the true behaviour. Whilst the bonding for some solids does closely resemble one or another of the models, many solids have bonding that is partway between the somewhat extreme cases described by the models. There are also solids that have different bonding mechanisms in different parts of their structures.

2.1.1   ELECTRONS IN ATOMS

Inside every atom there is a positively charged nucleus (containing protons and neutrons) and negatively charged electrons that can be considered to be orbiting around the nucleus in atomic shells. Electrons roughly the same distance from the nucleus and with similar energies will occupy the same atomic shell, but each of these shells can only hold a certain number of electrons. For example the K shell—which is the shell closest to the nucleus—has one s subshell that can hold a maximum of two electrons. Meanwhile the L shell—which is the next shell out—consists of an s subshell and a p subshell that can contain two and six electrons, respectively. (See Appendix D for a reminder of the Pauli exclusion principle, and Appendix C for a revision of electron shell notation.)

If an atomic shell contains the maximum possible quota of electrons, it is said to be "full" or "closed", and when the outermost shell of an atom is full, that atom is chemically stable and so will not react easily with other atoms. In fact the last statement is slightly misleading, as atoms in which both the s subshell and the p subshell of the outermost shell are full are also unreactive and described as having full outer

(a)

(b)

FIGURE 2.1 The apparatus shown in (a) was set up in 1927 by Prof. Thomas Parnell of The University of Queensland in Australia to demonstrate the fluidity of pitch. Pitch is made from tar, and was used in the past for sealing up and waterproofing joints on boats. It appears solid at room temperature, and is actually so brittle that it can be smashed into pieces with a hammer as (b) shows. However, during the first 70 years of the experiment, which continues today, eight drops of pitch gradually dripped out of the funnel, demonstrating that pitch does in fact have some of the properties of a liquid. (Courtesy of Professor John S. Mainstone, The University of Queensland, Australia.)

shells. (In discussions on bonding, whenever the outermost "shell" is referred to, it is a general term that, depending on the particular case, can mean either the entire outermost shell or the *s* and *p* subshells within the outermost shell.) If you look at the periodic table (see inside front cover) you will see that the noble gases—helium, neon, argon, krypton, xenon, and radon—which are all inert have full outer shells. Helium and neon have genuinely full outer shells, as He only has one *s* subshell and this contains two electrons, while Ne has just an *s* and a *p* subshell in its outer shell, and both of these are full. Each of the other noble gases has full *s* and *p* subshells in its outermost shell.

Another glance at the periodic table will reveal that for all of the other elements the outer shell is only partly filled. Electrons in the outermost occupied shell of an atom are known as *valence electrons*, and it is these electrons that take part in bonding processes between atoms. Atoms will always try to become more stable and so will attempt to acquire a full outermost shell either by losing electrons or gaining them in some way. If an atom is able to lose valence electrons, becoming not only a positive ion but also more stable in the process, it is said to be *electropositive*. By contrast, atoms that can accept more electrons into their outermost shell in order to fill it (or alternatively can share valence electrons) are described as *electronegative*. The *electronegativity* of an element is a measure of its ability to accept an additional electron, and in general the further to the right in the periodic table that an element is, the greater the value of its electronegativity. Electropositive elements such as the Group IA metals, for example, have much lower values of electronegativity than the halogens in Group VIIB.

2.1.2   IONIC BONDING

Ionic bonding involves the transfer of an electron from an electropositive element to an electronegative element, so is the bonding mechanism found in solids such as sodium chloride (NaCl) that are composed of metallic and nonmetallic elements. In NaCl it is the electropositive sodium atom, with an electronic configuration $1s^2 2s^2 2p^6 3s^1$, that donates the single electron in its outermost shell to chlorine. Since it has given up one of its electrons, the sodium atom now has a net positive charge and has become a positive ion. Chlorine, which has an electronic configuration of $1s^2 2s^2 2p^6 3s^2 3p^5$, is electronegative and uses the electron from the sodium to completely fill the $3p$ subshell, giving itself a closed outermost shell. As the chlorine has gained an electron, it now has a net negative charge and so becomes a negative ion. The resulting ions, Na^+ and Cl^-, now attract one another, as they have opposite electrostatic charges, and an ionic bond is formed as shown in Figure 2.2. Since each ion has a closed outer shell, the electrons are distributed evenly around the ion, giving it a spherically symmetric electronic charge distribution. Bonding in ionic solids is therefore nondirectional, as any given ion can attract an oppositely charged neighbouring ion no matter what the relative positions of the two ions are.

In an ionic solid, as Figure 2.3 illustrates, every positively charged ion is surrounded by negatively charged ions and vice versa, so that each ion in the solid ends up with ions that it is electrostatically attracted to as its nearest neighbours. Ionic solids are brittle, which means they fracture relatively easily and do not deform much

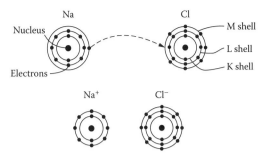

FIGURE 2.2 Ionic bonding in sodium chloride (NaCl). Sodium has only one electron in its outermost electron shell. This electron is transferred to the chlorine atom, which has space for just one more electron in its outer shell. The ions that this transfer creates are oppositely charged, so they attract one another.

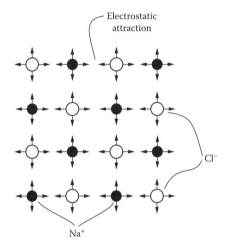

FIGURE 2.3 Ionic bonds between the ions in NaCl.

EXAMPLE QUESTION 2.1   IONIC BONDING

What is the electronic configuration of (a) the potassium ions and (b) the iodine ions in a crystal of potassium iodide? For each of your answers state which inert element has an identical electronic configuration. (The electronic configurations of the elements are listed in Appendix C.)

Answer on page 10.

before they break. This is because deformation involves rows of atoms slipping past other rows—a feat which is impossible in ionic solids, since it would bring atoms with the same charge near enough to each other to cause repulsion.

As well as making ionic solids brittle, ionic bonds make solids hard because they are very strong, and bond strength affects the hardness of materials. The

ANSWER TO QUESTION 2.1 ON IONIC BONDING

(a) The K^+ ions in a crystal of KI have an electronic configuration of $1s^2 2s^2 2p^6 3s^2 3p^6$, which is the same configuration as argon.

(b) The I^- ions in a crystal of KI have an electronic configuration of $1s^2 2s^2 2p^6 3s^2 3p^6 3d^{10} 4s^2 4p^6 4d^{10} 5s^2 5p^6$, which is the same configuration as xenon.

stronger the bonding, the stronger and harder the material (see Chapter 4). So ionic solids like sodium chloride and the other alkali halides (solids composed of a Group IA metal and a Group VIIB halogen), calcium carbonate, copper sulphate, metal oxides, and metal sulphides tend to be hard. In general they also have high melting temperatures because the stronger a bond is, the larger the amount of energy needed—and consequently the higher the temperature required—to break it so that a liquid can form. Ionic solids made up from two different types of ion where one ion is much larger than the other are, however, more strongly bonded and so have higher melting temperatures than ionic solids with similarly sized ions and hence slightly weaker ionic bonds. (The greater the distance between the ions in an ionic bond the weaker the bond is, so ionic bonds between different sized ions are strong because the smaller ions can get much nearer the larger ions than similarly large ions can.) By contrast, all types of ionic solid are good electrical and thermal insulators because almost all of the electrons are bound closely to particular ions, leaving very few "free" electrons available for conduction. However, if an ionic solid is melted, the atoms remain ions, and these ions can then conduct electricity.

Attraction and Repulsion

We have just seen that it is electrostatic attraction that holds individual ionic bonds together, and holds the various ions in ionic solids together too. But is an attractive force the only force an ion in an ionic solid feels?

The answer is no. There is also a repulsive force that occurs when two ions become so close that their electron shells start to overlap. In very simple terms, this can be thought of as a combination of an electrostatic repulsion of like charges and the effects of the *Pauli exclusion principle* (see Appendix D). This states that no two identical particles can occupy the same quantum state, and so does not allow identical electrons from each of the two ions to occupy the same overlapped electron shell. The only way for the ions to be so close together is for one of the identical electrons to be promoted to a higher electron shell. This increases the total energy of the system—as energy must be provided to promote the electron to the higher energy electron shell—and so creates a repulsive force. The repulsive force increases the nearer the ions get to one another, and stops them from coming so close together that their nuclei would fuse.

The upper and lower dashed curves in Figure 2.4 represent, respectively, the repulsive and attractive components of the total potential energy (solid curve) that an

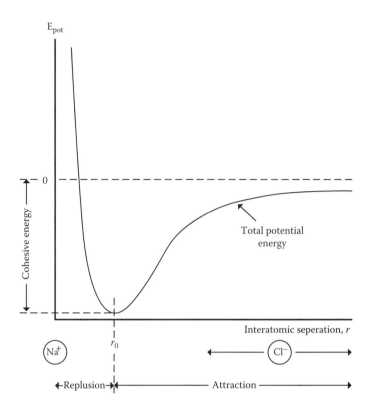

FIGURE 2.4   Potential energy as a function of interatomic separation for an Na^+ and Cl^- ion.

Na^+ ion and Cl^- ion feel as a function of their separation. Where the total potential energy has a minimum, the attractive and repulsive forces acting on the ions cancel each other out, so it is therefore at the separation distance that corresponds to this energy minimum—marked r_0 in Figure 2.4—that the two ions form a stable bond.

Although the attractive force is caused by a different mechanism in each different type of bonding, its potential energy curve has the same form as that shown in Figure 2.4 no matter what type of bonding is being considered. In the case of the repulsive force, not only does its potential energy curve look the same, but the same repulsion mechanism—which is equally applicable to atoms and ions—is also present in every type of bonding.

Cohesive Energy

If a solid were to be separated out into individual atoms (or ions), the energy that would be needed to do this is called the *cohesive energy*. The cohesive energy, which can range from around 0.1 eV/atom to approximately 10 eV/atom, is therefore a useful indicator of the strength of the bonding in different solids.

When any two atoms or ions are an infinite distance apart, the potential energy between them is taken to be zero—as the potential energy curve in Figure 2.4 suggests. Since the minimum in the potential energy curve occurs at the point where the atoms or ions are in a stable bond, the cohesive energy is equal to the depth of this energy minimum (as indicated in the diagram). Ionic crystals, which are held together by strong ionic bonds, have cohesive energies in the range 5–10 eV/atom.

2.1.3 COVALENT BONDING

In the last section we saw that in ionic bonding an electron is transferred from one atom to another. By contrast, *covalent bonding* occurs when valence electrons are shared between atoms. Covalent bonding is found in elements such as silicon, carbon (diamond) and germanium from Group IVB of the periodic table. As we will see in Section 2.1.7, it is also partially responsible for the bonding in III-V compounds (compounds comprising atoms from Group IIIB elements and from Group VB elements) like gallium arsenide (GaAs) and indium antimonide (InSb).

To see how covalent bonding works we will look at the relatively simple case of diamond. Any carbon atom can form four covalent bonds with other carbon atoms because, with its electronic configuration of $1s^2 2s^2 2p^2$, it has four valence electrons that can each be shared with another atom. Because a carbon atom can accommodate four more electrons—to give a total of eight electrons—in its outer shell, it can also share one of the valence electrons of each of its four neighbours. Figure 2.5(a) shows a simplified two-dimensional (2-D) diagram of the bonding between carbon atoms.

In a real diamond, each of the four outer atoms shown in the diagram also bonds with a total of four atoms (the central atom shown plus three further atoms not illustrated) to give a continuous three-dimensional (3-D) structure. This means that every carbon atom in the solid will have a closed outer shell, as it will not only be sharing its four valence electrons with its neighbours, but will also be sharing one of each of its four neighbours' electrons. (See Appendix C for a revision of electron shell notation.)

Since the outermost shells of atoms that take part in covalent bonding are not full, the electronic charge distribution is not spherically symmetric but is instead stronger in the areas in which the electrons tend to reside. This means that covalent bonds, which involve the sharing of these valence electrons, can only form along directions that encompass the electrons. The result is that every atom in a covalent crystal is bonded to four other atoms in such a way that the five atoms form a tetrahedral arrangement (see Figure 2.5[b]). For this arrangement the angle between the bonds, known as the *tetrahedral angle*, is 109°28′.

When the atoms in a covalent solid get close enough together for covalent bonds to form between them, their outer shells start to overlap. The details of the interaction between the atoms are very complicated, but in simple terms the overlap of the unfilled shells reduces the total energy of the system to a minimum value (like that shown in the potential energy curve of Figure 2.4), so the crystal holds together.

The cohesive energies of covalent crystals are between 1–5 eV/atom. Covalent solids like diamond, which have cohesive energies at the higher end of this range, have high melting temperatures (in this case above 3550°C) and are hard, like ionic solids. However, covalent crystals with lower values of cohesive energy, like grey tin, are softer and have much lower melting temperatures (grey tin melts at 232°C). Although diamond with its particularly strong covalent bonding is a very good insulator, many covalent solids are not such good insulators. In fact semiconductors (which will be

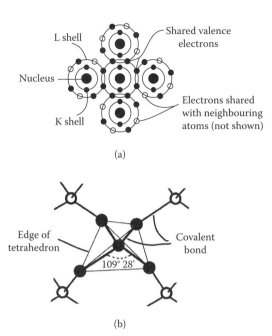

FIGURE 2.5 Two representations of covalent bonding between carbon atoms. In (a), a simplified 2-D schematic shows the details of the bonding. The central atom is sharing a single electron from each of its neighbours, giving it a full outermost shell of eight electrons. It is also sharing its own four valence electrons with the surrounding atoms. In a real solid, each of the four outer atoms shown here would be covalently bonded to a further three atoms, so each atom would be bonded to a total of four atoms. In (b), a 3-D view of the situation in a diamond is illustrated. Each atom and its four neighbours form a tetrahedral arrangement because covalent bonds can only occur along directions in which the valence electrons tend to be situated.

discussed in detail in Chapters 6 and 7) are all covalently bonded. This difference in behaviour from that of insulating ionic solids, in which the valence electrons are tightly bound to the ions, is because the valence electrons in covalent solids move to a certain extent as they are shared between atoms. As we saw in Section 2.1.2, ionic solids can, however, conduct electricity when the solid is melted. By contrast, covalent solids do not conduct electricity when molten because their constituent atoms are not ionised.

The atoms of a covalent solid are much less densely packed together than those of an ionic solid because covalent bonds are directional, while ionic bonds are not. This results in covalent solids having a more "open" structure than ionic solids, which will be discussed further in Section 2.2.

2.1.4 METALLIC BONDING

Looking at the periodic table (see inside front cover) will reveal that the majority of the known elements are metals. Bonding in metals is not described by either a covalent or by an ionic model. Instead, the atoms of metals are held together by a type of bonding known as *metallic bonding*.

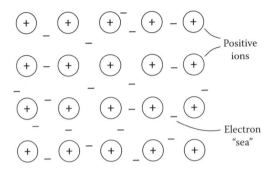

FIGURE 2.6 In a metal there is a "sea" of electrons surrounding positively charged ions.

Metals cannot form covalently bonded structures because their atoms do not have enough electrons in the outermost shell. For example, the metals in Groups IA (which include lithium, sodium, and potassium) and IIA (which include magnesium and calcium) of the periodic table have one and two valence electrons respectively, and four valence electrons are needed for a 3-D covalent solid to be formed. However, metallic bonding does involve the sharing of the valence electrons, so can be thought of as a form of covalent bonding in which all the valence electrons from every atom in the solid are shared between all of these atoms. One way to visualise this is to imagine (as depicted in Figure 2.6) a sea of valence electrons surrounding islands of positive ions that are created as a result of the valence electrons leaving their respective atoms. The attraction between these positive ions and the sea of negatively charged electrons forms the metallic bond which holds the solid together.

Metals are good electrical and thermal conductors because the valence electrons are not bound to the positive ions but are instead free to move around the solid and take part in conduction. (Thermal conduction is explained in Chapter 5, while electrical conduction is discussed in detail in Chapter 6.) Because the electrons in metals do not remain in the local environment of a single bond, or indeed of their parent atom, they are said to be *delocalized*. However, despite the fact that the electrons in solid metals are not bound to any particular atom, they are bound to the metal as a whole. They can only escape from the metal if they receive enough energy from, for example, a beam of light or X-rays, to overcome their attraction to the metal. (The ejection of electrons from metals by electromagnetic radiation is of course the well-known photoelectric effect.)

Unlike covalent bonding, metallic bonding is nondirectional. This, coupled with the fact that unlike ionic crystals metals have no oppositely charged ions side by side preventing deformation, makes metals ductile—which means they can be drawn out into a wire without breaking. Metals have cohesive energies in the range of approximately 1–5 eV/atom, so while some metals have relatively weak bonds and are therefore soft, others with stronger bonds are harder. Weaker bonding also means lower melting temperatures. Mercury, for example, has a low value of cohesive energy and a melting temperature of −39°C, so is liquid at room temperature, while tungsten with its stronger bonding melts at 3410°C.

2.1.5 VAN DER WAALS BONDING

Another way in which atoms can be held together in a solid is via *van der Waals forces*. These are weak intermolecular or interatomic electrostatic forces named after the Dutch physicist Johannes Diderik van der Waals (1837–1923), who won the 1910 Nobel Prize for physics for developing a new equation to describe gases which took account of the attraction between molecules.

In any atom, the negatively charged electrons are moving around the positively charged nucleus. So at any given moment in time the amount of shielding of the nucleus that occurs along any particular direction will depend on the position of the electrons at that instant. This means the atom has an *electric dipole moment* that is constantly changing as the electrons move around the nucleus. Any dipole consists of two equal and opposite charges slightly separated. The electric dipole moment (see Chapter 7) is the product of one of the electric charges and the distance between them. In the case of an atom the value of the negative electric charge will be the sum of the charges of all of the electrons, and similarly the value of the positive charge will be equal to the number of protons in the nucleus. A van der Waals bond is formed when the instantaneous dipole in one atom induces a temporary dipole in a neighbouring atom and causes it to become attracted (see Figure 2.7).

This weak attractive van der Waals force allows the inert gases with their closed outer shells to form solids, and can also cause attraction between molecules that have permanent electric dipole moments. Although van der Waals forces contribute to the bonding in all solids, they only have a very small effect in ionic and covalent solids, and are considered in this case to be a form of secondary bonding. However, for solids of the inert gases, including helium and neon, and solids of molecular gases, including ammonia and chlorine, van der Waals bonding is the primary, and only, bonding mechanism.

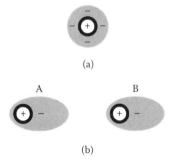

(a)

(b)

FIGURE 2.7 If the electrons in atoms remained still, they could be represented by a spherically symmetric "cloud" surrounding the nucleus, as shown in (a), and they would completely cancel out the positive charge of the nucleus. However, as these electrons in fact move around the nucleus, this electron cloud effectively alters shape all the time, producing a constantly changing electric dipole moment. In (b), the instantaneous electric dipole moment in atom A has induced a dipole in atom B. Atoms A and B are now electrostatically attracted to one another, and a van der Waals bond is formed.

Most van der Waals solids, which have cohesive energies of only around 0.1 eV/atom, have low melting points and are soft because the van der Waals force holding them together is so weak. However, some organic molecules bond primarily via van der Waals forces, including polymers and biological tissue like muscle and skin. Although these materials are made flexible and elastic by the weakness of the bonds, the large numbers of bonds between the relatively large molecules mean they remain solid up to fairly high temperatures. The cumulative effect of large numbers of van der Waals forces also allows a type of house lizard called a gecko to run vertically up walls and upside-down along ceilings—see Box 2.1: Sticky fingers.

2.1.6 HYDROGEN BONDING

A hydrogen atom, with its electronic configuration of $1s^1$, contains just a single electron, so its only option for bonding is to covalently bond with one other atom. When it covalently bonds to an atom of either oxygen, fluorine, or nitrogen, its valence electron spends most of its time nearer the atom of the other element than its own nucleus. This is because oxygen, fluorine, and nitrogen are three of the most electronegative (most able to attract electrons) of all the elements. Since its nucleus is left relatively unshielded, the hydrogen atom effectively becomes positively charged and can then attract a negatively charged part of either another molecule or part of its own molecule if that molecule is very large. The bond this attraction causes is known as a *hydrogen bond*. Whilst hydrogen bonds with their cohesive energies of 0.1–0.5 eV/atom are weaker than ionic, metallic, and covalent bonds, they are stronger than van der Waals bonds.

The familiar helix that the DNA molecule forms into is due to hydrogen bonding between different parts of the molecule. Hydrogen bonds are also found in many other organic molecules, including proteins, and are responsible for connecting the water molecules in ice together. Because the hydrogen bonding forces the oxygen atoms to be further away from the hydrogen atoms than they would be in an isolated water molecule, ice has a more "open" structure than water and hence a lower density. Several species of pond life owe much to hydrogen bonding, as they would freeze to death in cold weather if ice did not float on top of water.

EXAMPLE QUESTION 2.2   BONDING MECHANISMS

Which type of bonding is likely to be present in the following solids and why?

  (a) xenon (Xe)
  (b) caesium bromide (CsBr)
  (c) copper (Cu)
  (d) silicon (Si)

Answer on page 20.

BOX 2.1   STICKY FINGERS

"I was on a Hawaiian vacation and saw a gecko run across the ceiling to attack a large spider. I wondered, 'How can they do that?'" says Prof. Kellar Autumn, a biologist from Lewis and Clark College in the United States.

To answer his question, Prof. Autumn assembled a team of biologists and engineers from the University of California at Berkeley, the University of California at Santa Barbara, and Stanford University in the United States. The researchers eventually revealed that it is van der Waals forces—which cause molecules and atoms to become attracted to one another when they are very close together—that give geckos their sticking power. As well as increasing our understanding of biology, this discovery could lead to the development of new adhesives based on the same principles.

The bottom of a gecko's foot (see Plate 2.1) is covered in about half a million tiny hairs called setae. The end of each seta divides into hundreds of separate

PLATE 2.1   This photograph shows a Tokay gecko walking upside down on polished glass. Geckos detach their feet from whatever surface they are walking on by peeling their toes away. To attach their feet again, they uncurl their toes so that the ends of each toe—which are covered in tiny hairs called setae—can stick to the surface. It is an intermolecular attractive force known as the van der Waals force that allows these setae to stick to almost any surface, including the polished glass seen here. (Kellar Autumn, Lewis & Clark College, Portland, Oregon.)

PLATE 2.2 This scanning electron microscope image shows the ends of five gecko foot hairs (setae). Setae are around 100 millionths of a metre long, while the hundreds of separate tips called spatulae (that the setae can be seen dividing into) are only 200 billionths of a metre wide. Although the van der Waals forces which allow these spatulae to stick to almost any surface are very small, they produce a strong over-all attraction because their effect is multiplied many times over by the millions of spatulae on a gecko's foot. This adhesive force is so powerful that the gecko can support its entire body weight with a single toe. (Kellar Autumn, Lewis & Clark College, Portland, Oregon.)

tips called spatulae (see Plate 2.2) that are just 200 billionths of a metre wide and come into contact with whatever the gecko is walking on.

Autumn's team measured the adhesive force of a single seta early in 2000 via a tiny structure of silicon beams. The beams were covered in "piezoresistive crystals" whose electrical resistance alters if the beam bends. Attaching the seta in the same way a gecko would caused the beams to bend, so from the change in resistance this produced, the researchers were able to reveal how strongly the seta stuck.

Their results ruled out suction and friction as possible explanations for the geckos' climbing ability. Instead, their force values suggested that either capillary adhesion—where water molecules that cling to a surface form bonds with that surface and any other surface brought near to them, causing the two surfaces to stick together—or van der Waals forces were responsible. (Previous research had already discounted various other theories, including glue being secreted by the geckos' feet.)

If the capillary adhesion theory was true, setae would attach more strongly to hydrophilic (water loving) surfaces than to hydrophobic (water hating) surfaces. By using a force sensor first connected to a hydrophobic gallium arsenide (GaAs) semiconductor wafer and then to a hydrophilic silicon dioxide (SiO_2) wafer, Autumn's team showed that a gecko's toe attached just as strongly to both types of surface (see Plate 2.3). This, coupled with their discovery that

(c) Kellar Autumn 11/8/2000

PLATE 2.3 This photograph shows the foot of a Tokay gecko stuck firmly to a gallium arsenide (GaAs) wafer. Because GaAs is a hydrophobic (water hating) semiconductor and because gecko feet are hydrophobic as well, this confirms that it is intermolecular van der Waals forces rather than capillary adhesion of water molecules that allows geckos to climb walls and run across ceilings. (Kellar Autumn, Lewis & Clark College, Portland, Oregon.)

single setae adhered equally strongly to hydrophilic and hydrophobic surfaces, disproved the capillary adhesion theory.

Before announcing their results in the summer of 2002, however, the researchers provided additional confirmation that geckos stick to walls and ceilings by van der Waals forces. They made artificial setae from two different materials and, by measuring their adhesive forces, showed that their sticking power is only dependent on shape and size and not on material type, just as van der Waals theory predicts.

Speaking in 2002, while his team was trying to design a dry, self-cleaning glue from artificial setae, Prof. Autumn said he felt a commercially viable adhesive product could be on the market by 2004. The applications could range from easy-to-peel medical plasters to tape that would work under water and in space, while single setae could be used to pick up semiconductor wafers without marking them. "I would personally like to see a legged gecko robot walk on Mars someday", added Prof. Autumn.

ANSWER TO QUESTION 2.2 ON BONDING MECHANISMS

(a) Van der Waals bonding holds solid xenon together, since xenon has a closed outer shell and therefore no valence electrons to take part in ionic, covalent, or metallic bonding.

(b) Caesium bromide (CsBr) is an ionic solid because caesium, like sodium, has just one valence electron and is therefore strongly electropositive, while bromine, like chlorine, has seven valence electrons, making it strongly electronegative. The caesium donates its single valence electron to bromine to form the ionic bond.

(c) As copper is an element, it cannot contain ionic bonding, as this mechanism involves the transfer of an electron from one type of atom to another. It cannot be covalently bonded either because four valence electrons are needed for a covalent solid to be formed, and copper only has one valence electron. Instead copper, which is of course a metal, is held together via metallic bonds.

(d) Silicon, like carbon, has four valence electrons, so it bonds together covalently. Each silicon atom shares its valence electrons with its four neighbours, and also shares one valence electron from each of its neighbours, giving it a closed outer shell of eight electrons.

2.1.7 MIXED BONDING

So far we have looked at distinct types of bonding, but, as the introduction to this chapter mentions, some solids contain a mixture of different types of bonding. When the bonding in a solid is described as being "mixed", this can mean that different bonding mechanisms are present in different parts of the solid's structure. Alternatively, it can mean that there are the same types of bonds throughout the structure, but these bonds are a combination (or mixture) of two particular types of bonding.

Graphite (see Figure 2.8) is a good example of a solid with two distinct types of bonding present in different areas. Whilst van der Waals forces hold the individual layers of the structure together, the carbon atoms within each of the layers are covalently bonded. Because van der Waals bonding is much weaker than covalent bonding, it is much easier to shear (see Figure 4.1) a graphite crystal parallel to its hexagonal layers than perpendicular to them.

By contrast, gallium arsenide (GaAs) is a solid with the other type of "mixed" bonding, as its bonds can be considered to be partially covalent and partially ionic. Covalent bonds and ionic bonds can be thought of as two extremes of the same thing, and although there are many solids that have one type of bonding or the other, there are also plenty in which the bonds are mixed. Figure 2.9 shows bonding ranging from the extreme of purely covalent bonding to the other extreme of a purely ionic bond via two intermediate cases of mixed bonding. For compounds with mixed bonding, the extent to which their bonds are ionic in nature—known as the "ionicity"—depends on the relative electronegativities of the elements involved. The greater the difference in electronegativity, the more ionic in character the bond is.

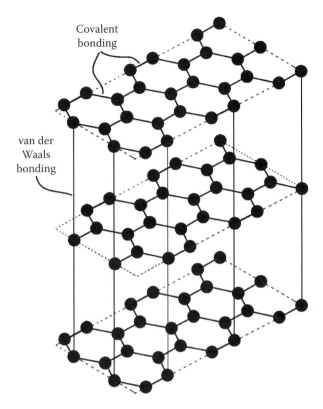

Covalent
bonding

van der
Waals
bonding

FIGURE 2.8 A schematic diagram of the structure of graphite. The horizontal layers of carbon atoms are held together in a hexagonal formation by covalent bonds, while van der Waals forces bond these layers to each other. Because van der Waals bonding is much weaker than covalent bonding, it is much easier to shear a graphite crystal parallel to its hexagonal layers than perpendicular to them.

It is also possible for bonds to form that are partially metallic and partially covalent (the transition metals, which include iron and nickel, contain this type of mixed bond), but in all cases it is important to remember that the various bonding models are just that: models. The real physical situation is almost always more complicated.

2.2 CRYSTALLINE SOLIDS

In the previous section we looked at the various ways in which atoms can bond together to form solids. Although many of the properties of solids are a consequence of the type of bonding that they have, some properties depend on the geometric arrangement of the atoms inside the solid.

The different structures that solids possess can be divided into two main types. *Crystalline* solids, which are covered in this section, are composed of a regularly repeating pattern of atoms, while as Chapter 3 will reveal, *amorphous* solids have structures in which the arrangement of atoms is much more random. The physics

EXAMPLE QUESTION 2.3   BONDING IN GaAs

Gallium arsenide (GaAs) is an important semiconductor material. It has a structure in which the atoms are at the same positions as in silicon, but half of the atoms are Ga and half are As. Each Ga atom has four (symmetrically placed) As atoms as nearest neighbours, and each As atom has four Ga atoms as its nearest neighbours. However, while silicon is covalently bonded, the bonds in GaAs are partially ionic and partially covalent. Explain why neither purely ionic nor purely covalent bonding is possible in GaAs.

ANSWER

Each Ga atom, with its electronic configuration of $1s^2 2s^2 2p^6 3s^2 3p^6 3d^{10} 4s^2 4p^1$, has three valence electrons and four nearest neighbours. It is therefore only possible for it to bond covalently with three of its neighbours. If it did so, it would still only have effectively six outer electrons, not the eight required to make a closed shell, and of course it would not be bonded with the fourth As atom.

Similarly each As atom, which has an electronic configuration of $1s^2 2s^2 2p^6$ $3s^2 3p^6 3d^{10} 4s^2 4p^3$, has five valence electrons and so could bond covalently with its four surrounding Ga atoms but would then have one too many electrons (nine in total) for its outermost shell. To make sure it only had eight electrons in its outermost shell, As could alternatively bond with just three of the Ga atoms, but it would not then be bonded with the fourth Ga atom. Pure covalent bonding is therefore not possible.

By contrast, pure ionic bonding is possible, in principle, if the three outer electrons of a Ga atom are transferred to an As atom to give both a closed outer shell. However, transfer of three electrons involves a large increase in energy (the second and third electrons must be removed from an increasingly positive Ga ion core). So pure ionic bonding is energetically unlikely.

The real situation is that some transfer of electrons takes place to provide some ionic bonding, *and* some sharing of electrons also occurs to give some covalent bonding.

of amorphous materials is less well understood, and more complicated, than that of crystalline materials. As a consequence, we will look in more detail at crystalline solids even though there are plenty of solids with amorphous structures, and even structures partway between amorphous and crystalline.

It is worth being aware from the start that although we will be considering the idealised case of perfect crystalline solids throughout this section, in reality they all contain defects. In fact, specialist techniques (see Chapter 7) need to be used in order to grow "perfect crystals". (Defects are discussed in detail in Chapter 3.)

2.2.1   Describing Crystal Structures

Crystalline solids are made up of atoms arranged in a 3-D pattern that repeats itself again and again throughout the entire solid. The way in which the atoms are arranged

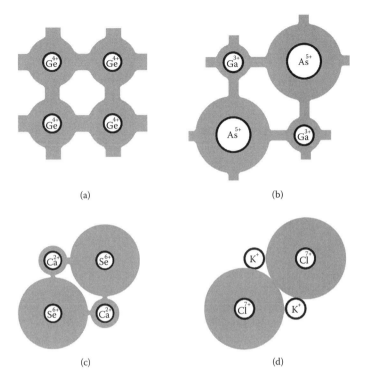

(a)   (b)

(c)   (d)

FIGURE 2.9 Covalent and ionic bonding can be thought of as two extreme cases of the same type of bonding. In (a), a purely covalent crystal of germanium (Ge) is illustrated. Each Ge atom has four valence electrons which are separated out in space by equal distances within the valence shell. The electronic density, shown as grey shading, is only large in the directions between the atoms in which these electrons have formed covalent bonds. In (b) a crystal of gallium arsenide (GaAs) is depicted. Although GaAs is covalently bonded, it can be seen that not only is the electronic density between the atoms slightly less than it was in (a), but it is larger around the arsenic than around the gallium, making the crystal slightly ionic in character. GaAs is therefore an example of a crystal with mixed bonding. (See Example question 2.3 on bonding in GaAs.) Another example is shown in (c), which illustrates the mixed bonding in calcium selenide (CaSe). In a perfectly ionically bonded crystal like potassium chloride (KCl), which is shown in (d), the atom donating its single valence electron is left with no valence electrons surrounding it at all, hence the lack of grey shading around the K^+ ions in (d). The valence electron from the electropositive atom is transferred to the electronegative atom, which in the case of KCl is chlorine. (Cl has seven valence electrons and uses the electron from the K to give itself a full outermost shell.) However, in ionic crystals where the electropositive atom has two (or three) valence electrons and the electronegative atom lacks two (or three) electrons in its outer shell, the difference in electronegativities of the two constituent elements is not as large as in a crystal like KCl. So for these crystals, the ionic bonding is less than perfect, as complete charge transfer does not take place. For example, in CaSe, the two valence electrons from the calcium are not transferred completely to the selenium. This means that although CaSe is mainly ionically bonded, the bonding is also partly covalent, as the small electronic density between the ions and surrounding the Ca ions shows. (The original version of this diagram was published in *Solid State Physics*, International Edition, Neil W. Ashcroft and N. David Mermin, p. 388, copyright Elsevier [W. B. Saunders Company], 1976. Reproduced with permission.)

is known as the *crystal structure*, and as we will see in Section 2.2.2, there are many different types of crystal structure.

Lattice and Basis

Although physical models of crystal structures can be built in which the atoms are represented by solid balls and the bonds between them by solid spokes, it is often more convenient to depict crystal structures in terms of a *lattice*. A lattice is an infinite array of points arranged in such a way that if you were able to stand on one of these points, no matter which of the points you chose to step onto, your surroundings would look exactly the same.

If one lattice point is arbitrarily chosen as the origin, the vector **r** between this origin and any other lattice point is given by

$$\mathbf{r} = u_1\mathbf{a} + u_2\mathbf{b} + u_3\mathbf{c} \tag{2.1}$$

where u_1, u_2, and u_3 are integers that can take any value—positive or negative— and **a**, **b**, and **c** are lattice vectors (also known as lattice constants) pointing in three different directions in space, respectively, so that the lattice is three-dimensional (3-D). (Appendix A includes a revision of vectors.)

As every single lattice point can be reached by putting all possible values of u_1, u_2, and u_3 into equation 2.1, this equation helps describe a lattice mathematically. To completely describe a lattice, however, not only do we need to know the lengths of the lattice vectors, but the angles α, β, and γ between the lattice vectors also need to be taken into account. (The lattice vectors and angles are collectively known as the *lattice parameters*.) Although we will look in depth at these angles in the next subsection, we will ignore them for the moment so we can concentrate on how equation 2.1 enables lattice points to be plotted from scratch from just one point, as shown in Figure 2.10.

Of course the lattice on its own cannot tell us what the structure of a particular crystal looks like. In order to represent a crystal structure, an identical *basis* of atoms must be added to every lattice point. The basis consists of either a single atom (or ion) or a group of atoms (or ions) with a fixed spacing and orientation to each other— determined by the type of bonding between them—and of a specific composition. For example, the caesium chloride structure has a basis of two ions—one caesium ion and one chlorine ion—that must appear at every lattice point in the same relative orienta- tion for every atom in the crystal to be represented. It is never the case that the basis varies and consists of, for example, two chlorine ions at one lattice point, a single chlorine ion together with one caesium ion at another lattice point, and two caesium ions at yet another point. If this occurred, the caesium chloride structure would not be produced. (In fact this arrangement would not produce a crystalline solid of any sort, as the order in which the atoms appeared in the structure would be randomised.)

Figure 2.11 shows a hypothetical lattice, basis, and crystal structure for a 2-D crystalline solid composed of two different types of atoms. (Sometimes the centre of one of the atoms of the basis coincides with the lattice point, as in this diagram, but this does not have to happen.) If this lattice was being used to depict a crystal structure with a basis of one atom, there would only be a single atom associated with

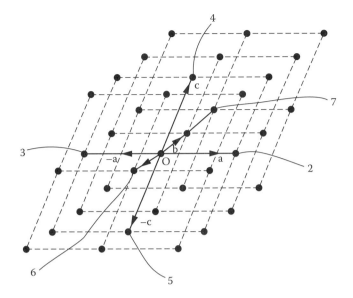

FIGURE 2.10 Part of a 3-D lattice constructed by plugging a range of different values for u_1, u_2, and u_3 into equation 2.1. For the purposes of this simplified demonstration, all of the lattice vectors have a magnitude of 1. So if the first point drawn is the point marked 0, and this is taken to be the origin, point 2 is the result of putting the values $u_1 = 1$, $u_2 = 0$, and $u_3 = 0$ into equation 2.1, while point 3 occurs when $u_1 = -1$ and $u_2 = u_3 = 0$. Points 4 and 5 are the result when $u_1 = u_2 = 0$ and $u_3 = 1$ and -1, respectively. Meanwhile point 6 is obtained with a value of -1 for u_2 and zero for u_1 and u_3, and point 7 corresponds to the values $u_1 = 0$, $u_2 = 2$, and $u_3 = 0$. The dashed lines connect these points to further lattice points that result from inserting other values of u_1, u_2, and u_3 into equation 2.1. It is clear from the 3-D shape that these dashed lines reveal that **a**, **b**, and **c** are not mutually perpendicular in this example, but as we will see in the next subsection there are lattices, known as cubic lattices, in which α, β, and γ are all equal to 90°.

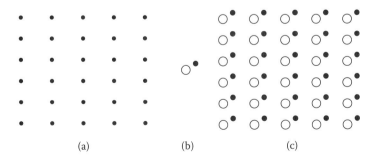

(a)   (b)   (c)

FIGURE 2.11 Crystal structures can be represented in terms of a lattice with a basis. In this figure a hypothetical 2-D crystal is shown. (a) is the lattice, while (b) is the basis of atoms which needs to be associated with each lattice point to generate the crystal structure shown in (c).

each lattice point, and so on for bases containing different numbers of atoms. Crystal structures can have a basis containing anything from one to thousands of atoms, but no matter how many atoms there are in the basis, the crystal can still be represented in this way. (Although if a crystal did have a basis of, say, 2000 atoms, the corresponding diagram would be somewhat more complicated than Figure 2.11[c]!)

Unit Cells

We have just seen that a crystal structure can be represented by a lattice and basis. But because there is a periodic repetition of groups of atoms throughout a crystal, we would not have to illustrate it with a diagram containing an infinite lattice, even if it were possible to do so. Instead crystalline solids can be depicted by a small "building block" which, when it is repeated again and again, builds up the entire structure of the crystal. This building block is known as a *unit cell*.

There is always more than one possible unit cell for any given lattice, and Figure 2.12 illustrates some of the different types of unit cell that a hypothetical 2-D lattice can have. Unit cells are created by choosing an arbitrary lattice point as the origin and connecting this point to the three points nearest to it in three different directions in 3-D, or to the two nearest points in two different directions in 2-D. The lines produced are then closed up to create a parallelogram in 2-D or a parallelopiped in 3-D. If the lattice point in the bottom left-hand corner is taken to be the origin for all of the cells shown in Figure 2.12, it can be seen that they have been created by first drawing lines joining this origin to two neighbouring points in two different directions.

The unit cells labelled A and B each contain a total of one lattice point, since there is only actually one quarter of each of the four points at the corners of the cell within the area of the cell. Any unit cell that contains just one lattice point is known as a *primitive unit cell* (sometimes shortened to "primitive cell"). This is the

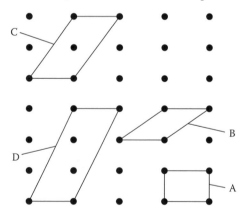

FIGURE 2.12   Several different unit cells for a hypothetical 2-D lattice. Cells A and B are primitive cells because they contain a single lattice point and have the smallest area that will still enable the entire lattice to be generated by repeated translations of the cell. By contrast, unit cells C and D are nonprimitive because they have a larger area and contain more than one lattice point.

minimum volume (in 3-D) or area (in 2-D) that any unit cell can have and still be able to build up the entire crystal structure if it is repeated over and over again by translating it along one of its edges.

There are also two other choices for the unit cell shown in the left of the picture. Both these cells still connect the origin in the bottom left-hand corner to two neighbouring lattice points in different directions, but it is clear that the area of these cells is larger than that of the primitive unit cells A and B. They also contain more lattice points. The cell marked C contains a total of two lattice points: one from within the cell and a total of one from each of the corners. Similarly, the cell labelled D contains three lattice points. Cells like C and D, which contain more than one lattice point, are known as *nonprimitive unit cells*. Like primitive cells, these cells will still replicate the pattern of the entire crystal structure when they are repeated over and over again.

Although a nonprimitive unit cell is more complicated than a primitive cell—it will contain more atoms, since each lattice point will have a basis of atoms associated with it—it is sometimes chosen to represent a crystal structure because its symmetry reflects the overall symmetry of the crystal better. We will see a good example of this later when we consider the face-centred cubic lattice, which has a cubic nonprimitive unit cell but a primitive cell of a different shape.

Of course real crystal structures have three dimensions, so a 3-D lattice and hence a 3-D unit cell is needed to represent them. Figure 2.13 shows a possible unit cell for a 3-D lattice. Referring back to Figure 2.10, it can be seen that the edges of any 3-D unit cell are in fact the lattice vectors **a**, **b**, and **c** which were defined in Equation 2.1. So the magnitudes of the lattice vectors are the lengths of the respective sides of the unit cell.

It is not enough, however, to know the physical size of a unit cell. The angles α, β, and γ between the lattice vectors—and hence between the sides of the unit cell—also need to be known. The crystal axes x, y, and z will be set at different directions

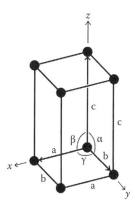

FIGURE 2.13 A 3-D unit cell showing the crystal axes x, y, and z, and the angles α, β, and γ between the sides of the cell. Since the distance between any two neighbouring lattice points in a given direction is equal to the magnitude of the lattice vector in that direction, the sides of the unit cell are in fact the lattice vectors **a**, **b**, and **c** defined in equation 2.1 The lengths of the respective sides are therefore equal to the magnitudes of the lattice vectors.

to each other depending on the values of α, β, and γ for a given crystal structure. For example, for a cubic lattice one possible unit cell could involve connecting an origin to the nearest lattice point in the same mutually perpendicular x, y, and z directions that form the Cartesian frame of reference. (See Appendix A for a revision of Cartesian coordinates.)

Bravais Lattices and Crystal Systems

In 1845, French physicist Auguste Bravais (1811–1863) realised that all the different primitive and nonprimitive unit cells that could be chosen for a 3-D lattice which satisfies the conditions in equation 2.1 could be classified into 14 types. These are known as "Bravais lattices" and are listed in Table 2.1. The Bravais lattices can in turn be grouped into seven different crystal systems, which as this table shows each have a different set of crystal axes. This means that seven different sets of crystal axes, each with a different combination of lengths for **a**, **b**, and **c** and with different angles between them, need to be used in order to generate the 14 Bravais lattices.

Any primitive Bravais lattice is just a 3-D version of the primitive unit cells A or B from Figure 2.12. They all contain one single lattice point, but in the case of a primitive Bravais lattice, the point is made up from 1/8th of each of the eight atoms at the cell corners (as we will see later). By contrast, as its name suggests a body-centred Bravais lattice has a lattice point at the centre of the cell, which means it is nonprimitive and is the 3-D equivalent of cell C in Figure 2.12. It therefore contains a total of two lattice points—one from each of the corners plus the one in the middle. Base-centred and face-centred Bravais lattices do not have straightforward

TABLE 2.1
The 14 Bravais Lattices and 7 Crystal Systems

Crystal System	Crystal Axes	Bravais Lattices	Materials with this Structure
Cubic	$\alpha = \beta = \gamma = 90°$ $a = b = c$	P (primitive) F (face-centred) I (body-centred)	Po, Cu, Au, Ag, Li, K, Fe, Diamond
Tetragonal	$\alpha = \beta = \gamma = 90°$ $a = b \neq c$	P (primitive) I (body-centred)	Rutile (TiO_2), Zircon ($ZrSiO_4$)
Orthorhombic	$\alpha = \beta = \gamma = 90°$ $a \neq b \neq c$	P (primitive) I (body-centred) F (face-centred) C (base-centred)	S, Ga, Cl (at 113K), Topaz
Hexagonal	$\alpha = \beta = 90°; \gamma = 120°$ $a = b \neq c$	P (primitive)	Be, Mg, Ti, Zn, Cd, Emerald
Trigonal	$\alpha = \beta = \gamma \neq 90°$ $a = b = c$	P (primitive)	Hg, Sb, Bi, As, Amethyst
Monoclinic	$\alpha = \beta = 90° \neq \gamma$ $a \neq b \neq c$	P (primitive) C (base-centred)	Gypsum, Orthoclase
Triclinic	$\alpha \neq \beta \neq \gamma \neq 90°$ $a \neq b \neq c$	P (primitive)	Axinite, Turquoise

2-D analogies. This is because as the next section (which describes each type of Bravais lattice in detail) will reveal, they have lattice points in the centre of the top and bottom faces of the cell, and on each of the six faces, respectively.

2.2.2 CRYSTALLINE STRUCTURES

Simple Cubic Structure

One of the most basic structures a crystalline solid can have is the simple cubic structure (sometimes shortened to "sc"). As its name suggests, the simple cubic lattice shown in Figure 2.14(a) is a cube with $\alpha = \beta = \gamma = 90°$ and a = b = c. (When the term "lattice" is used in the context of a Bravais lattice, it refers to a unit cell rather than an infinite array of points. So the unit cell shown in Figure 2.14[a] can be described as the simple cubic "lattice".) The simple cubic structure is not a popular structure for elements; in fact the rare metallic element polonium (Po) is the only one with this structure. As we will see later, there are, however, several compounds with structures based on the sc structure, including caesium chloride (CsCl) and ammonium chloride (NH_4Cl).

Elements with the sc structure have a basis consisting of just one atom, whereas the basis for sc compounds will consist of two or more atoms. We saw in Section 2.1.1 that the centre of the atom (or one of the atoms) of any basis does not have to coincide with each lattice point, but for convenience it is usually assumed that this is the case. This means that in general each corner of a unit cell will coincide with the centre of an atom, and in the particular case of an element with the simple cubic structure the centre of every atom will coincide with a cube corner.

In a real polonium crystal, you would find lots of the simple cubes shown in Figure 2.14(a) stacked up on top of each other and around each other to form the complete solid as Figure 2.14(b) illustrates. So if you chose to look at any atom within that structure, as long as it was not an atom on the surface, you would find it had the same number of *nearest neighbours* (atoms closest to it). The number of nearest neighbours is therefore a property of the lattice, and is known as the *coordination number*. Looking at Figure 2.14(b), if you consider the atom centred on the lattice

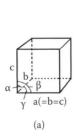
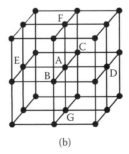

(a)   (b)

FIGURE 2.14   (a) The simple cubic lattice with the angles α, β, and γ, which are all equal to 90°, marked. (b) A schematic representation of part of a polonium crystal. In a real crystal the unit cell shown in (a) would repeat again and again to make up the whole structure. Each atom in the body of the solid will therefore have the same number of nearest neighbours.

point marked A, you can see that each of the atoms centred on the lattice points B, C, D, E, F, and G are its nearest neighbours. So the coordination number for the simple cubic lattice is 6.

Each atom throughout any crystal structure is bonded to its nearest neighbours, and since metallic bonding is nondirectional (see Section 2.1.4), most metals—as we will see in the next few subsections—have relatively large coordination numbers and closely packed atoms. By contrast, the atoms of covalently bonded solids cannot be so densely packed together because it is only possible for each atom to have four nearest neighbours, and these neighbouring atoms can only be set at fixed angles to the atom they surround (see Figure 2.5). (In fact, the "atoms" of a metallic solid are actually ions because of the metallic bonding between them, as are the "atoms" in ionically bonded solids. So whenever the "atoms" of a particular structure are referred to, this can be taken to mean either "atoms" or "ions", depending on the type of bonding holding the crystal together.)

The simple cubic lattice has just one lattice point at each corner, and only 1/8th of the volume of each of these eight lattice points is within the volume of the cell. As $8 \times 1/8 = 1$, there is therefore a total of one lattice point within the volume of an sc unit cell. This means it is a primitive unit cell. In fact although it is generally known as the "simple" cubic lattice, it can also be referred to as the "primitive" cubic lattice or Cubic P (as in Table 2.1).

Many diagrams of unit cells (both in this book and others) tend to show either the lattice points that atoms are centred on, or very small atoms fairly far apart from one another. In reality, however, atoms do not sit far apart from one another. For the purposes of discussions on the structure of elemental metals (and all of the solids of the Group VIII elements except helium), the atoms can in fact be assumed to be hard spheres just touching each other. As the simple cubic example in Figure 2.15(a) shows, this means the length of the lattice parameter **a** is able to give an indication of how tightly packed the atoms are in these solids. (Whenever the term "lattice constant" or "lattice parameter" is used on its own, it is more likely to mean **a** than **b** or **c**.) It also means that the atomic radius, r, can be calculated for elemental metals. For example, it can be seen from Figure 2.15(a) that $2r = a$,

$$\therefore \ r = \frac{a}{2} \ \text{for simple cubic} \tag{2.2}$$

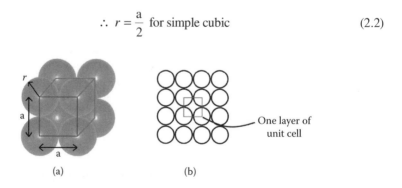

(a)   (b)

FIGURE 2.15   The simple cubic structure (a) is formed by layers of atoms like those in (b) being stacked directly on top of one another.

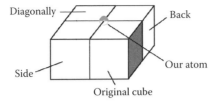

FIGURE 2.16 Four neighbouring cubic unit cells and one of the atoms that they share. In a real crystal the unit cell repeats again and again to build up the entire structure, so each of the corner atoms is in fact shared by eight unit cells because any given cubic unit cell will have seven other unit cells surrounding it. The "original cube" in the diagram represents a hypothetical cubic unit cell with just one atom at one corner. It has unit cells butted onto its back and side, and slotted in diagonally, so this single atom (shown in the back top left-hand corner position of the original cube) has half of its total volume shared by the cells shown. The other half of its volume is shared between the four unit cells that would sit on top of the unit cells shown in the diagram if this was a real crystal. As the total volume of the atom is shared between eight unit cells, each of the cells in the diagram contains 1/8th of an atom. If we now imagine the unit cells in our example are simple cubic unit cells with an atom at each corner, they will each contain the equivalent of one atom. This is because there are eight atoms in total (one at every corner), which are each shared by eight cubes, so only 1/8th of each atom is in a particular unit cell, and $8 \times 1/8 = 1$.

Looking at Figure 2.15(a) again, we can see that it shows how layers of atoms are stacked up directly on top of one another to make the simple cubic structure. This picture gives us another way of visualising the number of nearest neighbours because, as Figure 2.15(b) shows, any atom from one of the layers of atoms that the unit cell has been formed from has four atoms touching it in north, south, east, and west directions. Since layers like this are stacked directly above one another to form the sc structure, there will also be an atom on top of the chosen atom and another below it, giving a total of six nearest neighbours.

We have already seen that there is a basis of one atom for Po, the only element to have the simple cubic structure; in other words, for Po each lattice point has a single atom associated with it. So there is only a total of one atom within the volume of the simple cubic lattice, as Figure 2.16 illustrates. (If this basis of one atom was not in fact centred on each lattice point, but was arranged so that it sat in the middle of the sc unit cell, there would still only be a total of one atom per unit cell. This is because each atom would remain in the same position relative to its neighbours, and so if one atom was shifted into the centre of a unit cell, all the other atoms would end up sitting at the centre of unit cells too.)

The number of atoms per unit volume is known as the *atomic density*, and for elements with the simple cubic structure is as follows:

$$\text{Atomic density of simple cubic crystal} = \left(\frac{1}{a^3} \right) \text{atoms/m}^3 \qquad (2.3)$$

where a is the length of the lattice parameter, and so a^3 is the volume of the unit cell.

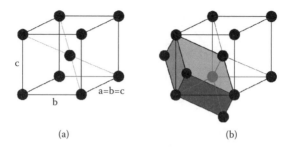

(a)   (b)

FIGURE 2.17 (a) The body-centred cubic lattice and (b) one possible primitive cell for the bcc structure shown in relation to the unit cell in (a). (Note that this primitive cell is not cubic, although the three atoms that have been added to part [b] compared with part [a] are body-centred atoms in adjacent cells.)

Body-Centred Cubic Structure

Atoms will always have a natural preference to pack closely together when they crystallise into solids because energy is minimised in that way. This means that close-packed structures are more stable than more open structures and is the reason why hardly anything has the simple cubic structure—the atoms are just too far apart. However, with an additional lattice point in the middle of a simple cubic lattice—as in Figure 2.17(a)—there is much less free space. This structure is known as body-centred cubic (often shortened to "bcc") and is the choice of many metals, including lithium (Li), sodium (Na), potassium (K), rubidium (Rb), and caesium (Cs) from Group IA of the periodic table; chromium (Cr), molybdenum (Mo), and tungsten (W) from Group VIA; and iron (Fe).

The body-centred cubic unit cell shown in Figure 2.17(a) is the 3-D analogue of cell C in Figure 2.12, and like this 2-D cell it contains two lattice points, so is not a primitive cell. By contrast, it can be seen that the primitive cell shown in Figure 2.17(b) only contains a total of one lattice point. This is only one of the possible primitive cells that could have been chosen for this structure, but any bcc primitive cell would have to have half the volume of the unit cell because by definition it can only contain one lattice point. Despite being more complicated, in the respect of containing more lattice points, the unit cell shown in Figure 2.17(a) is usually used to represent the bcc structure because its symmetry reflects the overall symmetry of real-life bcc crystals better than any of the possible primitive cells.

If we consider the elemental metals that have the bcc structure, which have a basis of one atom at each lattice point, we can derive an expression for the atomic density of the body-centred cubic structure in the same way that we did for the sc structure. In this case not only is there an atom centred on each of the cube corners, but there is also an atom in the middle. This gives a total of two atoms per unit volume, as the contribution of the corner atoms combines to give a total volume of 1 (as in the case of the simple cubic structure), and the body-centred atom is entirely within the unit cell. So for an element

$$\text{Atomic density of body-centred cubic crystal} = \left(\frac{2}{a^3}\right) \text{atoms/m}^3 \qquad (2.4)$$

EXAMPLE QUESTION 2.4   MASS DENSITY

Calculate the mass density of lithium (Li), which has a lattice parameter of 0.350 nm.

ANSWER

Looking up the crystal structure of lithium will reveal that it is bcc, and as we have just seen, the atomic density of a body-centred cubic crystal is $2/a^3$ atoms/m^3. If you are asked to work out the mass density rather than the atomic density of an element, all you have to do is multiply the number of atoms in a unit cell by the mass of a single atom of whatever element you are dealing with. So in this case the atomic weight of Li needs to be multiplied by 2.

$$\therefore \text{ The mass density of a bcc crystal} = \frac{2 \times M_{atom}}{a^3}$$

The atomic weight of Li is listed on the periodic table as 6.94 u; in other words, it is $6.94 \times (1.6604 \times 10^{-27}) = 1.152 \times 10^{-26}$ kg, although you may get a slightly different answer if you use the preset physical-constant button on your calculator. (See Appendix C for a reminder of atomic mass units.)

$$\therefore \text{Mass density of Li} = \frac{2 \times 1.152 \times 10^{-26}}{\left(0.350 \times 10^{-9}\right)^3} = 537.376 \text{ kg m}^{-3}$$

(Remember that a is given in nm, and you always need to convert to SI units for calculations.)

Before deriving an expression for the radius of atoms in an elemental crystal with the bcc structure, it is useful to see how layers of atoms need to be stacked in order to form the bcc structure. The first layer of atoms is arranged like those shown in Figure 2.18(a). The second layer of atoms then covers the spaces between the atoms of the first layer as in Figure 2.18(b). Finally the third layer of atoms (Figure 2.18[c]) occupies positions directly overhead those of the first layer of atoms.

The bcc structure has a coordination number of eight because there are four atoms from layer 3 (which as we have just seen is a mirror of layer 1) plus four atoms from layer 1 acting as nearest neighbours to each atom in layer 2. Equally, every atom in layer 1 has four nearest neighbours from the layer 2 above it (see Figure 2.18[b]) together with a similar group of four nearest neighbours from the layer below it, and as this layered structure repeats over and over in the pattern 1, 2, 1, 2, 1, 2, ..., 1, 2, every atom (except those on the surface) will have eight nearest neighbours.

If we now think more about how the atoms are packed in the body-centred cubic structure, we can see (as Figure 2.19 shows) that, unlike the simple cubic structure

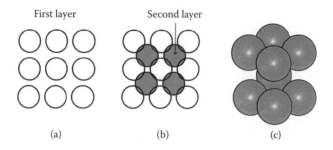

First layer   Second layer

(a)   (b)   (c)

FIGURE 2.18 Successive layers of atoms stacked up in the order shown in this series of diagrams creates the bcc structure.

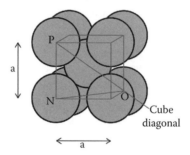

FIGURE 2.19 In an elemental bcc crystal, the atoms "touch" along the [111] cube diagonal. (The notation for directions and planes in unit cells is discussed in Section 2.2.7.)

where the atoms touch along the edge of the cubic unit cell, here the atoms touch along the cube diagonal instead.

To find the relationship between the atomic radius r and the length of the lattice parameter **a** for a bcc crystal, we first need to work out the length of the cube diagonal. We can use Pythagoras's theorem to find this. If we label the bottom rear right-hand corner of the unit cell as O and the top front left-hand corner as P, the length of the diagonal will be OP. If we also label the bottom front left-hand corner as N, we can see that the distance $ON^2 = a^2 + a^2 = 2a^2$,

$$\therefore OP^2 = ON^2 + a^2 = 2a^2 + a^2 = 3a^2$$

Since there are three atoms touching along the cube diagonal, OP = 4r (with 2r coming from the centre atom and 1r from each of the two corner atoms.)

$$\therefore (4r)^2 = 3a^2$$

and so

$$r = \frac{a\sqrt{3}}{4} \quad \text{for body-centred cubic} \tag{2.5}$$

Although atoms are not, of course, hard spheres touching each other, the rough estimates of atomic density and radius that equations like these give are useful when calculations need to be done on properties that are affected by structure. As we will see in Chapter 4, ductility and brittleness are among the mechanical properties affected by crystal structure. Also, as Box 4.1 reveals, strain is inherent in quantum dots because of differences in either interatomic spacing or atomic sizes between the two different types of semiconductor that the dots are made from. The distances between the atoms in solids (and therefore an estimate of the atomic radius) can be obtained experimentally via X-ray diffraction, which will be discussed in detail in Chapter 5.

Face-Centred Cubic Structure

Another way for atoms to be closely packed together is to have the face-centred cubic (fcc) structure shown in Figure 2.20(a). Instead of having an extra lattice point at the centre of each cube like the bcc structure, there is a lattice point at the centre of each face. So the fcc lattice looks rather like a die with a 1 on every face. ("Die" is the singular of the word "dice.") The metals copper (Cu), silver (Ag), and gold (Au) that make up Group 1B of the periodic table have the fcc structure, as do several transition metals, including nickel (Ni) and platinum (Pt). Other elements crystallising in the fcc structure include the metals aluminium (Al), strontium (Sr), and calcium (Ca) and the van der Waals-bonded solids of neon (Ne), argon (Ar), krypton (Kr), xenon (Xe), and radon (Rn).

An fcc unit cell contains a total of four lattice points, and so elements with this structure (that have a basis of one) have four atoms per unit cell. The volume of three of these atoms is obtained from the fact that half an atom is inside the unit cell from each of the face centres, while the remaining atom is made up from 1/8th of each corner atom (as with the other cubic structures). So, for elements,

$$\text{Atomic density of face-centred cubic crystal} = \left(\frac{4}{a^3}\right) \text{atoms/m}^3 \qquad (2.6)$$

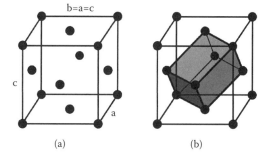

(a)   (b)

FIGURE 2.20 (a) The face-centred cubic lattice and (b) one possible primitive cell for the fcc structure shown in relation to the unit cell in (a).

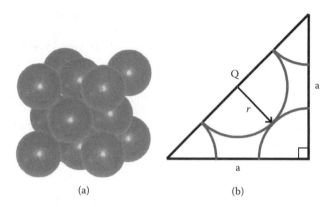

(a)   (b)

FIGURE 2.21 If the atoms of an elemental metal with the fcc structure are assumed to be hard spheres just touching one another as shown in (a), the atoms will come into contact along the face diagonals as indicated in (b), which illustrates half of one face.

In order to contain just one lattice point, any primitive cell for the fcc structure, such as that shown in Figure 2.20(b), must have only a quarter of the volume of the unit cell shown in Figure 2.20(a).

If we now imagine the atoms of an elemental metal (or Group VIII element) centred about the lattice points shown in Figure 2.20(a) as hard spheres just touching one another as in Figure 2.21(a), it becomes clear that the atoms touch along the face diagonals for the fcc structure. Figure 2.21(b) shows a schematic diagram of half of one face.

From Pythagoras, $Q^2 = a^2 + a^2$, and we can see that $Q = 4r$

$$\therefore (4r)^2 = 2a^2$$

and

$$\therefore r = \frac{a\sqrt{2}}{4} \text{ for face-centred cubic} \tag{2.7}$$

As Figure 2.21(a) suggests, the fcc structure is a more closely packed structure than either the simple cubic or body-centred cubic structures. In fact in the fcc structure, 74% of the available volume is occupied by atoms, while in the bcc structure 68% is occupied. The simple cubic structure has much more free space, with only 52.4% of the available volume taken up by its atoms. These figures may also be expressed as a *packing fraction*. In this case the maximum available volume is taken to be 1, and the numbers show what fraction of this available volume is filled by atoms. So for sc the packing fraction is 0.524, for bcc it is 0.68, and for fcc it is equal to 0.74. Box 2.2: "A menace for mathematicians" looks at how mathematicians have attempted to understand close packing. However, if you are asked to work out the packing fraction, you will need to calculate what proportion of the unit cell is occupied by atoms, as shown in the following worked example.

EXAMPLE QUESTION 2.5   PACKING FRACTIONS

Calculate the packing fraction of the body-centred cubic structure.

ANSWER

In the absence of any information to the contrary, it would be safe to assume that you were being asked to consider an elemental solid with a basis of one. Having assumed this, the first thing to do is to work out the number of atoms per unit volume for whatever structure you are dealing with. For a bcc structure there is a total of 2, made up from one atom from the cube corners plus one atom at the centre.

Because we are treating atoms as hard spheres for the purposes of all these calculations, the volume of each of these two atoms can be taken to be $4/3\pi r^3$. So the total volume occupied by atoms within the unit cell is the number of atoms per unit cell multiplied by the volume of each atom, i.e.,

$$2 \times \frac{4}{3}\pi r^3 = \frac{8}{3}\pi r^3$$

For bcc,

$$r = \frac{a\sqrt{3}}{4}$$

and to get the packing *fraction* this volume must be divided by the volume of the unit cell a^3. So,

$$\frac{\frac{8}{3}\pi \left(\frac{a\sqrt{3}}{4}\right)^3}{a^3} = \text{packing fraction}$$

$$= \frac{8}{3}\pi \frac{\sqrt{3}^3}{64}$$

$$= 0.68$$

Looking at the photographs accompanying Box 2.2, it can be seen that each atom in the 2-D array of layer 1 has six nearest neighbours arranged hexagonally around it. Because each atom in layer 2 fits into the hollows formed by three atoms below it, each atom in layer 1 will also have three atoms from the layer above it as its nearest neighbours. In addition, because of the way the structure repeats with layer 1 from one unit cell sitting on top of layer 3 from another, each atom of

BOX 2.2   A MENACE FOR MATHEMATICIANS

In the late 16th century, the British explorer Sir Walter Raleigh asked his mathematical assistant Thomas Harriot if he could come up with a formula for working out the number of cannonballs in a stack. Little did he know that this seemingly simple question would lead to a problem that ended up taking mathematicians nearly 400 years to solve.

Raleigh was soon provided with a formula, but studying the cannonballs had set Harriot—who was a firm believer that matter was composed of atoms—thinking about all the various arrangements in which spheres could be stacked. He supposed that these spheres could just as easily represent atoms as cannonballs, and that learning how they could be arranged would reveal how matter was formed. Harriot corresponded regularly with the German astronomer and scientist Johannes Kepler, and in 1606 he suggested that an atomic theory could explain the results of Kepler's optics experiments.

Although initially sceptical, Kepler started to come round to this idea and began his own studies, which included trying to find the best way to stack spheres so there was the least possible space between them. In 1611 he described his work in a booklet called "The Six-Cornered Snowflake". In this he claimed that the face-centred cubic arrangement (see Figure 2.20) was "the tightest possible, so that in no other arrangement could more pellets be stuffed into the same container". This statement became known as the Kepler conjecture.

To anyone who has ever stacked a pile of oranges, the Kepler conjecture is intuitively obvious. They will naturally use the face-centred cubic arrangement, in which the layers are built up by placing oranges in the hollows left between the oranges in the layer below (see Plates 2.4–2.7). Experiments on crystals have also revealed to physicists that the face-centred cubic (fcc) arrangement, and the equally closely packed and mathematically equivalent close-packed hexagonal (cph) structure, are the most densely packed arrangements that atoms can have. But proving this mathematically is anything but easy.

However, after centuries of attempts by various mathematicians, Prof. Thomas Hales, formerly from the University of Michigan and now at the University of Pittsburgh in the United States, became the first person to provide a rigorous proof of the conjecture. "I started working on the problem as a hobby in 1988. The solution finally came in 1998," recalls Prof. Hales.

Hales based his work on that of a Hungarian mathematician called László Fejes-Tóth, who worked on Kepler's conjecture in the 1950s. Using this approach involved studying up to 50 spheres instead of the infinite number that some other mathematicians—who had failed to solve the problem—had tackled. Even so, the packing density of every conceivable arrangement of those spheres had to be estimated to see if it was less than that of the fcc arrangement. This meant using powerful computers to solve inequalities involving hundreds of variables. Worse still, "There are many candidates for packings of clusters of spheres that are nearly as good as the optimal ones," says Hales.

Plate 2.4   Plate 2.5

Plate 2.6   Plate 2.7

These photographs show how the face-centred cubic (fcc) structure is built up. The plastic balls represent the atoms of an elemental fcc crystal. The first layer of the face-centred cubic structure (Plate 2.4), with its hexagonal arrangement, is more closely packed than that of either the bcc or sc structure, and so the entire structure contains more atoms per unit volume. As Plate 2.5 shows, the second layer of atoms covers some of the spaces between the atoms of the first layer, while the third layer (Plate 2.6) covers the spaces between the atoms of the second layer. Since none of the atoms in the third layer occupy positions directly overhead those of the first or second layer, these layers are repeated in the sequence 1, 2, 3, 1, 2, 3, 1,…, etc., to build up the entire structure. As it was impossible to include an unlimited number of balls in these photographs, the repetition of these layers results in a pyramidal shape (Plate 2.7).

Eventually, however, Hales had studied the packing density of every possible arrangement of spheres. He found there were no packing densities greater than $\pi/\sqrt{18}$—which is approximately equal to 0.74, and is the packing density of the fcc and cph structures—which proved the Kepler conjecture.

Since then, Hales has made use of the techniques he developed for solving the Kepler conjecture. "Many of the same methods were used later in my proof of the Honeycomb conjecture, which states that the hexagonal honeycomb is the optimal way to partition a region into cells of equal area," he says.

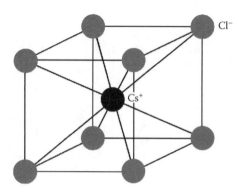

FIGURE 2.22   A unit cell for the CsCl structure.

layer 1 will also be filling the hollows between three atoms in the layer 3 below it. This means there are three nearest neighbours from the layer below layer 1 as well, giving a total number of 6 + 3 + 3 = 12 nearest neighbours for the fcc structure.

Caesium Chloride Structure

The caesium chloride structure—and the next three structures that we will consider— are not themselves Bravais lattices, but are instead based on the Bravais lattices we have just been looking at. (Whilst there are many different crystal structures that are produced by adding a basis of two or more atoms to one of the Bravais lattices, the four cubic structures described here have been chosen because they are relatively simple and are shared by a fairly large number of solids.)

The caesium chloride structure shown in Figure 2.22 looks similar to the bcc structure, but unlike the bcc structure contains two different types of atoms. While one type of atom sits in the body-centred position in the middle of the cube, the other type occupies the cube corners. In the case of CsCl itself, it is a Cs^+ ion that is found in the centre, while Cl^- ions reside at the cube corners. A more accurate way of looking at the CsCl structure is as a simple cubic structure with two atoms at each lattice point as opposed to one.

It is clear from the diagram that each atom in the caesium chloride structure is surrounded by eight atoms of the opposite type, so the coordination number of the CsCl lattice is 8. (Although the unit cell shown in Figure 2.22 has the Cs^+ ion in the middle and Cl^- ions at the corners, the CsCl crystal structure would still be produced if it had a Cl^- ion in the centre and Cs^+ ions at the corners instead.) While most of the alkali halides (ionically bonded compounds composed of a Group I metal and a Group VIIb halogen) have the sodium chloride structure that we will look at next, caesium chloride (CsCl), caesium bromide (CsBr), and caesium iodide (CsI) crystallise in the caesium chloride structure. Ammonium chloride (NH_4Cl) and some types of brass (CuZn) are among the other substances that have the caesium chloride structure.

Sodium Chloride Structure

Two crystal structures based on the face-centred cubic structure are the sodium chloride structure (which is also known as the "rock-salt" or the "halite" structure)

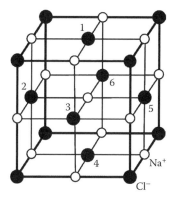

FIGURE 2.23 Many ionic compounds crystallise in the sodium chloride structure. Each ion has six nearest neighbours. For the Na^+ ion in the centre of this diagram, the Cl^- ions marked 1–6 are its nearest neighbours.

and the zincblende structure. The sodium chloride structure, shown in Figure 2.23, can be thought of as two interlinked fcc lattices, with one fcc lattice containing entirely Na^+ ions slid along half a cube diagonal with respect to the other lattice, which is composed of Cl^- ions. (The Cl^- ions are shown in black in Figure 2.23, while the Na^+ ions are depicted in white.) In fact NaCl has a face-centred cubic lattice with a basis of two atoms, Na^+ and Cl^-, separated by half a cube edge.

Each Cl^- ion has six nearest neighbours of Na^+ ions, and each Na^+ ion has six nearest neighbours of Cl^-, so the coordination number is 6. This is shown in the diagram by the Cl^- ions marked 1 through 6, which surround the Na^+ ion in the centre of the picture. Lithium fluoride (LiF), magnesium oxide (MgO), manganese oxide (MnO), silver bromide (AgBr), lead sulphide (PbS), and potassium bromide (KBr) are some of the well-known ionic compounds that crystallise in the sodium chloride structure.

Zincblende Structure

Like the sodium chloride structure, the zincblende structure can also be considered to consist of two interpenetrating fcc lattices (see Figure 2.24), but this time one fcc lattice is slid along one quarter of a cube diagonal with respect to the other. So in reality the zincblende structure consists of an fcc lattice with a basis of two atoms which are separated by quarter of a cube diagonal.

Several compounds have this structure, including zinc sulphide (ZnS)—which as well as having the cubic zincblende form can also have a hexagonal crystal structure known as wurtzite (polymorphic compounds like this are discussed in Section 2.2.5)—zinc selenide (ZnSe), aluminium phosphide (AlP), and gallium arsenide (GaAs). In every case, one of the two interpenetrating fcc lattices is made up entirely from one of the elements of the compound, while the other lattice is composed of atoms of the other constituent element. To help illustrate this, the zinc atoms are shown in black in Figure 2.24, while the sulphur atoms are white, but exactly the same structure would be reproduced if the atoms were interchanged so that the sulphur atoms occupied the corner and face-centred positions.

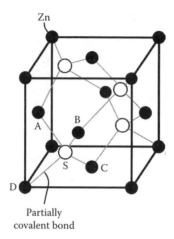

Partially
covalent bond

FIGURE 2.24   Zinc sulphide (ZnS) is one example of a compound with the zincblende structure. It has mixed bonding that is partially ionic but predominantly covalent. The highly covalent nature of the bonding means a tetrahedral arrangement of atoms forms, as indicated by the Zn atoms marked A, B, C, and D, which are bonded to the annotated sulphur atom. Other compounds with the zincblende structure include aluminium arsenide (AlAs), copper chloride (CuCl), cadmium sulphide (CdS), silver iodide (AgI), silicon carbide (SiC), boron nitride (BN), gallium arsenide (GaAs), and gallium phosphide (GaP).

Since four atoms of one element surround every atom of the other element, the coordination number for the zincblende structure is 4. For the sulphur atom singled out in the diagram, the zinc atoms marked A, B, C, and D are its nearest neighbours. Many of the compounds with the zincblende structure have a type of mixed bonding (see Section 2.1.7), which is a combination of ionic and covalent bonding, and it is the covalent part of this mixed bonding that gives each group of five atoms a tetrahedral arrangement.

Diamond Structure

The diamond structure (shown in Figure 2.25) is like the zincblende structure, but in this case all the atoms are the same type. As its name suggests, diamond (carbon atoms) has this structure, as does silicon, grey tin, and germanium. In a diamond, each carbon atom has four nearest neighbours, so the coordination number for the diamond structure is 4. It is the loosest packed of all the crystal structures, with a packing fraction of just 0.34. As with the zincblende structure, the tetrahedral arrangement of the nearest neighbours in the diamond structure comes about because of covalent bonding between the atoms (see Section 2.1.3). Diamond is an example of an element with a basis of more than one atom, as it has an fcc structure with a basis of two atoms.

Hexagonal Close-Packed Structure

We have now seen all of the cubic Bravais lattices (and some of the variations on them) that crystals can have; however, as Table 2.1 shows, there are six other shapes

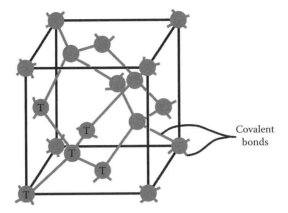

FIGURE 2.25 Each atom in the diamond structure is covalently bonded to its four nearest neighbours. The tetrahedral arrangement that this produces is shown in the diagram by the atoms marked with a *T*. Covalent bonding is discussed in detail in Section 2.1.3.

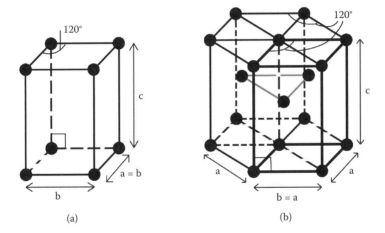

FIGURE 2.26 (a) The simple hexagonal lattice and (b) the hexagonal close-packed (hcp) structure. Many elements have the hcp structure, including scandium (Sc), titanium (Ti), zinc (Zn), cadmium (Cd), ruthenium (Ru), zircon (Zr), yttrium (Y), hafnium (Hf), rhenium (Re), osmium (Os), thallium (Tl), and the rare earths gadolinium (Gd), terbium (Tb), dysposium (Dy), holmium (Ho), erbium (Er), tellurium (Tm), and luthenium (Lu). The bold outlining reveals the simple hexagonal lattice that the hcp structure is based on.

of structure they can possess: tetragonal, orthorhombic, hexagonal, trigonal, mono-clinic, and triclinic. The Bravais lattices associated with these crystal systems will be discussed in the next few subsections, starting with the simple hexagonal struc-ture shown in Figure 2.26(a), which has $a = b \neq c$, $\alpha = \beta = 90°$, and $\gamma = 120°$.

As we saw earlier, there are many different crystal structures that can be pro-duced by adding a basis of two or more atoms to one of the Bravais lattices. The hexagonal close-packed (hcp) structure shown in Figure 2.26(b)—also known as the close-packed hexagonal (cph) structure—is the most common of the structures

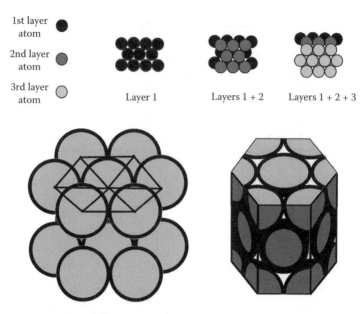

FIGURE 2.27   Building up the hcp structure layer by layer.

based on the simple hexagonal Bravais lattice (which is indicated by the bold outlining in the diagram). Elements with the hcp structure—which include beryllium (Be), magnesium (Mg), and titanium (Ti)—actually have the simple hexagonal lattice with a basis of two atoms. The hcp structure is the most favourable structure for the elements, which is not surprising, since the atoms are so closely packed their minimum energy is low. In fact the packing fraction for the hcp structure is the same as for the fcc structure, namely 0.74.

In order to make up the hcp structure by stacking layers of atoms, we can see from Figure 2.27, and by referring back to Box 2.2, that the first two layers of the hcp structure are the same as layers 1 and 2 for the fcc structure. But for hcp, the third layer of atoms sits directly overhead the positions of the first layer of atoms, giving a repeating pattern of layer 1 then layer 2 then layer 1 then layer 2 throughout any hcp solid. The basis of two atoms for elements with the hcp structure will therefore consist of one atom from layer 1 together with an adjacent atom from layer 2.

As with the fcc structure, the coordination number for the hcp structure is 12. This comes from the six atoms in contact with any one atom in layer 1, plus the three nearest-neighbour atoms from layer 2 above it, together with the three nearest neighbours from the layer 2 below the layer 1 shown in the diagram.

Tetragonal Structure

There are two types of tetragonal lattice: simple, with an atom at each corner of the packing case shape (see Figure 2.28[a]), and body-centred, with an additional atom in the middle (see Figure 2.28[b]). If you look at the simple tetragonal lattice shown in Figure 2.28(a), you can see that both the top and bottom faces of the tetragonal structure are squares (a = b ≠ c), and as with the cubic structures, $\alpha = \beta = \gamma = 90°$.

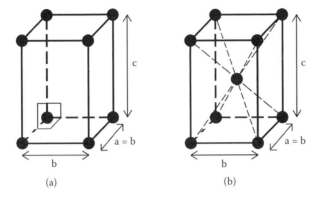

FIGURE 2.28 (a) The simple tetragonal lattice and (b) the body-centred tetragonal lattice.

(Figure 2.14, which illustrates the simple cubic lattice, shows the angles α, β, and γ.) Both white tin (β-tin) and indium (In) have tetragonal structures, as do some minerals including the gemstone zircon ($ZrSiO_4$) and rutile—otherwise known as titanium dioxide (TiO_2).

Orthorhombic Structure

In contrast to the tetragonal shape where two sides are the same length, the orthorhombic structure (see Figure 2.29) has three sides of differing lengths ($a \neq b \neq c$). As $\alpha = \beta = \gamma = 90°$, you could think of an orthorhombic unit cell as a cabinet that you want to put in a small space in a room. To make sure it fits, you would need to measure not only its height, but its depth and width as well.

As Figure 2.29 shows, there are four variants of the orthorhombic structure: simple, body-centred, face-centred, and an unusual structure known as orthorhombic C, which has an atom at the centre of the top and bottom faces, and is known as base-centred.

If elements with the simple orthorhombic structure are considered, there will be one atom at each corner. This means there is a total of one atom per unit cell, and as the sides are all different lengths, the volume of the unit cell is $a \times b \times c$. So for elements

$$\text{Atomic density of simple orthorhombic crystal} = \left(\frac{1}{a \times b \times c} \right) \text{atoms/m}^3 \quad (2.8)$$

A handful of elements crystallise in the orthorhombic structure, including gallium (Ga), sulphur (S), chlorine (Cl) at 113K, and bromine (Br) at 123K. Several minerals also have this structure, including barite, the gemstone topaz, and aragonite—a form of calcium carbonate ($CaCO_3$). In addition, some organic molecules crystallise in the orthorhombic structure, including one type of chitin found in lobster tendons.

The Trigonal, Triclinic, and Monoclinic Structures

In contrast to the orthorhombic structure, there is only one form of the trigonal structure, which is shown in Figure 2.30. This structure, which is sometimes called

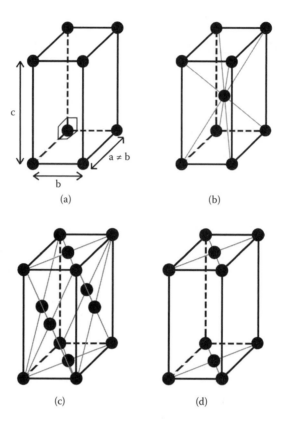

FIGURE 2.29   (a) The simple orthorhombic lattice, (b) the body-centred orthorhombic lattice, (c) the face-centred orthorhombic lattice, and (d) the base-centred orthorhombic lattice.

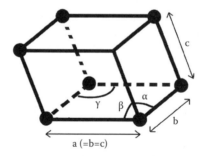

FIGURE 2.30   The trigonal lattice.

rhombohedral, can be thought of as a cube stretched diagonally and has $\alpha = \beta = \gamma \neq 90°$ and $a = b = c$, as the diagram shows. Since over 70% of elements crystallise in the fcc, bcc, hcp, or diamond structures, it is perhaps not surprising that the trigonal structure is not a very popular choice for the elements. Having said that, mercury (Hg), antimony (Sb), bismuth (Bi), samarium (Sm), and arsenic (As) do crystallise in this structure. Some minerals, including the gemstones smoky quartz, rose quartz, and

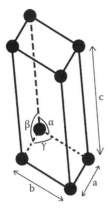

FIGURE 2.31 The triclinic lattice.

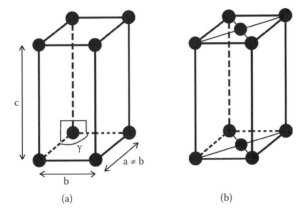

FIGURE 2.32 (a) The simple monoclinic lattice and (b) the base-centred monoclinic lattice.

amethyst (purple coloured quartz)—which all have the chemical formula SiO_2—and calcite (a form of calcium carbonate) also have the trigonal structure.

The final two crystalline structures—triclinic and monoclinic shown in Figures 2.31 and 2.32, respectively—are an even less popular choice for the elements, although many compounds crystallise in these structures. For the triclinic structure, $\alpha \neq \beta \neq \gamma \neq 90°$ and $a \neq b \neq c$. It only has the simple form shown in the diagram, whereas the monoclinic structure can exist either in simple form (see Figure 2.32[a]) or as a base-centred structure (shown in Figure 2.32[b]).

At a quick glance, the simple monoclinic structure shown in Figure 2.32 looks the same as the simple orthorhombic structure shown in Figure 2.29. But the difference is that both the top and bottom faces in the orthorhombic structure are rectangles, and they are parallelograms in the monoclinic structure, where $\alpha = \beta = 90° \neq \gamma$ and $a \neq b \neq c$. Just to confuse us even further, the tetragonal shape (Figure 2.28) also looks similar to both monoclinic and orthorhombic, but its top and bottom faces are squares.

The gemstone turquoise and the mineral axinite are among the solids that have the triclinic structure, while horse haemoglobin and the minerals azurite, orthoclase,

EXAMPLE QUESTION 2.6   CRYSTAL STRUCTURES

This diagram shows the cubic unit cell of what is known as the C15 structure, which contains two types of atom.

   (a) Is this a primitive cell?
   (b) What is the Bravais lattice of this structure?
   (c) How many atoms are there in this cell?
   (d) If you consider only the larger atoms, what structure do they show?
   (e) How many atoms are there in the basis?
   (f) If the larger grey atoms are of element A and the smaller black atoms are of element B, what is the chemical formula for the material?

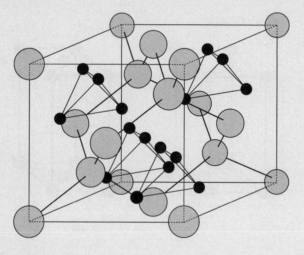

Answer on page 50.

and gypsum—which has a chemical formula of $CaSO_4 \cdot 2H_2O$ and is used as plaster for setting broken bones—are examples of materials that crystallise in the monoclinic structure.

2.2.3   QUASICRYSTALS

We saw in Section 2.2.1 that crystal structures can be depicted in terms of a lattice in which the vector **r** between any arbitrarily chosen lattice point and any other lattice point is given by equation 2.1. Lattices like this, and the crystals they describe, can only have 2-, 3-, 4-, or 6-fold rotational symmetries. Taking a crystal with 6-fold symmetry as an example, this means that in one complete 360° rotation, there are six occasions on which the crystal would appear exactly the same if it was viewed by someone who did not know it had been rotated. It would appear identical on these occasions because all the rotated lattice points would just "overwrite" the positions that they originally had before the rotation. (Similarly, a crystal with 4-fold symmetry would appear identical four times in any complete rotation, and so on.)

This is not the whole story, however. In 1984 an X-ray diffraction experiment revealed that a sample of an aluminium-manganese alloy had 5-fold symmetry. This material was called a *quasicrystal* as although, like a crystal, it produced a sharp diffraction pattern—which is obtained when X-rays scatter off crystals, and indicates the arrangement of their atoms (see Chapter 5)—it did not possess the expected symmetry characteristics of a crystal. Since then other alloys (mixtures of two or more metals; see Chapter 4) containing either two or three metals have been discovered that can form quasicrystals with 5-fold or 10-fold rotational symmetry.

Research into quasicrystals is at the time of writing this book a growing field internationally. There is a lot of interest in these unusual materials, not only because of the nature of their structure, but also because they have some useful properties. For example, quasicrystals have good wear-resistance and low-friction surfaces, which makes them ideal for coating things like machine parts and the heads of electric razors.

2.2.4   Liquid Crystals

There is also another type of crystal that is not really a crystal in the true sense of the word. This is a *liquid crystal*, which has a partly ordered structure, but can flow like a liquid. In fact the liquid crystal state is classed as a separate state of matter, and is not usually discussed in much detail in solid state physics courses. It is worth noting, however, that many organic materials become liquid crystals at certain temperatures.

2.2.5   Allotropes and Polymorphs

Some of the solids that were quoted in Table 2.1 and Section 2.2.2 as having particular crystal structures only have these structures within certain temperature ranges. Similarly, solids can have different structures at different pressures. When an element has more than one possible crystal structure it is known as *allotropic*, and each of its possible structures is termed an *allotrope*.

Tin is a good example of an allotropic solid. It has a body-centred tetragonal structure and is a metal at room temperature; however, below 13.2°C, tin exists as a covalently bonded semiconductor with the diamond structure. The room-temperature form of tin is known as white tin (β-tin), and since it is a metal is ductile. By contrast, the low-temperature grey tin (α-tin) is brittle. The slow change from ductility to brittleness at low temperatures, which results in the less-dense grey tin gradually crumbling into a powder, is known as "tin disease" or "tin plague" and can cause some bizarre problems. For example, church organists in Northern Europe are often indirect victims of tin disease, as it can cause old tin-based organ pipes to disintegrate during cold winters. Another element that is well known for being allotropic is carbon. Not only can carbon exist in the form of diamond and of graphite, it was discovered in 1985 that it also forms into a structure known as buckminsterfullerene (often called a "buckyball").

ANSWER TO QUESTION 2.6 ON CRYSTAL STRUCTURES

(a) No this is not a primitive cell. Any crystal structure is represented by a basis of atoms centred on each lattice point, and it is clear that there are several repeating groups of atoms that look identical. This means the basis is repeated more than once, and so there must be more than one lattice point in this unit cell.

(b) The Bravais lattice of this structure is fcc. This can be seen by ignoring the black atoms and concentrating on the grey atoms. Whilst there are additional grey atoms at other positions, there are grey atoms at each of the cube corners and at the centre of each face. Like the NaCl and zincblende structures, the C15 structure is therefore based on the fcc lattice and has a basis containing more than one atom.

(c) There are 24 atoms in this cell. 16 of these are the black atoms—which are entirely contained within the volume of the cell—and a further 4 are grey atoms also entirely within the volume of the cell. The 1 is the total contribution from the grey atoms at the cube corners, and the final 3 are the result of the fact that half of the volume of each of the face-centred grey atoms is within the cell.

(d) If the larger grey atoms were the only atoms shown in this diagram, it would be a picture of the diamond lattice shown in Figure 2.25.

(e) There are 24 atoms in total in this unit cell, which is based on the fcc structure. The fcc structure has 4 lattice points per unit cell, so there must be 6 atoms in the basis. This is made up from the 4 black atoms in a tetrahedral arrangement and 2 grey atoms, as indicated here.

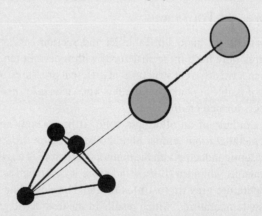

(f) Since there are 8 grey (A) atoms and 16 black (B) atoms in this unit cell, the chemical formula for the material must be AB_2.

FIGURE 2.33 Scanning electron microscope (SEM) image of some calcium carbonate crystals formed in a kettle. Two different polymorphs of CaCO₃ have formed: the needle-like aragonite, which is actually an orthorhombic crystal, and calcite, which has a trigonal structure. (Courtesy of Alan Piercy.)

When a compound has multiple structures it is described as *polymorphic*, while its different forms are called *polymorphs*. Polymorphism occurs in all sorts of compounds, ranging from fairly simple materials like calcium carbonate (see Figure 2.33) and zinc sulphide to more complex crystalline organic materials like pigments and medicinal drugs. The drug paracetamol has two common polymorphs for example. Like the allotropes of tin and carbon, different polymorphs of compounds can have very different physical properties. In the case of medicinal drugs, a patient can receive a different dose from the same amount of the same drug if they are given two different polymorphs which each have a different solubility.

2.2.6 SINGLE CRYSTALS AND POLYCRYSTALS

Any crystal made up from a perfect pattern of atoms, so that all the unit cells join together in the same way and point in the same direction, is known as a *single crystal*. Single crystals do grow naturally; in fact many gemstones are single crystals, but most single crystals are artificially created for use in the semiconductor industry. (The various growth methods are discussed in Chapter 7.)

Sometimes single crystals have an outer shape that reveals the shape of the underlying crystal structure. You can see this for yourself if you look closely at a grain of salt. You might just be able to make out that it is a tiny cube, which we would expect, as we know that the NaCl lattice is cubic.

The vast majority of crystalline materials, however, are not one huge "single" crystal but are instead *polycrystalline*. This means that they grow in the form of lots of little crystals known as "crystallites" or "grains", which vary in size and point in different directions with respect to each other. The boundaries separating these individual crystallites are known as *grain boundaries*. As we will see later, grain boundaries can affect some of the properties of solids. Sometimes their effects are negative, but often polycrystalline solids are deliberately grown if their properties are ideal for a certain application. For example, the polycrystalline diamond films shown in Figure 2.34 have been grown for use as radiation detectors, as the figure caption explains.

2.2.7 DIRECTIONS, PLANES, AND ATOMIC COORDINATES

Directions

It was mentioned in Section 2.2.2 that it is often important to know the atomic density and atomic radius when performing calculations on the properties of crystals. Some properties, including the electrical conductivity, can have different values along different directions in a crystal. For example, the electrical conductivity of graphite (see Figure 2.8) is 5000 times higher in a direction parallel to the horizontal layers of carbon atoms than in a direction perpendicular to them. Meanwhile, deformation in crystals is caused by one plane of atoms sliding over another plane (see Chapter 3), and it is planes of atoms in crystals that scatter X-rays and so enable X-ray diffraction pictures to be produced (see Chapter 5). So in order to understand more about effects and properties like this, a way of describing directions and planes within crystals is also needed.

We have already seen that axes can be added to a unit cell (see Figure 2.13). These axes give us a reference frame for defining directions and planes. Directions are encased in square brackets and are in general given by $[hkl]$, where h represents the distance along the x-axis, and k and l represent the distances along the y-axis and z-axis, respectively, that need to be moved in order to coincide with a vector pointing in that direction. For example, for cubic crystals the x direction is [100], y [010], and z [001], as illustrated in Figure 2.35. These numbers may look familiar; in fact they are the same as those used to define the Cartesian unit vectors (see Appendix A).

Other important directions for cubic crystals (and indeed other shapes of crystal) are the face diagonals and the body diagonals, and one of each is shown in Figure 2.35. The face diagonal that appears in the diagram is [110] because it intersects the x–y plane and does not go up the z-axis at all. By contrast, the [111] direction runs from the origin and up on through the corner diagonally opposite it. The set of cube edge directions (which are equivalent in the sense that they are all along or parallel to the cell axes) is written as <100>. Similarly, the set of face diagonals (all across faces) is <110>, and the set of body diagonals (all pointing diagonally across the unit cell) is <111>.

Obviously not all crystals are cubic, but this method of defining directions can be extended to cover every type of crystalline solid. For example, a fourth axis is used in addition to the three standard axes when describing hexagonal crystals. This four-axis coordinate system is known as the "Miller-Bravais" coordinate system and is illustrated in Figure 2.36. Describing directions in terms of this coordinate system is a bit more complicated than for cubic crystals, but the principles are the same. The notation for the z-direction [0001] is, however, straightforward and is marked on the diagram.

(a)

(b)

FIGURE 2.34 Natural diamonds are formed at enormous temperatures and pressures beneath the surface of the Earth, but synthetic diamond can be grown in a laboratory using a variation of a standard technique used in the electronics industry for making semiconductor devices and integrated circuits known as chemical vapour deposition (CVD). The SEM microscope images (a) and (b) show polycrystalline diamond films grown by the microwave plasma enhanced CVD process by Dr. Philippe Bergonzo and his team from CEA Saclay (the French Atomic Energy Commissariat) in 2001. Radiation detectors can be made from diamond films like these because, although visible light passes straight through diamond, it absorbs UV light, X-rays, neutrons, and alpha particles and produces an electric current when it does so. Because diamond also has a very high resistance to both radiation and chemicals, diamond detectors could work in harsh environments such as inside nuclear reprocessing plants where strong acids are used. The photograph (c) shows a prototype alpha particle detector produced by the CEA team in 2002 for monitoring radioactive waste. The diamond film can be seen in the centre of the cylindrical detector unit. By tailoring the CVD growth process, the researchers have shown that diamond films can be optimised for other applications, such as UV detection on solar exploration missions. (Courtesy of Philippe Bergonzo.)

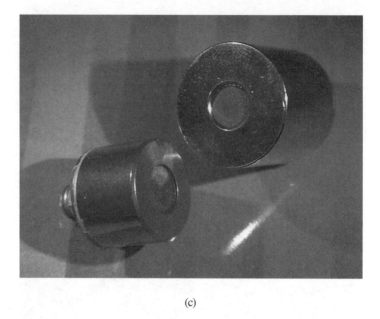

(c)

FIGURE 2.34 (continued)

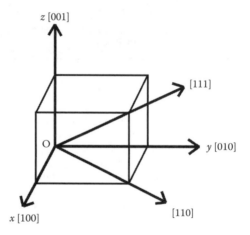

FIGURE 2.35   Some important directions in a crystal relative to the origin O.

Planes

Planes are given in general by the notation (hkl), where h, k, and l are known as the *Miller indices* of the plane (and by $(hkil)$ for hexagonal crystals). For cubic crystals, the (hkl) plane lies perpendicular to the direction $[hkl]$ as shown in Figure 2.37. Although the (0001) plane shown in Figure 2.36 happens to be perpendicular to the [0001] direction in a hexagonal crystal, there is not a simple relationship like this for all of the planes in either the hexagonal crystal system or the other noncubic crystal systems.

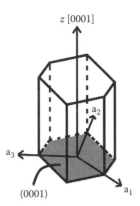

FIGURE 2.36 The Miller-Bravais coordinate system for hexagonal crystals, which consists of four axes, three of which (namely a_1, a_2, and a_3) are in the basal plane of the unit cell and are set at 120° to each other. The direction along the z-axis [0001] and the (0001) plane perpendicular to this direction are indicated.

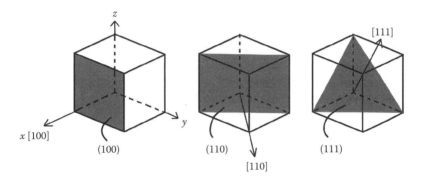

FIGURE 2.37 Some important planes in a cubic crystal. The (hkl) planes are at right angles to the [hkl] directions.

As with the system for defining directions, the basic method of labelling planes is, however, the same for all crystal systems. (We will look at the cubic system in the following example simply because it is particularly easy to visualise.) If a cubic unit cell with cube edges of length "a" is considered, the plane (hkl) cuts the x-axis at a/h, the y-axis at a/k, and the z-axis at a/l. For example, the (110) plane (see Figure 2.37) cuts the x-axis at a/h, which in this case is $a/1$ (or in other words "a"), and the y-axis at a/k, which as $k = 1$ is also a. However, the (110) plane does not cut the z-axis at all, as $a/0$ is ∞.

Any two planes will be equivalent if the arrangement of atoms within each plane is the same. So in many cases planes that are parallel to one another are equivalent, and will in fact have the same Miller indices. Sets of equivalent planes sit inside curly brackets {hkl} or {$hkil$}. For example, the set of planes (100), (010), (001) in a cubic crystal is {100}. In this case the planes are not parallel—they are in fact various cube faces—but because of the cubic symmetry, the arrangement of atoms on each face is identical.

EXAMPLE QUESTION 2.7   MILLER INDICES

What are the Miller indices of the plane shown in this diagram? Assume that the perspective of the illustration is being ignored in favour of the sides of the unit cell being in true proportion to one another.

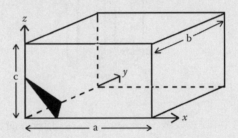

ANSWER

Having been told that the sides of the unit cell are in proportion to one another, the easiest approach is to measure the axes of the unit cell with a ruler. This should reveal that the plane cuts the x-axis at $a/4$, the y-axis at $b/2$, and the z-axis at $c/2$. Although this is an orthorhombic unit cell rather than a cubic one, its planes will be labelled in the same way. This means that any plane will intercept the unit cell axes at a/h, b/k, and c/l. The Miller indices of this plane are therefore 4, 2, and 2, so this is the (422) plane.

It is possible to view planes of atoms directly, and Figure 2.38 shows a scanning tunnelling microscope image of a plane of silicon atoms. Any other plane in the crystal that has exactly the same arrangement of atoms will be equivalent to this, and so would form part of the same set.

Atomic Coordinates

There are times, such as during investigations on crystal structures, when it is useful to describe exactly where individual atoms are. This can be indicated by a set of three coordinates (S_x, S_y, S_z), which are like a 3-D version of a map grid reference. Any given set of atomic coordinates describes how far that particular atom is from the origin, which has coordinates of 000, of the unit cell it is associated with.

The first number in the set reveals how far the atom is along the x-axis, while the second and third numbers indicate its distance along the y- and z-axes, respectively. Since atomic coordinates are always given in relation to a unit cell, the maximum value of any coordinate is 1, which is on the surface of the cell. Any fractions describe positions part way along any of the unit cell axes. For example, a position halfway along any axis will be denoted by a coordinate of ½. This means the atom at the centre of a bcc unit cell is at ½½½, and the atom at the front face-centre of

FIGURE 2.38 This image of silicon by Lloyd Whitman, Steven Erwin, and A. Baski was one of the winners in the U.S. Naval Research Laboratory's "Science as Art" competition. It is entitled "Sunrise over Si(114)" and is reproduced with kind permission of the Naval Research Laboratory (NRL). The following caption (also reproduced with permission) is the NRL description: An artistic view of the atomic-scale topography of Si(114) as revealed by scanning tunnelling microscopy (STM). (An artificial background and computer-simulated lighting have been added.) The tallest rows of atoms are 1.63 nm apart, with the atoms visible along these rows separated by 0.77 nm. Due to the rapidly shrinking size of electronic devices, an understanding of the atomic-scale structure of semiconductor surfaces is vital to the development of future Navy electronics. We have used STM and first-principles theoretical calculations to determine the structure of all Si surfaces oriented between the (001) and (111) planes. A number of new stable surfaces, including Si(114), have been discovered, and are being investigated as possible substrates for novel Si-based electronics.

the fcc lattice shown in Figure 2.20(a) will have coordinates of 1½½. Meanwhile, the atom in the corner diagonally opposite the origin of any cubic unit cell is at position 111.

Another way of thinking of atomic positions is in terms of a position vector, which has the same form as the vector defined in equation 2.1. If such a vector were indicating the location of an atom rather than the position of a lattice point, u_1, u_2, and u_3 would have to represent the integers 0 or 1 or any fractional real number in between, rather than any integer. (See Appendix A for a reminder of position vectors.)

FURTHER READING

Hogan, J. "Space station unlocks new world of crystals". *New Scientist* no. 2394 (10 May 2003): 17.
(More references for further reading, as well as Web links, are available on the following Web page: http://www.crcpress.com/product/isbn/9780750309721.)

EXAMPLE QUESTION 2.8   ATOMIC COORDINATES

Identify the cubic structure that has atomic coordinates of: 000, 100, 110, 010, 001, 101, 111, 011, ½½1, ½½0, ½0½, ½1½, 1½½, 0½½.

ANSWER

Sketching out a cube, adding axes, then plotting each of the atoms as follows will reveal the face-centred cubic structure.

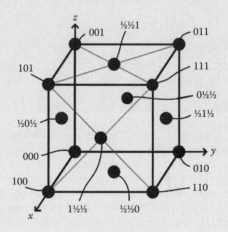

SELECTED QUESTIONS FROM
QUESTIONS AND ANSWERS MANUAL

Q2.1

(a) What structure does the compound caesium iodide crystallise in?
(b) What does the term "coordination number" mean, and what is its value for the lattice representing your answer to part (a)?
(c) What type of bonding does CsI have?
(d) What is the electronic configuration of the two constituents of the compound when they are in solid form?

Q2.5   The cohesive energy of a solid is 9 eV/atom. What does this tell you about the solid?

Q2.9   Assuming iron (Fe) has a lattice parameter, a, of 0.287 nm, what is the atomic radius and atomic density of an iron crystal?

Q2.11   Calculate the atomic density of a simple orthorhombic crystal with a = 1.046 nm, b = 1.288 nm, and c = 2.448 nm.

Q2.14 For a cubic crystal lattice, what do the following represent?

(a) <111>
(b) [010]
(c) (111)
(d) {100}

3 The Rejection of Perfection

Defects, Amorphous Materials, and Polymers

CONTENTS

3.1   DEFECTS

We tend to see defects as a negative thing. If something fails to work properly we say that it is "defective", and it is hard to open a magazine or newspaper without coming across an article encouraging us to remedy a range of supposed defects in either our characters, dress sense, handling of finances, or attitude towards our health. The *Oxford English Dictionary* reinforces this view when it describes a defect as a "lack of something essential or required". But the reputation defects seem to have gained is not well deserved. In physics, the opposite of the dictionary definition tends to be true—often the defect itself is in fact essential and required.

For example, it is impurities in semiconductors that give them the electrical properties needed for semiconductor devices such as transistors to work. Meanwhile the colour of some crystals—including the gemstone ruby—is created by defects in their structure. This does not mean, however, that defects always have desirable properties. One type of defect, for instance, makes single crystals many times weaker mechanically than they would be if they had a perfect structure.

There are two major classes of defect: *point defects* and *dislocations*. Point defects (which are discussed in the next section) are defects that involve either a single atomic site or a small number of sites. By contrast, dislocations (which we will look at in Section 3.1.2) are defects that involve a whole line of atoms.

3.1.1   POINT DEFECTS

Schottky Defects, Frenkel Defects, and Impurities

The simplest type of point defect is a *Schottky defect*. Schottky defects are more commonly known as vacancies, and as this name suggests are vacant lattice sites left by missing atoms (see Figure 3.1). (In discussions on defects, the terms "lattice site" and "atomic site" are both taken to mean atomic positions that would be occupied by an atom of a given crystal if that crystal had a perfect structure.)

The atoms around any vacancy tend to alter their positions slightly (as you will see if you look closely at Figure 3.1), so in general there will be some distortion of the crystal structure near Schottky defects. There is also another type of atomic movement that can occur when vacancies are present: Atoms can "diffuse", which means they can migrate through a crystal by moving from one atomic position to another. This is because very little energy is needed for an atom to leave its normal lattice site and move into a Schottky defect. Of course when it moves, an atom leaves behind its own vacancy, so the defect is effectively moving through the crystal in the opposite direction to the atom.

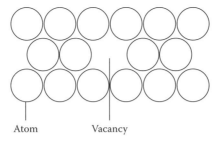

Atom   Vacancy

FIGURE 3.1   Part of an atomic plane in a crystal that contains Schottky defects, which are also known as vacancies.

There tends to be a stream of vacancy movement when there is a concentration gradient—in other words when there are more vacancies in one area of a crystal than another. As diffusion always occurs from regions of high concentration to those of low concentration, this movement means crystals with concentration gradients of vacancies eventually end up with a more even spread of vacancies throughout their volume.

Every crystal, whether it has grown naturally or been artificially created, contains Schottky defects, and the number of vacancies increases as the temperature increases. This is because the heat provides the energy needed to form more defects.

The equilibrium number of vacancies, n, for a crystal at a temperature, T, is given by

$$n = N \, \exp\left(-E_v/k_B T\right) \tag{3.1}$$

where N is the total number of lattice sites, E_v is the energy needed to create the defect in the particular solid under consideration, and k_B is Boltzmann's constant. For many metals there can be as many as one vacancy in every 10,000 atoms when they are close to their melting points.

Vacancies tend to be produced in pairs in ionic solids, with one positive ion vacancy and one negative ion vacancy per pair. This prevents the solid from ending up with an overall electrostatic charge, which would occur if one ion type had more vacancies than the other. (An overall charge cannot develop like this because it would increase the total electrostatic energy of the crystal and crystals always try to minimise their energy.) As we will see in Chapter 6, the presence of vacancies in ionic crystals increases their electrical conductivity.

A *Frenkel defect* also involves a vacant lattice site, but in this case there is an *interstitial* atom—that is, an atom sitting in a position between normal lattice sites—near the vacancy (see Figure 3.2). The interstitial atom in a Frenkel defect is known as a "self-interstitial", as it is one of the component atoms of the crystal. In fact it is the interstitial atom leaving its site in the lattice that causes the vacancy to form. Irradiating a crystal can create Frenkel defects, as can large amounts of deformation. The equilibrium number of Frenkel defects, n, for a crystal at a temperature T is given by

EXAMPLE QUESTION 3.1   VACANCY CONCENTRATIONS

If you have a 1 cm^3 sample of aluminium, how many vacancies would it contain at (a) 300K and (b) 200K if the energy needed to create the vacancy is 0.75 eV and there are 10^{22} atomic sites in every cubic centimetre of the sample?

ANSWER

The energy needed to create a vacancy is given as 0.75 eV = $0.75 \times 1.602 \times 10^{-19}$ = 1.2×10^{-19} J. With this value converted to SI units, we can now work out the number of vacancies for each of the two temperatures:

(a) at 300K

$$n = N \exp\left(-E_v / k_B T\right) = 1 \times 10^{22} \exp\left(-1.2 \times 10^{-19} / 1.381 \times 10^{-23} \times 300\right)$$

$$= 1 \times 10^{22} \exp(-28.96)$$

$$= 2.65 \times 10^9 \ \text{vacancies/cm}^3$$

(b) at 200K

$$n = N \exp\left(-E_v / k_B T\right) = 1 \times 10^{22} \exp\left(-1.2 \times 10^{-19} / 1.381 \times 10^{-23} \times 200\right)$$

$$= 1 \times 10^{22} \exp(-43.45)$$

$$= 1.35 \times 10^3 \ \text{vacancies/cm}^3$$

It is worth noting that in both cases the number of vacancies is a very small proportion of the number of atoms. In fact there is one vacancy for every 3.8×10^{12} atoms at 300K, while at 200K the number of vacancies is 1 in 7.4×10^{18}. (If you are ever asked a question like this in an exam, you can quickly check that you have not gone horribly wrong in your calculations by making sure the answer reveals the fact that the number of vacancies increases as the temperature increases.)

$$n = \sqrt{\left(NN'\right)} \exp(-E_I / 2 k_B T) \qquad (3.2)$$

where N is the total number of lattice sites, N' is the number of interstitial sites, E_I is the energy needed to move an atom away from its lattice site and into an interstitial position, and k_B is Boltzmann's constant. In contrast to Schottky defects, Frenkel defects do not have to exist as pairs in ionic crystals because the missing electric charge at the vacancy is simply sitting a bit further away, as it is the charge on the interstitial.

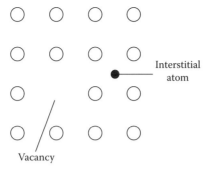

FIGURE 3.2 A Frenkel defect.

Self-interstitials are not the only atoms that can occupy interstitial sites, however. Impurity atoms can become incorporated into crystals either as interstitials or by replacing the "host" atoms. (Any crystal that impurities enter can be referred to as the "host" as it is effectively hosting the impurity atoms.) Impurities that replace atoms at normal lattice sites are known as *substitutional* impurities.

Like vacancies, interstitial atoms lead to a distortion of the part of the crystal structure that surrounds them. In fact self-interstitials cause such a large distortion that they rarely form. By contrast, it is relatively easy for impurity atoms to sit in interstitial positions in the lattice if they are smaller than the atoms of the host crystal. For example, carbon, which has an atomic radius of approximately 0.71 Å (0.071 nm), can be added to iron, which has an atomic radius of around 1.24 Å (0.124 nm), to make steel. The carbon sits in interstitial positions in the iron, and like other interstitial impurities in crystals is able to diffuse through it relatively easily by moving from one interstitial site to another.

Impurity atoms are often added to pure solids to enhance their properties. For example, carbon is added to iron to make it harder, while various different impurities are added to semiconductor materials like silicon and gallium arsenide to change their electrical conductivity (the "doping" of semiconductors with impurities will be discussed in detail in Chapters 6 and 7). Impurities also occur naturally; for instance, chromium ions in aluminium oxide (Al_2O_3) cause the otherwise transparent crystals to turn red and become known as rubies. The chromium ions substitute for aluminium ions, and although this causes a small amount of strain to the crystal structure, up to 1% chromium can be incorporated.

Colour Centres

It is not just impurity atoms that can change the colour of crystals. Point defects known as *colour centres* can cause transparent crystals to appear a different colour than they would be without these defects present. This is because colour centres can absorb some wavelengths of light and not others, and when we look at a solid the colour it appears is a combination of the wavelengths of visible light that it reflects and/or transmits and/or reemits after absorption. (Different processes occur depending on the solid in question and the conditions it is under.)

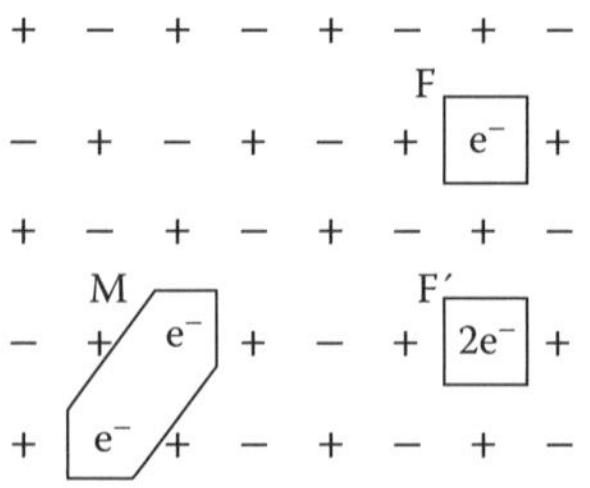

FIGURE 3.3 The F centre (top right), in which an electron becomes trapped at a negative ion vacancy; the F′ centre (bottom right), which consists of two electrons trapped at a negative ion vacancy; and M centre (left), which is composed of two adjacent F centres. (Adapted from Figure 7.1 in Townsend, P. D., and Kelly, J. C., *Colour Centres and Imperfections in Insulators and Semiconductors*, Chatto & Windus for Sussex University Press, London, 1973. Reproduced with permission.)

Colour centres can be produced in several ways. These include the addition of impurities and bombarding crystals with X-rays, gamma rays, neutrons, or electrons. In the latter case the radiation damages the crystal structure and causes defects of various sorts to form. For example, diamonds can become blue if they are irradiated with electrons. (A heat treatment may also be required to complete this colour change; heat treatments are discussed in the next chapter.) Naturally occurring colour centres also exist; for instance, a rare form of fluorite (CaF_2) called "Blue John" has a bluish/purple colour because of the presence of colour centres known as F centres.

There are various types of colour centre, and the F centre—named after the German word "Farbe", which means colour—was the first to be understood. It consists of an electron trapped at a negative ion vacancy (top right in Figure 3.3). The electron makes sure the ionic crystal remains electrostatically neutral. Other colour centres include the F′ centre (bottom right in Figure 3.3), which is like the F centre but has two electrons trapped at the vacancy, and the M centre (left in Figure 3.3), which consists of two adjacent F centres bound to each other. There is also the R centre, which is made up from three F centres, and the N centre, which is a cluster of four F centres.

Nonstoichiometry

When there is a simple ratio between the different elements of a compound, and the atoms of the compound are present in the correct numbers for this ratio to hold true, that compound is said to be "stoichiometric". For example, a perfect calcium fluoride crystal, which has a chemical formula of CaF_2, will have two fluorine atoms for every atom of calcium throughout its structure.

By contrast, "nonstoichiometric" compounds like iron oxide, which has an approximate chemical formula of $Fe_{1-x}O$ (where x is a fraction less than 1), have a more complex relationship between their constituent atoms. In this case, some of the iron ions within the compound can exist as Fe^{2+} while others are Fe^{3+}, as shown in Figure 3.4. If we consider a small region of an iron oxide crystal like this, we can see that if two of the Fe ions in this particular region have a valency

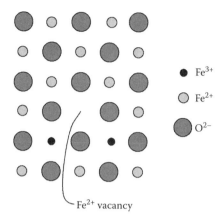

FIGURE 3.4 Part of the nonstoichiometric compound iron oxide. An Fe^{2+} vacancy must be formed for every two Fe^{3+} ions present in order to keep the crystal electrostatically neutral.

of +3, then an Fe^{2+} vacancy needs to be formed in order for the crystal to remain electrically neutral overall. In this case the two excess positive charges from the two Fe^{3+} ions are effectively cancelled out by the two missing positive charges from the Fe^{2+} vacancy.

As there is an ion missing from this part of the crystal, the chemical formula FeO does not accurately represent the structure because there is an additional oxygen ion in comparison with the number of iron ions. The x in the formula $Fe_{1-x}O$ therefore reduces the number of iron ions overall when there are Fe^{3+} ions present.

While some compounds are naturally nonstoichiometric, nonstoichiometric compounds can also be made from stoichiometric compounds like calcium fluoride by deliberately adding additional atoms of one element.

Other Defects

Grain boundaries and even the surfaces of solids can be classed as defects, and under certain circumstances, the fact that atoms vibrate about their usual positions in the lattice can be thought of as a defect. This is because the vibrations are displacing the atoms, and some of the properties of solids are dependent on the actual positions of the atoms in the lattice at any given time. In fact, the properties that will be affected are those in which the time scale for whatever interaction with the lattice gives rise to the macroscopic property is small compared with the period of the atomic vibration.

For example, an electron moving through a metal as part of an electric current will only take around 10^{-16} s to move between two atoms about 10^{-10} m apart if it is moving at approximately 10^6 ms^{-1}. Since the vibrations of atoms in a lattice at room temperature have a period of vibration of about 10^{-13} s, then the positions of the atoms will affect the passage of the electron through the solid. Because the size of the vibrations increases as the temperature increases, any property—such as electron transport—that is affected by lattice vibrations will be altered more when the crystal is at higher temperatures. In fact, we will see in Chapter 6 that electrons are scattered by these atomic vibrations, and so the electrical resistance of metals increases with increasing temperature.

Although we are concentrating in this section on defects on an atomic scale, it is worth bearing in mind that there are also much larger scale defects that solids can contain. These include cracks and pores. Macroscopic defects like this are often referred to as "bulk" or "volume" defects, and can have catastrophic effects on the mechanical properties of solids, as we will see in Chapter 4.

3.1.2 DISLOCATIONS

Edge Dislocations and Screw Dislocations

We have just seen that there are many different types of point defect. By contrast, there are only two separate kinds of dislocations: edge dislocations and screw dislocations. The edge dislocation (shown in Figure 3.5) is the easier of the two to visualise. It can be thought of as either the addition of an extra plane of atoms into one section of a crystal, or alternatively as the removal of part of a plane of atoms. Like point defects, dislocations (which are sometimes referred to as "line defects") cause distortion of the crystal structure. You can see this in Figure 3.5 where the planes of atoms either side of the dislocation effectively bend round the end of the dislocation and come back into a position that restores the usual structural arrangement of the crystal a short distance away from it.

The symbol ⊥ is sometimes used to represent an edge dislocation. The vertical line in this symbol indicates the *dislocation line*, which is the line running along parallel to the end of the extra part-plane of atoms, and for the example shown in Figure 3.5 runs perpendicular to the plane of the paper into the crystal. Crystals are not the only materials in which defects looking like edge dislocations can form. As Figure 3.6 shows, similar changes in pattern can occur on a larger scale. In this case

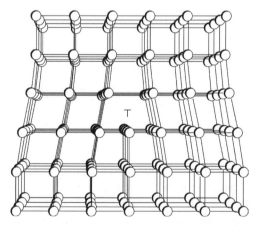

FIGURE 3.5   An edge dislocation in a simple cubic crystal. (Adapted from Figure 9.3, p 105, Saxby, G., *The Science of Imaging, An Introduction*, Institute of Physics Publishing, Bristol and Philadelphia, 2002. With permission.)

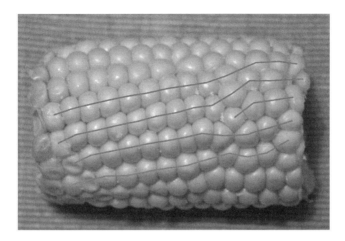

FIGURE 3.6 This picture of a corn-on-the-cob shows a larger scale equivalent of an edge dislocation. Like an extra plane of atoms forming in part of a crystal, part of an extra row of kernels has grown. (Courtesy of Alan Piercy.)

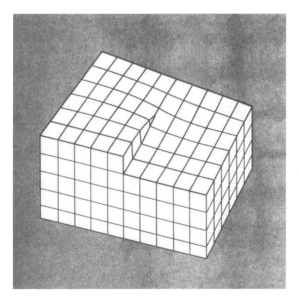

FIGURE 3.7 A screw dislocation. (This figure originally appeared as Figure 18.17, p. 433, Beiser, A., *Perspectives of Modern Physics*, 1981, McGraw-Hill, © The McGraw-Hill Companies, Inc. Reproduced with permission.)

part of an extra row of kernels has grown between two otherwise regularly spaced rows on this corn-on-the-cob.

The second type of dislocation, the screw dislocation, can be seen in Figure 3.7. A screw dislocation can be thought of as a cut part way into a crystal where the region of the crystal one side of the cut is shifted by one atomic spacing relative to the region of crystal on the other side. Screw dislocations get their name from the fact that this displacement causes the atomic planes near the dislocation to spiral around the dislocation line (which runs parallel to the original cut), just like the threads spiral around the body of a screw. The symbol C can be used to represent a screw dislocation.

Although both distinct types of dislocation can exist, real dislocations in crystals are often "mixed dislocations", which means they are a combination of edge and screw dislocations. Sometimes dislocations end on the surface of a crystal, but they can also form closed loops within crystals. As we will see in Chapter 4, dislocations have a major effect on the mechanical properties of many solids.

Dislocations and Crystal Growth

In order to grow a crystal, something needs to be used as a "seed" for the atoms that will make up the crystal to attach themselves to and gradually build up on. For example, to produce a large single crystal, a much smaller piece of seed crystal can be exposed to a vapour containing the same type of atoms as itself. The atoms from the vapour can condense onto a perfect atomic plane, as shown in Figure 3.8(a), and so build up another plane of atoms directly on top of it. However, they will condense much more readily at atomic sites on the seed crystal that already have some nearest neighbours present because they will be more strongly attracted to these nearest neighbour atoms. This situation is shown in Figure 3.8(b), where a screw dislocation has reached the crystal surface and new planes of atoms spiral around the screw dislocation as more and more atoms are deposited, producing growth patterns like those shown in Figure 3.9.

The growth rates of some crystals are much faster than would be expected if they were perfect and were therefore growing via the relatively slow mechanism illustrated in Figure 3.8(a). In these cases the crystals are likely to contain screw dislocations, which are allowing the new atomic planes to spiral continuously upwards.

Dislocation Density

Every crystalline solid will contain dislocations, and the quantity of dislocations present can be expressed in terms of the *dislocation density*. This is defined as the

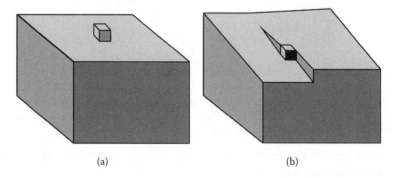

(a)   (b)

FIGURE 3.8 Crystal growth, with an atom (a) weakly attracted to a perfect crystal plane, and (b) strongly attracted to a screw dislocation. (This illustration was published in *Solid State Physics*, International Edition, Neil W. Ashcroft and N. David Mermin, p. 636, copyright W. B. Saunders Company (Elsevier), 1976. Reproduced with permission.)

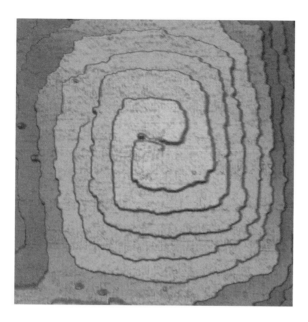

FIGURE 3.9   This image by P. M. Thibado, B. R. Bennett, B. V. Shanabrook, and L. J. Whitman was one of the winners in the U.S. Naval Research Laboratory's "Science as Art" competition. It is entitled "Double Spiral of GaSb on GaAs" and is reproduced with the kind permission of the Naval Research Laboratory (NRL). The following caption (also reproduced with permission) is the NRL description: A scanning tunneling microscopy image of the surface of a 4 μm-thick GaSb film grown by molecular beam epitaxy on GaAs (001). The image, 0.5 μm × 0.5 μm, was acquired in situ following growth of the film. Spiral-like structures, such as the one seen here, grow around threading dislocations in the film caused by the 7% lattice mismatch with the substrate. Note the pair of dislocations emerging from the surface at the top of the spiral, along with a third emerging half-way down on the side; each one creates a 0.3-nm-height "step". In situ atomic-scale characterization is an important component in the development of growth procedures that minimize the dislocation density in heteroepitaxial layers. (See Section 7.3.2 for more on epitaxial growth methods.)

number of dislocation lines intersecting a randomly chosen and therefore randomly oriented unit area of a crystal. In good-quality silicon and germanium crystals that have been grown under carefully controlled conditions for use in the semiconductor industry, the dislocation density is much less than 10^2 cm^{-2}. By contrast, the dislocation density of metal crystals that have been heavily deformed—which occurs when metal objects are made by hammering or machining metal sheets into specific shapes—can be as large as 10^{12} cm^{-2}. In general, the dislocation density is higher in metal crystals than in nonmetallic crystals. Various experimental techniques including electron microscopy can be used to detect dislocations, and therefore enable dislocation densities to be estimated.

We have already seen that dislocations cause distortion of the crystal structure, and crystals will always seek to minimise any distortions because they increase the energy of the crystal. The tendency towards minimum energy means that dislocations make themselves as short as possible, and that large dislocations are likely to

EXAMPLE QUESTION 3.2   DEFECTS

If the spheres in the pictures below represent the atoms in a crystal plane, which type of defect does each image reveal, and where is it?

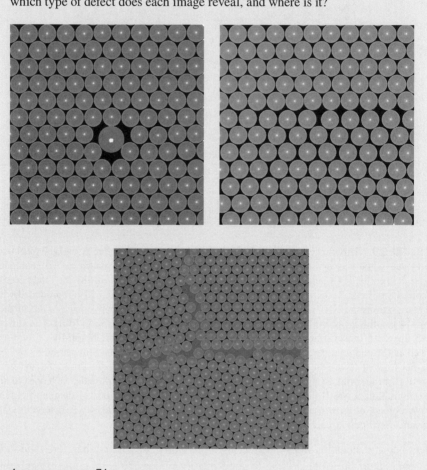

Answer on page 74.

break up into several smaller dislocations. It also means that dislocations that have formed as closed loops will contract whenever they can.

Stacking Faults

In Chapter 2 we saw how various crystal structures can be built up layer by layer. For example, Box 2.2 shows how layers of atoms need to be stacked to form the face-centred cubic (fcc) structure. In a perfect fcc crystal, the pattern of layers will repeat in the sequence layer 1, layer 2, layer 3, layer 1, layer 2, layer 3, layer 1, etc., throughout the entire crystal. However, real fcc crystals can contain a defect in this pattern, which occurs when either a layer in the sequence is missing (as illustrated

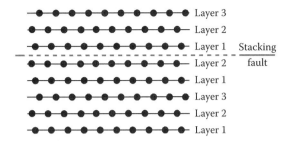

FIGURE 3.10   One type of stacking fault which can occur in an fcc crystal.

in Figure 3.10) or an additional layer is introduced. This type of defect is known as a "stacking fault".

The crystal structure depicted in Figure 3.10 is a mixture of the regular fcc layering of layer 1, layer 2, layer 3, layer 1, etc., and some hexagonal close-packed (hcp) stacking of layer 1, layer 2, layer 1, etc. (see Figure 2.2.7). Hexagonal close-packed structures themselves can also contain stacking faults. In this case an fcc layer 3 can appear instead of the hcp layer 2 or alternatively be introduced as an additional layer.

In general, for close-packed structures which have a hexagonal array of atoms as layer 1, there is no geometrical reason why the next layer of atoms could not sit equally well in the positions of layer 3 of an fcc lattice as in the positions of layer 2 of an fcc lattice. By contrast, you would not expect to find stacking faults in bcc structures because there are no alternative sites for the atoms in the layer resting on top of layer 2 (see Figure 2.18).

3.2   AMORPHOUS MATERIALS

3.2.1   STRUCTURE OF AMORPHOUS MATERIALS

Although many solids are crystalline, there are also plenty that are amorphous. The word "amorphous" comes from both the Latin and Greek words for "shapeless", and is generally taken to mean something without an organized structure. Despite having this label, the atoms are not in fact completely randomly arranged in amorphous materials. However, they are arranged in a more random way than the atoms of a crystalline solid.

Amorphous materials can either be naturally occurring—like the volcanic mineral obsidian—or artificially created from crystalline solids. In fact almost all crystalline solids can be made amorphous, and the amorphous version of a particular material can have very different properties to the crystalline material it was produced from, despite sharing the same composition. Amorphous polymers for instance tend to be transparent, whereas polymers with a mainly crystalline structure are generally opaque or translucent.

In amorphous solids, the interatomic forces and bonds between atoms are very similar to those in crystals. This means that the number and positioning of nearest neighbours for any given atom in an amorphous solid is on average the same as that of another atom in the solid. However, because there are slight differences in

ANSWER TO QUESTION 3.2 ON DEFECTS

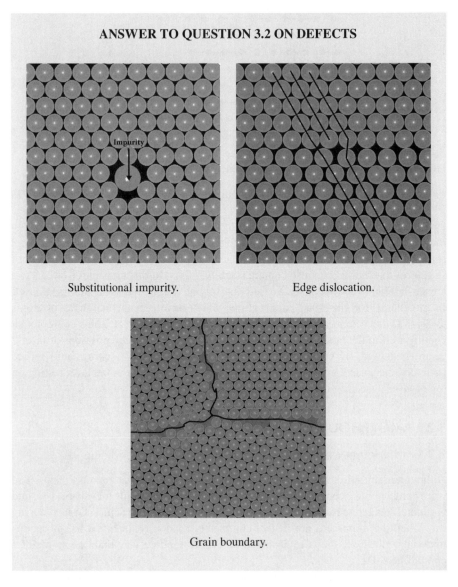

Substitutional impurity.   Edge dislocation.

Grain boundary.

bond angles and atomic spacings throughout the solid, it is impossible to predict the position of distant atoms from any given point in the solid. Because the immediate environment of each atom is the same, we say that amorphous materials possess "short range order" (SRO), but because of the changes in atomic spacing and orientation at a larger scale throughout amorphous materials, they are said to lack "long range order" (LRO).

The theoretical understanding of amorphous solids is much less developed than that of crystalline solids. However, the next section should at least give some idea of the structure of this interesting class of materials, which already have a variety of important technological applications—and look set to find more in the future.

FIGURE 3.11 A dense random packing of hard spheres can be achieved by pushing toy balls randomly into a plastic sack until no more can be squeezed in.

3.2.2 Models of Amorphous Structures

Dense Random Packing of Hard Spheres Model

Because of the absence of long-range order, the structure of an amorphous material cannot be represented by a lattice. Instead, two models are commonly used to represent the positions of the atoms in amorphous structures: the *dense random packing (DRP) model* and the *continuous random network (CRN) model*.

As with crystalline solids, the structure of an amorphous solid is mainly governed by the type of bonding that holds its atoms together. For amorphous materials in which the bonding is nondirectional—that is, for amorphous metals, ionic solids, and molecular solids held together by van der Waals bonding—the DRP model (see Figure 3.11) provides the best representation of the structure. By contrast, for covalently bonded materials like silicon, which have directional bonding, the CRN model (see next subsection) is used to represent the structure.

The DRP model uses hard spheres to represent atoms. This means that the "atoms" do not become distorted as they are pushed together. Early DRP models (which were first made in the 1950s) consisted of ball bearings placed in a bag which was squeezed until the balls could be pushed no closer. If the packing this process achieved was noncrystalline, the balls would then be glued into place to form the DRP model. Nowadays, computer simulations tend to be used instead, and Figure 3.12 shows how the simulated models are built up from a "seed" of three spheres.

As each new sphere is added, it has to touch three neighbouring spheres and be so close to them that there is not enough space left to fit another sphere into any gaps that are left in the structure. These criteria produce fairly closely packed structures,

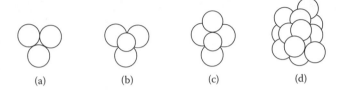

(a)   (b)   (c)   (d)

FIGURE 3.12 Computer simulations of the DRP model are built up from a "seed" of three spheres as shown in (a). When a new sphere is added, it has to satisfy the criteria that it touches three spheres already in the model (see [b] and [c]), and leaves no space between itself and its neighbours that is large enough to be filled up by another sphere, as illustrated in (d).

TABLE 3.1

Comparison of Packing Fractions for Some Common Crystalline and Amorphous Structures

Structure	Packing Fraction
fcc	0.74
bcc	0.68
DRP	0.61–0.64
sc	0.524

---

with packing fractions (see Section 2.2.2) of around 0.61–0.64. Table 3.1 lists the packing fractions for some common crystalline and amorphous structures.

Each of the spheres representing the atoms in the DRP model has a certain volume, so short-range order—in which the inter-atomic distances are approximately the same as for a corresponding crystalline solid—is produced. This is simply because there are only a limited number of ways in which a very small handful of spheres can arrange themselves and still fulfil the two conditions previously mentioned. However, there is no long-range order, as the bond angles are not consistently the same throughout the solid.

Continuous Random Network Model

Like the DRP structure, the continuous random network (CRN) structure has inter-atomic distances very similar to those in a crystalline solid. In this case, however, the bonds between the atoms are directional, but the interbond angles are not the same as those in a crystalline solid of the same composition, and are different in different parts of the amorphous structure, leading to the required absence of long-range order. Despite this variation, the interbond angles usually remain within ±10% of the values they would have in a crystalline version of the same solid. This leads to the bonds between some atoms bending, and a range of different bond lengths.

As with the DRP structure, CRN models may be constructed physically—using balls to represent the atoms and spokes between them to represent the bonds—or produced by computer simulation.

It is not possible to describe all types of amorphous materials using either the DRP or CRN model, but they are two of the most widely used models. Whatever model is used, however, it can only ever be an approximation of the true structure. This is because in order to build up any type of model of an amorphous structure, scientists need to have experimental data on the structure of amorphous materials to use as a starting point, and this is not as complete as the information that can be obtained on crystalline solids. As we saw in Chapter 2, knowing the positions of the atoms that make up the unit cell of a crystal is enough to determine the positions of the atoms throughout the whole solid, as the pattern of atoms in the unit cell repeats again and again. By contrast, as we have just seen, the atomic positions in a localised area of an amorphous solid bear little relation to the structure as a whole, since amorphous solids lack long-range order.

Defects in Amorphous Solids

Like crystalline solids, amorphous solids can contain defects. In some ways the idea of defects in an amorphous structure is a strange one because the entire structure can be thought of as being "defective" in itself. However, if the CRN or DRP models are taken to represent an "ideal" amorphous solid, then any departure from the structures they predict can be thought of as a defect.

The simplest types of defect in amorphous structures are vacancies, which both the CRN and DRP structures can contain, and "dangling bonds", which can only occur in materials with the CRN structure.

Dangling bonds are produced when there is a vacancy. The missing atom would normally bond with a neighbouring atom, but because there is no atom there, the neighbouring atom is left with an incomplete bond containing fewer electrons than are needed to fill an electron shell and so complete the bond. Dangling bonds are therefore lacking in electrons, and so tend to trap any impurity atoms in the amorphous material that contain the extra electrons they require.

Impurity atoms can also be found in amorphous materials, and can even be deliberately added to amorphous semiconductors—just like they can be added to crystalline semiconductors—to alter their electrical properties. Amorphous selenium, for example, is the basis of the photocopier, while solar cells are often made from amorphous rather than crystalline silicon because it is cheaper to produce and can be made into thin films. These films are larger than single crystals and therefore enable bigger devices to be built, which can capture more light. (Semiconductor applications will be discussed in detail in Chapter 7.)

3.2.3 GLASSES

One of the most commonly used amorphous materials is glass. When we use the word "glass" like this in everyday language, we mean the silica glass that windows and bottles are made out of. However, in physics, the word "glass" has a much more general meaning. It is used to describe any amorphous solid that has been made by cooling a liquid so quickly that instead of becoming a crystalline solid, the liquid becomes an amorphous solid as the *glass transition temperature* (T_g) is passed. So glasses are all amorphous materials, but not all amorphous materials are glasses.

EXAMPLE QUESTION 3.3   AMORPHOUS POLYMERS

You have a sample of polythene with a density of 970 kg m^{-3} that you know is part crystalline and part amorphous. What fraction of the total mass of the sample is crystalline, given that crystalline polythene has a density of 1000 kg m^{-3} and amorphous polythene has a density of 865 kg m^{-3}?

ANSWER

The first step is to work out a formula for mass fractions of components in a mixture of known density as follows:

$$\text{Density of mixture} = \rho_M = \frac{\text{Total mass}}{\text{Total volume}} = \frac{M_M}{V_M}$$

If the mass fraction of component A in the mixture is F_A, then

$$\text{Mass of component A in the mixture} = M_A = F_A M_M$$

and

$$\text{Mass of component B in the mixture} = M_B = F_B M_M = \left(1 - F_A\right) M_M$$

Hence the volumes of components A and B in the mixture are

$$V_A = \frac{M_A}{\rho_A} = \left(\frac{F_A}{\rho_A}\right) M_M \text{ and } V_B = \frac{M_B}{\rho_B} = \left(\frac{F_B}{\rho_B}\right) M_M$$

so that the total volume is

$$V_M = V_A + V_B = M_M \left[\left(\frac{F_A}{\rho_A}\right) + \left(\frac{F_B}{\rho_B}\right)\right] = M_M \left[\frac{F_A \rho_B + F_B \rho_A}{\rho_A \rho_B}\right]$$

and the density of the mixture is

$$\rho_M = \frac{M_M}{V_M} = \left[\frac{\rho_A \rho_B}{F_A \rho_B + F_B \rho_A}\right] \tag{1}$$

but $F_B = (1 - F_A)$, so that the denominator of equation (1) becomes

$$F_A \rho_B + \left(1 - F_A\right) \rho_A$$

$$= F_A \rho_B + \rho_A - F_A \rho_A$$

$$= F_A \left(\rho_B - \rho_A\right) + \rho_A$$

Equation (1) becomes

$$\rho_M = \left[\frac{\rho_A \rho_B}{\rho_A + F_A \left(\rho_B - \rho_A \right)} \right] \quad (2)$$

Equation (2) allows the density of a mixture to be determined for a given mass-fraction of component A. It may also be used to determine the mass-fraction from the known density of the mixture—which is what we need to do in order to answer this particular question.

This is achieved by cross-multiplying equation (2), which gives

$$\rho_M \rho_A + F_A \rho_M \left(\rho_B - \rho_A \right) = \rho_A \rho_B$$

$$\therefore F_A \rho_M \left(\rho_B - \rho_A \right) = \rho_A \rho_B - \rho_M \rho_A$$

and

$$F_A = \frac{\rho_A \left(\rho_B - \rho_M \right)}{\rho_M \left(\rho_B - \rho_A \right)} \quad (3)$$

In this case, component A is the crystalline polythene and component B the amorphous polythene. If we now add in the numbers, we have

$$F_A = \frac{1000 \left(865 - 970 \right)}{970 \left(865 - 1000 \right)} = 0.802$$

In other words, 80.2% of the polythene sample is crystalline.

To form a glass, the cooling rate must actually be much faster than it would be if a crystalline solid were to be formed. This is because glasses can only form from *supercooled liquids*, which are liquids that have been cooled so quickly that they are still liquid at a temperature below their normal freezing points because there has not been enough time for crystallization into a solid to occur. Figure 3.13 shows the phase diagram (diagram showing under what conditions—such as temperature and pressure, or temperature and volume—different phases of a material exist) for a glassy material.

The lower curve shows what happens when a liquid is cooled slowly. It changes phase into a crystalline solid when it reaches the melting point (T_m), then remains in the same state as the temperature decreases further. By contrast, the upper curve shows what happens when the liquid is cooled rapidly. When it reaches T_m, the liquid is unable to change phase, as it is being cooled so rapidly that there is no chance for crystallization to occur. So instead it becomes a supercooled liquid below its melting

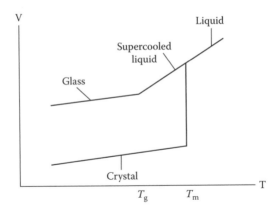

FIGURE 3.13   Phase diagram of a glassy material. If the liquid is cooled slowly, it will change phase into a crystalline solid when it reaches the melting point (T_m). This is illustrated by the lower curve. If, however, the liquid is cooled quickly, it will not have time to undergo a phase transition at T_m and will become a supercooled liquid. On further cooling, this supercooled liquid will form a glass as it is cooled through the glass transition temperature T_g.

point, which in turn changes into a glass as it is cooled through the glass transition temperature T_g.

As the diagram shows, there is not an abrupt change of phase at the glass transition temperature like there is at the melting point. This is because the glass transition temperature is not in fact a single temperature like T_m. Instead it is a range of temperature over which the supercooled liquid gradually changes into a glass. Despite the fact that T_g is not a single temperature, it is still used to represent the transition from a liquid to a glass. So for temperatures below T_g the material is a glass, whereas above T_g it is a liquid.

Although the cooling rate must be fast enough to prevent crystallization occurring, there are still a range of cooling rates (from 10^{-5} to 10^8 K s^{-1}) that can be used to form glasses, depending on the material. Within this range, the slower the cooling rate, the larger the supercooled region and the lower the value of T_g. In fact T_g can vary by up to 20% for different cooling rates.

The process of forming amorphous materials by rapid cooling is known as *melt quenching*, and was the first technique ever used to produce amorphous materials. As we will see in the next section, this is only one of the techniques that can be used to create amorphous materials.

3.2.4   PREPARATION OF AMORPHOUS MATERIALS

As with other types of amorphous materials, glasses can have all the same types of bonding as crystalline solids. But whilst glasses can only be formed by cooling from a molten form of the material, there are several techniques that can be used to prepare other types of amorphous materials. However, only five methods—thermal evaporation, sputtering, glow-discharge composition, chemical vapour deposition, and melt quenching—are regularly used in industry and academia.

Thermal Evaporation

Thermal evaporation involves vaporizing a starting compound (crystalline version of the material required) and collecting a thin amorphous film (with a width in the order of tens of micrometres) of the evaporated compound. It takes place inside a vacuum chamber to reduce the chance of the amorphous material being contaminated with other atoms. There are two ways to heat the compound to its vaporization point.

If the compound has a fairly low melting point (and hence a relatively low vaporization point), it is made into a powder and then placed in a "boat" made from a metal with a high melting point such as tungsten or molybdenum. An electrical circuit is then made by connecting electrodes—which are in turn connected to a DC current source—to each end of the boat. A large DC current is passed through the circuit and the boat acts like a resistor, becoming so hot that the compound in it is vaporized.

However, the boat does not become hot enough to vaporize compounds with high melting points, which have to be heated to their vaporization points by a beam of high-energy electrons from an electron gun instead. Whichever method is used, the vaporized compound is then collected on a cold substrate. The arrangement of the boat and substrate in a typical thermal evaporation chamber is shown in Figure 3.14.

The substrate is much colder than the vapour of the compound, so when atoms from the vapour arrive on the surface they are "frozen" into the random positions in which they hit the substrate. Since the atoms are too cold to move into the positions needed to form a crystalline solid, an amorphous solid is produced instead. In fact, to produce amorphous semiconductors—such as amorphous selenium (which can be written as a-Se for short), amorphous germanium (a-Ge), and amorphous gallium arsenide (a-GaAs)—the substrate is often held at about room temperature. These

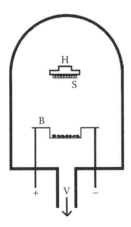

FIGURE 3.14   In this diagram of a thermal evaporation chamber, the substrate (S) is at the top of the chamber above the boat (B), which contains the compound to be vaporized. The temperature of the substrate is controlled by the attached heater (H), while the electrodes that enable a DC current to be passed through the boat, and so heat it, are at the bottom of the chamber. This diagram originally appeared as figure 1.5, p. 9 in Elliott, S. R., *Physics of Amorphous Materials*, 2nd ed., Longman Group UK Limited, 1990. Reproduced with kind permission of S. R. Elliott.

BOX 3.1   REGENERATION SCHEME

If you were asked to list materials that can be used inside the human body as medical implants, it is unlikely that glass fibres would be one of your choices. That is unless you were Prof. Jonathan Knowles from the Eastman Dental Institute at University College London in the U.K.

"During my PhD I made some glass fibres and it struck me that they could be used for growing tissue. Many years later I had the chance to pursue this idea," recalls Prof. Knowles, whose team is developing glass fibres that could one day be routinely used as a scaffold to help bone, muscles, and ligaments regrow after surgery or accidents. The fibres may also be able to play a part in the gene therapy treatments of the future.

The Biomaterials and Tissue Engineering Division headed by Prof. Knowles—working in collaboration with the Hammersmith Hospital in London for this project—produce the glass for their fibres using a technique known as melt quenching. This involves cooling a molten mixture of the compounds the glass is made from so quickly that it cannot form a crystalline solid. The amorphous glass produced is then melted in a furnace that has a 1-mm-wide hole at the bottom. Once the glass has been in this furnace for about an hour—enough time to remove any bubbles from it—the plug is removed from the hole and the molten glass flows through it. Finally a rotating drum draws the stream of glass out into a fibre. This can vary in width from 5 to 200 µm—the faster the glass is pulled out, the thinner the fibre produced.

Since the ultimate aim is for the fibres to be implanted inside the body, the glass is made from many of the same constituents as our bones. This should reduce the chance of rejection and therefore helps towards making the fibres suitable for a range of potential applications. "We envisage the fibres being used for growing any tissues with orientation, such as muscle and ligament. In dentistry for example we could use them for regeneration of muscle that has been lost either due to trauma or congenital abnormalities. So far we have successfully grown muscle cells (see Plate 3.1), ligament cells and bone cells on them," said Prof. Knowles in 2003, shortly after publishing the results of these studies.

As well as being biocompatible, the glass fibres can be made into any shape required, and can dissolve over a period of time, which would remove the need for a second operation to take out any implanted scaffolds that replacement tissue was growing on. The solubility of the glass fibres means that they could not only aid soft tissue regrowth, but could also help fractured bones to re-form. "They could be used to reinforce a degradable polymer to produce a completely degradable composite with enhanced mechanical properties for load bearing use," explained Prof. Knowles, who was also hopeful the fibres

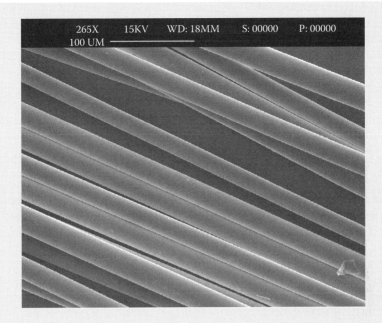

265X   15KV   WD: 18MM   S: 00000   P: 00000
100 UM ─────────────

PLATE 3.1   This micrograph shows muscle cells growing on soluble, biocompatible glass fibres as part of a series of experiments published between 2000 and 2002 that also revealed that ligament cells and bone cells could be grown on this type of fibre. At the time these results were announced, none of the researchers involved knew exactly why the fibres were promoting cell growth, but they believed the process was linked to both the chemistry of the glass and the shape and size of the fibres. Their hope was that if they could eventually understand and control the tissue regeneration, these fibres could allow muscles, ligaments, and bones to regrow after surgery or accidents without leaving the patient with the scarring normally associated with tissue regrowth. (Courtesy of Jonathan Knowles, Eastman Dental Institute.)

could provide a substrate (base) for growing genetically modified cells for gene therapy treatments.

Various genetic diseases including muscular dystrophy—in which muscles waste and become weaker as muscle fibres degenerate—could potentially be treated by gene therapy. The treatment consists of removing diseased cells from a patient and modifying their genetic structure so they behave like normal healthy cells. More of these "normal" cells are then grown on a substrate, which is implanted into the body—complete with its covering of cells—in the hope that the new cells will continue to reproduce and cure the condition by eventually outnumbering the abnormal cells responsible for disease.

semiconductors are covalently bonded solids, but although thermal evaporation can also be used to form the amorphous form of some metals, in this case the substrate needs to be held at around 4K to stop polycrystalline films from forming.

The substrate temperature can also be altered in order to reduce the number of defects present in the amorphous film deposited on it. The higher the substrate temperature, the easier it is for atoms to move around on its surface and so the less structural defects are formed. However, the substrate cannot be held at too high a temperature without crystallization occurring, so the temperature has to be chosen carefully for each material.

Sputtering

The sputtering process also takes place inside a vacuum chamber, and produces a thin film on a substrate. Instead of using a powdered starting compound, however, the atoms that will form the amorphous film are obtained from a solid target. This target is bombarded with ions from a plasma that is formed inside the chamber by introducing an inert gas such as argon and then applying an r.f. (radio frequency) field between the target and substrate. The electric field ionises the gas molecules, creating the plasma in the same way that the starter motor in a fluorescent light provides the spark that ignites the gas in the tube and causes it to glow. It also causes positive ions from the plasma to be attracted to the target during the negative part of its cycle, and electrons to become attracted to the target in the positive half of the cycle.

The ions are so energetic that they knock out either single atoms or clusters of atoms from the target material. The freed atoms then become deposited on the substrate, and an amorphous film is built up as the target is gradually eroded away by the ions from the plasma. Figure 3.15 shows the main features of the sputtering process.

Glow-Discharge Decomposition

Like the sputtering technique, glow-discharge decomposition uses a plasma to produce thin amorphous films. However, in this case the plasma is formed from a gas of the source material which decomposes, allowing its component ions to be collected

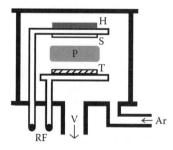

FIGURE 3.15   Apparatus and setup for the sputtering process. A plasma (P) is formed after an inert gas such as argon (Ar) is introduced into the chamber and a radio frequency (RF) field is applied between the target (T) and substrate (S), the temperature of which can be controlled by the heater (H). A vacuum system (V) pumps the argon away. This diagram originally appeared as figure 1.7, p. 11 in Elliott, S. R., *Physics of Amorphous Materials*, 2nd ed., Longman Group UK Limited, 1990. Reproduced with kind permission of S. R. Elliott.

on a substrate. The substrate is therefore positioned in the plasma so that the ions can become deposited on it.

Chemical Vapour Deposition

Another technique that relies on the chemical decomposition of a gas of the source materials is the chemical vapour deposition (CVD) technique. Unlike the glow-discharge decomposition technique, however, heat is used to decompose the gas rather than the creation of a plasma causing the decomposition. This technique is not only used to produce amorphous materials; it is widely used in the electronics industry for manufacturing integrated circuits and semiconductor devices, and will be discussed in more detail in Chapter 7.

3.3 POLYMERS

Polymers are solids composed of extremely large molecules, which are in turn made up from lots of small molecular units known as *monomers*. Figure 3.16 shows a two-dimensional representation of the polymer polyethylene (more commonly called polythene) and its monomer ethylene.

Polythene is one of a large selection of artificially created polymers, which includes polyvinyl chloride (PVC), nylon, and polystyrene. Artificial polymers like this are often referred to as "plastics", and they have many diverse uses ranging from food and drinks packaging to medicine. For example, transdermal patches are made from various polymer layers, one of which is impregnated either with medicinal drugs or—for people trying to give up smoking—nicotine, which the user sticks onto their skin. The drug molecules reach the bloodstream of the patch wearer by diffusing through the bottom of the patch and upper layers of the skin, then entering the capillaries.

Natural polymers, including rubber, cotton, wool, and leather, are also widely used. The most abundant natural polymer is cellulose, which forms the main component of the cell walls of most plants and trees and is made into viscose by the textile industry. Proteins and enzymes are among the other natural polymers.

Artificial polymers have been replacing more traditional materials for several decades, and at the time of writing this book there is much interest in using polymers to make electronic and optoelectronic devices and in further development of electroactive polymers (EAPs). EAPs are plastics that can move or change their size and

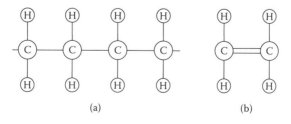

(a)   (b)

FIGURE 3.16 The polymer polyethylene is shown in (a), while its monomer ethylene is depicted in (b).

shape when they are stimulated by electricity, and could find many uses, including sensors in fabrics—for example, for measuring the amount of tension in a car seat belt—and artificial muscles for robots and prosthetic limbs.

3.3.1 STRUCTURE OF POLYMERS

Bonding in Polymers

We have already seen that polymers are made up from long chain-like molecules, which are in turn composed of a series of smaller molecular units called monomers. Most polymers are organic. In other words, they are compounds containing carbon. In fact many of them, including polythene, are hydrocarbons, which means they are composed of hydrogen and carbon. (When discussing polymers, the term "polymer" can either mean the long chain molecule itself, or the actual solid which is in effect composed of several polymers.)

The atoms in polymer chains are covalently bonded to one another. These bonds are indicated by the lines in Figure 3.16, where the double line between the carbon atoms of the monomer represents a "double" covalent bond in which two pairs of electrons are shared between two atoms. As we will see later, some polymer solids effectively show a form of mixed bonding as hydrogen bonds and van der Waals forces hold their various covalently bonded polymer chains together.

Polymerization

Polymers are created artificially in a process known as *polymerization*, and there are two basic ways in which the polymer chains can be formed. In "chain growth polymerization" each monomer becomes part of the polymer in turn, so the polymer grows one monomer at a time like a necklace that is made longer by continually adding a single link to it. By contrast, in "step growth polymerization", growing chains consisting of several monomers already linked together can join other growing chains just as easily as they can join a monomer.

Within each of these types of polymerization, there are two possibilities for how the growth progresses. If the entire monomer joins to either another monomer or to a growing chain, the reaction is known as "addition polymerization". (As Figure 3.16 suggests, polythene grows by addition polymerization, as all parts of the monomer become part of the final polymer.) If, however, only part of the monomer molecule ends up as a section of the final polymer, the process is known as "condensation polymerization". In this case the unused part of the molecule is given off as a by-product of polymer production; for example, hydrogen chloride (HCl) gas is given off during the manufacture of a type of nylon known as nylon 6,6.

Crystallinity

Although all polymers consist of monomers joined together to form long chains, these chains can have different structures. The three main forms of chains—linear, branched, and cross-linked—are illustrated in Figure 3.17.

Linear polymers contain linear polymer chains, which as Figure 3.17(a) shows are long chains with nothing attached to their sides. It is easy for linear chains to slide past

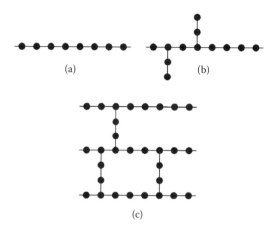

FIGURE 3.17 Three possible structural arrangements for polymer chains.

one another, as they are arranged randomly like strands of cooked spaghetti in a bowl and are only held together by relatively weak van der Waals or hydrogen bonds. This makes polymers with linear structures—like PVC, nylon, and polystyrene—much more flexible than polymers with a branched structure (see Figure 3.17[b]) like polypropylene.

The side branches, which are part of the main polymer chain and therefore consist of covalently bonded atoms, make it more difficult for the polymer chains to slide past each other. This in turn makes branched polymers stiffer and stronger than linear polymers. They also have a lower density because branched chains cannot pack as closely together as linear chains.

Polymers consisting of cross-linked polymer chains (Figure 3.17[c]) are even stiffer than branched polymers because the chains cannot slide past one another at all. The cross-links are made up of covalently bonded atoms that then link the chains together with covalent bonds. Because covalent bonds are much stronger than van der Waals or hydrogen bonds (see Section 2.1), more heat is needed to break the bonds and so melt a cross-linked polymer than a linear or branched polymer. The differences in the different types of polymer structure can be exploited during manufacturing processes.

Although we have just looked at each structure in turn, real polymers often contain a mixture of different types of structure. For example, a mainly linear polymer could have a certain amount of cross-links or side branches. This can affect how crystalline it can be.

Whilst no polymer is completely crystalline, linear polymers can have a high degree of crystallinity if the polymer chains line up in such an orderly fashion that the atoms repeat with the same regularity that would be found in an ordinary crystal. For example, one form of polythene can be 95% crystalline, and it is possible to define unit cells for polymers with structures as crystalline as this. There cannot be such a high degree of crystallinity in a branched polymer, however, as the side branches prevent the chains from lining up in a regular way throughout the structure. It is possible for regions of branched polymers to be crystalline though.

Both linear and branched polymers can also be totally amorphous, and cross-linked polymers are almost always amorphous. As well as being denser than amorphous polymers, crystalline polymers tend to have higher melting points. This is because the atoms are more closely packed together and the van der Waals bonds between them are therefore stronger, making them harder to break. For the same reason, crystalline polymers tend to be mechanically stiffer than their amorphous counterparts.

Like other types of solid, polymers are not always perfect. Point defects such as vacancies and interstitials can be found in crystalline areas of polymers, while the boundaries between different crystalline regions within a polymer can also be thought of as defects.

3.3.2 THERMOPLASTICS

When some polymers are heated they become soft, and only harden again if they are allowed to cool. These polymers are known as *thermoplastics* or thermosoftening plastics. Most thermoplastics are flexible and fairly soft at room temperature, but become even softer and more flexible when they are hot. They consist of either linear polymer chains or branched polymer chains that can move past each other. Because no covalent bonds are broken when thermoplastics are heated, they can be softened and reshaped several times, and so can be recycled. (If the temperature is increased to an extremely high level, the covalent bonds between the atoms in the polymer chains will break, and at this point the thermoplastic will irreversibly break down. This occurs at a far higher temperature than the ordinary melting point, however, so does not affect the recycling process.)

PVC and polythene are two well-known examples of thermoplastics. There are two types of polythene: high-density polythene and low-density polythene. As Figure 3.18 shows, high-density polythene consists entirely of linear polymer chains that can pack together easily. By contrast, low-density polythene has branches that stop the chains from packing so closely together and hence make this polymer a relatively low-density material. Pipes, toys, and household objects such as buckets can be made from high-density polythene, while low-density polythene is mainly made into sheets or thin films so it can be used for plastic bags, tubing for ball point pens, and drinks bottles.

There are a range of uses for PVC as well, depending on how much *plasticiser* is added to the PVC. (Some of these applications are revealed in Example question 3.4.) Plasticiser is a substance—usually a liquid—which gets in between the polymer chains and pushes them further apart. This reduces the strength of the bonding between the chains and so increases the flexibility of the polymer (see Section 3.3.5 for more on additives). Without plasticiser, PVC is hard and rigid because, although it is a linear polymer, its bulky chlorine atoms are arranged randomly on either side of the polymer chains and so prevent them from sliding easily past one another.

Polystyrene and PTFE (polytetrafluoroethylene) are two more widely used thermoplastics. Polystyrene can be expanded by a gas to produce a lightweight rigid white substance used for packaging and insulation, while PTFE is a tough material that is resistant to chemicals and is used to coat cooking utensils. There are also variations

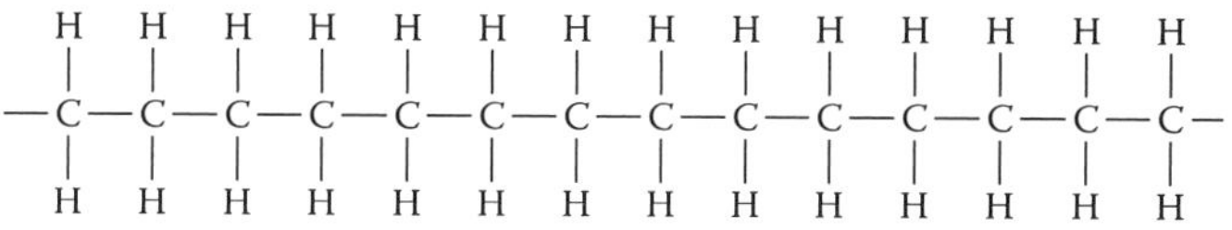

(a)

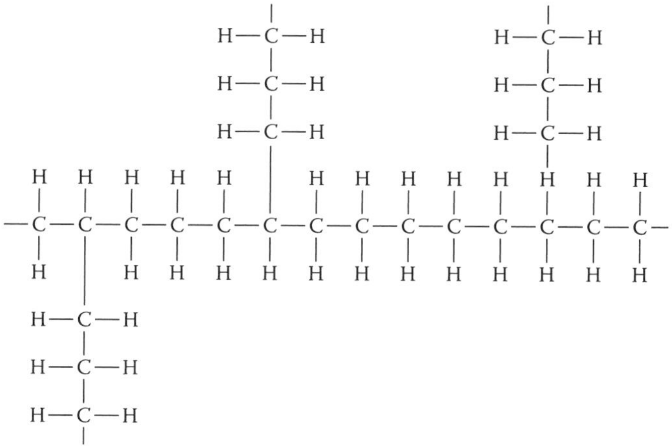

(b)

FIGURE 3.18 The two forms of polythene. High-density polythene (a) consists of linear polymer chains that can pack together easily and thus produce a material with a high density. By contrast, low-density polythene (b) has side branches that prevent the polymer chains from packing so closely together and so reduce the overall density of the material.

EXAMPLE QUESTION 3.4   APPLICATIONS OF PLASTICS

You run a small business manufacturing plastic goods and have three different batches of PVC available to you. Batch 1 consists of PVC without any plasticiser added; Batch 2 contains PVC with a low percentage of added plasticiser; while Batch 3 has a high plasticiser content. Which batch would you use to manufacture the following goods:

(a) a raincoat
(b) a drain pipe
(c) a soft-drink bottle

Answer on page 90.

ANSWER TO QUESTION 3.4 ON APPLICATIONS OF PLASTICS

(a) No one will want a raincoat which is stiff and uncomfortable to wear, so to make them as flexible as possible, Batch 3 with the high plasticiser content should be chosen.

(b) Without plasticiser, PVC is hard and rigid, so Batch 1 would be the most suitable batch to manufacture drain pipes out of.

(c) Soft-drinks bottles need to be more flexible than drain pipes, but not so flexible that they are unable to hold fluids. Batch 2 with a low percentage of added plasticiser is therefore the most sensible choice for manufacturing this product.

on standard thermoplastics, such as thermoplastic elastomers which combine the properties of thermoplastics and elastomers (see Figure 3.19).

3.3.3 THERMOSETS

Unlike thermoplastics, *thermosets* or thermosetting plastics char and start to decompose when they are heated, so cannot be reshaped. They are cross-linked polymers, so are rigid. The cross-linked structure prevents thermosets from being recycled, as heat causes the covalent bonds in the cross-links to break. Once broken, these bonds cannot be re-formed, so the plastic becomes unusable. However, since it takes a reasonable amount of heat energy to break these bonds, and so allow the plastic to flow, thermosets have higher melting temperatures than thermoplastics, enabling them to be used in hotter environments. They also tend to be stronger and harder than thermoplastics.

Bakelite and certain polyesters are thermosets. Most goods made from these materials and other thermosets are manufactured by moulding so that the cross-linking occurs while the polymer is in the mould. This is because other than cutting the thermoset, there is no other way of changing its shape once it has formed.

Another type of widely used thermosets are the epoxies. These include some well-known brands of glue, as well as mortars and high-strength glues for use in the construction industry. The epoxy adhesives used by the building trade are strong enough to stick concrete to concrete, or concrete to steel. Polyester-based mortars and putties can also be used in conjunction with concrete or brickwork.

3.3.4 ELASTOMERS

Both natural and synthetic rubbers belong to a class of polymers known as *elastomers*. Elastomers are elastic materials, that is, materials that return to their original shape and size when a force stretching or compressing them is removed. (Elasticity will be discussed in detail in Chapter 4.)

They have a mainly amorphous structure consisting of curled-up chains of linear molecules that have a certain amount of cross-linking between them. It is these

FIGURE 3.19   The handle of this measuring jug is covered in a layer (approximately 1.5 mm thick) of a thermoplastic elastomer called Santoprene®. Thermoplastic elastomers behave like rubbers, but have the advantage of being able to be melted and moulded like thermoplastics. This not only makes them recyclable, but also means goods can be manufactured from them more cheaply and quickly than from conventional rubbers, as it is easier to mould thermoplastics than to process rubbers. The textured finish of the Santoprene® covering, together with its inherent properties of softness and a high coefficient of friction, result in a handle which is easy for the user to grip. (Courtesy of Lakeland Limited.)

cross-links that make elastomers elastic by making sure they return to their original shape and size after they have been uncurled by stretching. Without the cross-links, the polymer chains could just slide into new positions when a force was applied and remain there, producing a permanent shape change.

Rubbers can become hard when they are very cold and even if they are stretched by a large amount. This is because in both cases they have become partly crystalline. It is the crystallization of rubber that causes old balloons to go wrinkly as they deflate.

While synthetic rubber is manufactured from petroleum, natural rubber is the latex of the rubber tree (*Hevea brasiliensis*) from South America. Although it can be used in its natural state, rubber can be made much more elastic and less sticky by adding sulphur to it. The process of adding sulphur to rubber is known as *vulcanisation*, and was invented by Charles Goodyear in 1839. Some of the sulphur atoms form cross-links between chains, which then make the rubber harder. If more and more sulphur is added, rubber eventually becomes a hard solid known as ebonite.

As well as being flexible and easy to mould, rubber is waterproof and very durable. These properties make it ideal for one of its best-known applications—vehicle tyres. In fact a large percentage of the rubber used world-wide is made into tyres for cars and other road vehicles, although it does find other applications; for example, waterproof boots and clothing, and gaskets, seals, and valves are commonly made from rubber.

3.3.5 ADDITIVES

There are not many polymers that have exactly the right properties for the uses they are put to, and various substances can be added to the polymer structure to give it the qualities it needs. For example, tensile strength, stiffness, and abrasion resistance can be improved by adding a *filler* to the polymer. Various substances can be used as fillers, and as well as generally being cheaper than the polymer, and thus reducing the overall cost of the material, each different filler can change the properties of the polymer in a different way, as Table 3.2 shows. This allows polymers to be tailored for specified applications. (Some of the mechanical properties of solids mentioned in this table are discussed in Chapter 4.)

Fillers are not the only types of additives. If the substance added to a polymer helps to prevent it being damaged by its working environment, it is known as a *stabiliser*. For example, some plastics can change colour if they are regularly exposed to sunlight, and can even suffer a reduction in their mechanical properties. These effects can be reduced by adding a stabiliser such as carbon black, which can absorb UV light.

Other additives include plasticisers, which were discussed in Section 3.3.2 and are used to increase the flexibility of certain plastics, zeolite which can absorb a variety of substances including water molecules, and flame retardants (see Example question 3.5).

TABLE 3.2
Effect that Different Fillers Have on Polymers

Filler	Effect on Properties
Cotton flock	Increases the impact strength, but reduces water resistance
Cellulose fibres	Increases both tensile and impact strength
Glass fibres	Increases the tensile strength and stiffness, but lowers the ductility
Mica	Improves electrical resistance
Graphite	Reduces friction

Note: The filler can make up as much as 80% of a plastic.

EXAMPLE QUESTION 3.5   ADDITIVES

Which of these additives—glass fibres, titanium dioxide, or trisbromoneopentyl alcohol flame retardant—would make the following polymers more suitable for the stated uses and why?

(a) polyurethane foam mattress
(b) polyester boat hull
(c) rigid PVC window frame

Answer on page 94.

FURTHER READING

Ashley, S. "Artificial muscles". *Scientific American* (Oct. 2003): 35–41.

Elliott, S. R. *Physics of Amorphous Materials.* 2nd ed. Longman Group UK Limited; copublished in the United States with John Wiley & Sons, Inc., New York, 1990, Sections 1.3, 2.2, and 3.5.

Hull, D., and Bacon, D. J. *Introduction to Dislocations.* 4th ed. Oxford: Butterworth-Heinemann, 2001.

Townsend, P. D., and Kelly, J. C., *Colour Centres and Imperfections in Insulators and Semiconductors.* London: Chatto & Windus for Sussex University Press, 1973, Chapters 3 and 7.

(More references for further reading, as well as Web links, are available on the following Web page: http://www.crcpress.com/product/isbn/9780750309721.)

SELECTED QUESTIONS FROM
QUESTIONS AND ANSWERS MANUAL

Q3.1 If you have a 1 cm^3 sample of aluminium, how many vacancies would it contain at 250K if the energy needed to create the vacancy is 0.75 eV and there are 10^{22} atomic sites in every cubic centimetre of the sample?

Q3.3

(a) What is a colour centre?
(b) What type of bonding do crystals containing colour centres have?
(c) Are colour centres naturally occurring or artificially made?

Q3.8 What is a supercooled liquid?

Q3.9 What will the supercooled liquid in question Q3.8 form if it is cooled even further?

ANSWERS TO QUESTION 3.5 ON ADDITIVES

(a) If the trisbromoneopentyl alcohol flame retardant was added to the foam mattress, it would make it safer for anyone sleeping on it in the event of a fire. In general, flame retardants work either by releasing a gas which interferes with combustion or by initiating a chemical reaction which cools any burning areas.

(b) Glass fibres would help make a polyester boat hull stronger and stiffer.

(c) Titanium dioxide is the main active ingredient in many suncreams as it absorbs UV light. This also makes it a useful additive in plastic products—like PVC window frames—that will be subjected to sunlight during use.

Q3.12   What is the main difference in structure between high-density and low-density polythene, and how does this affect the density?

Q3.13   Which of the following would you add to rubber and why? (a) Carbon, (b) Sulphur, or (c) Mica?

4 Stressed Out
The Mechanical Properties of Solids

CONTENTS

4.1   INTRODUCTION TO MECHANICAL PROPERTIES OF SOLIDS

Given the choice of using rubber or steel to construct a permanent road bridge, we are all likely to opt for steel. After all, we know it would be strong enough to support the weight of cars and people, whereas the rubber would bend and possibly tear. The metal is an obvious choice. But what if you were asked to choose between steel and reinforced concrete?

Now the decision is not so clear cut. Both materials can be—and indeed are—used to make bridges, and the material chosen will be the one which can best satisfy a number of requirements, including the size and position of the bridge. Someone will have made a choice like this for many of the products and engineering structures that we use, and different solids have many different types of properties which enable them to be used in a wide variety of applications.

For example, plastic drinks bottles are made from lightweight, slightly flexible, transparent and recyclable polymers, while saucepans are made out of metals with good thermal conductivity and corrosion resistance. Some properties, such as translucency or even electrical conductivity, can make no difference for certain applications, but be vital for others. It doesn't matter for instance whether the metal wire used to make a piece of jewellery can conduct electricity particularly well or not, but the metal wire used to make an electrical cable must be a good conductor. The mechanical properties of materials, however, such as hardness and elasticity, are generally important for any application, as materials are usually subjected to some sort of force during their working lives.

In this chapter we will see how several of the main mechanical properties are measured, and why they are of interest. We will also look at ways in which the mechanical properties of solids can be altered or supplemented for particular applications.

4.1.1 STRESS AND STRAIN

The mechanical properties of solids reveal how they behave when a *stress* is applied to them. Stress is defined as the force per unit cross-sectional area and is measured in pascals (Pa), where 1 Pa is equivalent to 1 Nm^{-2} (one newton per square metre). In its simplest form, stress can be expressed by the following equation:

$$\sigma = \frac{F}{A} \tag{4.1}$$

where σ is the stress, F is the force, and A is the cross-sectional area the force is applied to.

Whenever stress is applied to a material, the material will deform in some way and so change shape. Figure 4.1 shows the three different types of stress—*tensile stress*, *compressive stress*, and shear stress—and the type of deformation they produce.

The change in volume or shape that the stress acting on an object produces is known as *strain*. Strain is dimensionless, and is defined as

$$\varepsilon = \frac{l - l_0}{l_0} \tag{4.2}$$

where ε is the strain, l_0 is the length of the object before any stress is applied to it, and l is the length of the object whilst it is under stress.

Strain is proportional to stress in most materials for very small stresses. However, as we are about to see, larger amounts of stress produce very different amounts of strain in different types of solids.

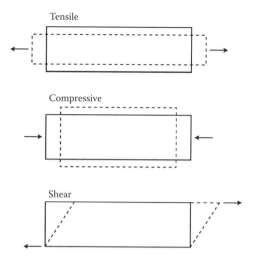

FIGURE 4.1 The three different types of stress—tensile, compressive, and shear—and the changes in shape they cause. In each case the bold lines represent the shape of the original object, while the dotted lines show the change in shape produced by the stress. As this shows, a tensile stress pulls on an object and stretches it; a compressive stress squeezes an object and squashes it; and a shear stress distorts an object. (Modified from Types of Stress Diagram (P. 404) from *The Penguin Dictionary of Mathematics*, 2nd ed., edited by David Nelson (Penguin Books, 1998). Copyright © Penguin Books Ltd. 1989, 1998.)

Stress-Strain Curves

The relationship between stress and strain for a particular material can be plotted as a stress-strain curve like that shown in Figure 4.2, which shows a hypothetical stress-strain curve for a metal being subjected to a tensile stress.

Up to point A, which can be known either as the *elastic limit* or "limit of proportionality", the material in the figure obeys Hooke's law, named after the English

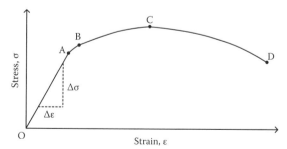

FIGURE 4.2 A typical stress-strain curve for a hypothetical metal subjected to a tensile stress. In the elastic region between the origin and the yield stress at point B, the metal will return to its original shape and size when the stress is removed. In the region O to A, the stress is proportional to the strain and the slope of this linear region is known as Young's modulus. Beyond point B, permanent "plastic" deformation occurs until the material fractures at the breaking stress marked as point D. Point C represents the tensile strength, which is the greatest tensile stress the material can withstand before breaking.

physicist Robert Hooke (1635–1703). He discovered that beneath the elastic limit, stress is directly proportional to strain, and that if the stress is removed the material returns to its original shape.

The slope of the linear section of the curve from O to A is *Young's modulus*. Young's modulus is sometimes known as the tensile modulus, and can even be called the modulus of elasticity. This last term is somewhat misleading, however, as there are in fact three commonly used moduli of elasticity—Young's modulus, the bulk modulus, and the rigidity modulus, which correspond to tensile, compressive, or shear applied stress, respectively.

Hooke's law can be expressed mathematically by the following equation in which $\Delta\sigma$ is the change in stress, $\Delta\varepsilon$ is the change in strain, and the constant of proportionality E is Young's modulus.

$$\frac{\Delta\sigma}{\Delta\varepsilon} = E \tag{4.3}$$

Young's modulus, E, is named after the English physicist, physician, and Egyptologist Thomas Young (1773–1829), who discovered it in 1807. It can be thought of as a measure of the *stiffness* of a material, that is, how much the material resists *elastic deformation* when it is under stress. The larger the value of Young's modulus, the stiffer the material is. Or putting it another way, the greater the value of Young's modulus, the smaller the amount of strain caused in a material by a given stress.

Beyond the elastic limit, in the region between A and B, although the stress is no longer proportional to the strain, the material will still return to its original shape if the stress is removed. The region of the curve between the origin and point B is therefore known as the elastic region. However, if stress continues to be applied to a material after it has exceeded point B—which is called the "yield stress", "yield strength", or "yield point"—it will become permanently deformed. This permanent *deformation* is known as *plastic deformation*.

The *tensile strength* indicated by point C (which can also be called the "ultimate strength" or even the "ultimate tensile stress") is the largest amount of tensile stress that a solid can withstand without breaking. Although this particular graph is illustrating the behaviour of a hypothetical solid under tensile stress, in general the amount of stress a material can withstand before fracturing is known as its *strength*. Some materials can be stronger if one type of stress is applied rather than another. Concrete, for example, is relatively weak under tension (when stretched), but very strong under compression (when squashed). This makes concrete pillars a good choice for supporting bridges because they are so strong under compression.

Finally, if the material remains under stress it fractures, and the point at which this happens is known as the "breaking stress", "breaking strength", or "break point".

Stress-strain curves look very different for different types of materials, depending on whether those materials are *brittle* or *ductile*. Brittle materials like glass, stone, thermosets, and ceramics tend to shatter if they are hit with a hammer, and little or no plastic deformation takes place before they break. They will not show much

EXAMPLE QUESTION 4.1   STRESS AND STRAIN

How much longer will a 20 cm long piece of titanium (which has a Young's modulus of 107 GPa) become if it is subjected to a tensile stress of 250 MPa? Assume the deformation is elastic.

ANSWER

Having been told that the deformation is elastic, in the absence of any further details, you would be justified in assuming that the stress (σ) is proportional to the strain (ε) and that prior to the stress being applied the titanium was not strained or under stress, so that

$$\frac{\sigma}{\varepsilon} = E$$

where E is Young's modulus. You have been given the value of E for titanium, and you also know how much stress is being applied. This means you can easily work out the strain, but in this case you have been asked to calculate the increase in length. This is quite straightforward, however, as the increase in length is related to the strain as follows:

$$\varepsilon = \frac{l - l_0}{l_0}$$

where l_0 is the length of the object before any stress is applied to it, and l is the length of the object whilst it is under stress, so $(l - l_0)$ = the change in length, Δl.

Combining these two expressions and putting in the numbers reveals the extension:

$$\frac{l_0 \sigma}{E} = \Delta l = \frac{0.2 \times 250 \times 10^6}{107 \times 10^9} = 0.467 \text{ mm}$$

extension before reaching their elastic limit either, and will fracture before their yield point is reached.

By contrast, ductile materials like metals, wood, and some thermoplastics show a large amount of plastic deformation before they break. This allows them to be machined in various ways, drawn out into wires, and hammered into new shapes relatively easily. They also tend to extend by a reasonable amount before their elastic limit is reached. Elastic materials such as rubber show a large amount of elastic deformation, but stress is only proportional to strain for a fraction of their elastic region (see Example question 4.2).

EXAMPLE QUESTION 4.2   STRESS-STRAIN CURVES

Which of these stress-strain curves is most likely to represent (a) a rubber, (b) a metal, and (c) a diamond?

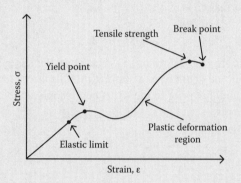

Curve A

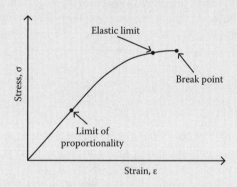

Curve B

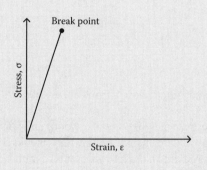

Curve C

Answer on page 102.

TABLE 4.1

The Effect of Different Types of Bonding on the Macroscopic Properties of Solids

Type of Bonding	Cohesive Energy Range (eV/atom)	Macroscopic Properties
Ionic	5–10	Hard, brittle, high melting temperatures. Good electrical and thermal insulators, but can conduct electricity when molten.
Covalent	1–5	High melting temperatures and hard for solids with cohesive energies at the highest end of this range; softer and lower melting temperatures for covalent solids with lower values of cohesive energy.
Metallic	1–5	Ductile, high electrical and thermal conductivity. Metals with cohesive energies at the bottom of this range are softer and have lower melting temperatures than those with higher values of cohesive energy.
van der Waals	0.1	Soft, flexible, elastic, low melting point.

How Bonding Affects Mechanical Properties

Although elastic strain is a macroscopic property that we can easily observe, for example when we stretch an elastic band, it occurs because the atoms the material is made from move slightly farther apart and the bonds between them are stretched. In fact the larger the value of Young's modulus, the more resistance the atoms have to being separated. So materials with stronger bonds will have larger values of Young's modulus (see Example question 4.3).

This is just one example of many in physics where behaviour at a microscopic level affects behaviour on a macroscopic scale (size comparable with objects we can see without the aid of a microscope). Table 4.1 summarises the effect that different types of bonding have on some of the properties of solids. As this reveals, and as we saw throughout Section 2.1, the type of bonding that holds the atoms together in a solid influences not only its mechanical properties—for example, the stronger the bonding, the harder the material is—but its thermal and electrical properties as well. (These will be discussed in detail in Chapters 5 and 6, respectively.)

It is worth remembering that different solids with the same type of bonding can have very different values of cohesive energy (which is a good indicator of the strength of the bonding, and was discussed in Section 2.1.2). For instance, diamond with its very high value of cohesive energy is much harder and has a much higher melting temperature than grey tin, which is another covalently bonded solid but which has a lower value of cohesive energy. It is also worth recalling that many solids contain a mixture of different types of bonding (see Section 2.1.7), and in some cases this can produce some unusual mechanical effects. A graphite crystal, for example, will be easier to shear parallel to its horizontal layers than perpendicular to them (see Figure 2.8). This is because the van der Waals forces bonding the layers vertically

ANSWER TO QUESTION 4.2 ON STRESS-STRAIN CURVES

(a) Curve B represents a rubber as it has a large elastic region, but only a small linear elastic region, and shows relatively little plastic deformation before breaking.

(b) Curve A represents a metal, and shows all the same characteristics as the curve shown in Figure 4.2 and discussed earlier.

(c) Curve C represents a brittle material like diamond, which shows no plastic deformation before reaching its break point.

to each other are much weaker than the covalent bonds holding the carbon atoms together within each horizontal layer. (It is not only bonding that affects the mechanical properties of solids, however. As we will see in the next section, defects play a large part in determining how a solid reacts to external forces.)

EXAMPLE QUESTION 4.3   STIFFNESS OF MATERIALS

In your haste to prepare your research results for a presentation, you have accidentally typed the information in the second and third columns of this table in the wrong order. How should it read?

Material	Young's modulus (GPa)	Bonding Mechanism
Diamond	196	Metallic
Iron	1,000	Covalent
Silicon	3	van der Waals (mainly)
Nylon	107	Covalent

Answer on page 104.

4.1.2 PLASTIC DEFORMATION

Whilst elastic deformation occurs on the atomic level by the bonds connecting atoms stretching as the atoms move further apart from one another, plastic deformation involves the bonds between atoms breaking. New bonds are then formed when the atoms that have split with their original neighbours have moved to a new position within the solid. This makes the deformation permanent, and different processes lead to bonds breaking depending on whether the material is crystalline or amorphous. We will look first at what happens inside crystals once their elastic limit has been passed.

Slip

Dislocations (see Section 3.1.2) were originally proposed in the 1930s to explain what happens when crystals are plastically deformed in some way. By this time

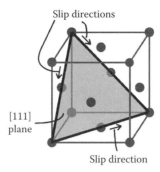

Slip directions

[111]
plane

Slip direction

FIGURE 4.3 Three of the slip systems for an fcc crystal. Slip will occur in the plane of closest atomic packing for any crystal, which in this case will be in any of the set of four {111} planes. Within each of these planes there are three different directions in which slip can occur, which are each lines of closest atomic packing, giving fcc crystals a total of 12 slip systems.

researchers had observed the plastic deformation of metal crystals under a microscope, although they had not yet seen dislocations. However, the pioneers could see that the deformation was occurring by one plane of atoms sliding over the top of another. This process is known as *slip*.

For metal crystals, slip occurs most easily along a particular crystallographic plane known as the "slip plane", and along a particular direction known as the "slip direction". (As there can be more than one possible slip direction for a single slip plane, the term "slip system" is often used to denote a particular combination of slip plane and slip direction.) Slip planes will usually be the planes of closest packing. In the case of fcc metals, this means the four equivalent {111} planes, whereas for bcc metals, displacement will take place along the {110}, {112}, and {321} sets of planes. The slip direction is any line of closest atomic packing within the slip plane. So for fcc metals the slip direction is the set of face diagonals <110> (see Figure 4.3), while for bcc metals it is the set of <111> directions. (See Section 2.2.7 for an explanation of the notation for atomic planes and directions.) Metals with the fcc or bcc structure are generally ductile because there are a large number of planes and directions that slip can occur along. By contrast, metals with the hcp structure tend to be brittle because there are fewer slip systems.

To produce slip, a shear stress (see Figure 4.1) must be applied to a crystal. In this case this means a stress applied parallel to the slip plane that acts in the slip direction. The elastic limit marks the level of stress above which the atomic planes will begin to slip, producing a permanent or plastic deformation. If calculations are carried out to determine the elastic limit for crystals by taking into account the stress needed to break bonds and so allow slip to occur, the results are two to four orders of magnitude higher than experimentally recorded values of the elastic limit. This is because the calculations only apply to perfect crystals, which contain no defects. In reality most crystals contain anything from 10^2 to 10^{12} dislocations per square centimetre, and it is these dislocations that so dramatically reduce the elastic limit—and the more dislocations that are present in a crystal, the lower the value of the elastic limit.

ANSWER TO QUESTION 4.3 ON STIFFNESS OF MATERIALS

The table should read as follows:

Material	Young's Modulus (GPa)	Bonding Mechanism
Diamond	1,000	Covalent
Iron	196	Metallic
Silicon	107	Covalent
Nylon	3	van der Waals (mainly)

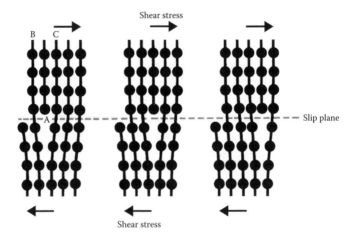

FIGURE 4.4   Movement of an edge dislocation.

Figure 4.4 helps illustrate how dislocations lead to low values of elastic limit. The line of atoms marked A represent the atoms at the end of an edge dislocation. The dislocation line runs perpendicular to the page in the same way a pin would point if you pushed it through the paper. (The dislocation line also runs perpendicular to the slip direction in an edge dislocation, whereas in a screw dislocation the dislocation line is parallel to the direction of slip.)

The planes of atoms marked B and C are distorted in the region around the end of the dislocation. So if a shear stress is applied as indicated, it is relatively easy for the bonds between any two atoms in the C plane that are respectively above and below the slip plane to break. The bottom half of the original C plane then bonds with the half-plane of A atoms, forming a new complete plane. The effect of these original C bonds breaking and then re-forming as the atoms bond with new neighbours is to shift the dislocation to the right by one whole atomic spacing.

With the shearing stress still applied, this process of the bonds breaking in the plane of atoms to the right of the dislocation, and new bonds forming in their place with the atoms in the dislocation, continues until the dislocation can move no further.

This occurs either when it reaches the crystal surface, or if it becomes tangled up with another dislocation.

When a dislocation travels along a slip plane in this way, its movement amounts to the same thing as a slip displacement of the top part of the crystal, and therefore causes a plastic deformation of the crystal. In perfect crystals, all the bonds between the layers above and below the slip plane would have to break simultaneously for slip to occur. However, since dislocations allow slip to take place by the atomic bonds breaking one line at a time, the shear stress needed for crystals with dislocations to plastically deform is much less than that needed to permanently deform a perfect crystal.

One way to think of the difference in shear stress needed to produce slip in perfect crystals and in crystals with dislocations is to imagine moving a large heavy carpet over a floor. If we were to try to shift the carpet all at once by dragging it, we would require a large force to do so—a situation comparable to producing slip in a perfect crystal. However, if we made a ruck in one side of the carpet and encouraged it to travel across to the other side by using a flicking motion, the carpet would still move, but we would have applied a much lower force to it. This mirrors the situation for a dislocation moving through a crystal—the end result is the same, but much less force has been applied.

Although we have mainly been considering metal crystals so far, there are other crystalline solids whose plastic deformation is governed by the motion of dislocations. These include ceramics, which are compounds like $CsCl$ containing one metallic element together with one nonmetallic element. When the bonding in a ceramic is predominantly ionic there is very little plastic deformation because there are very few slip systems which would not bring oppositely charged ions near enough to one another to cause repulsion. Even covalently bonded ceramics show less deformation than metals because they have very strong bonding and relatively few slip systems. This means ceramics tend to be more brittle than metals.

Of course dislocations are not the only defects that can affect the mechanical properties of solids. Figure 4.5 for example shows the result of a small crack being allowed to grow into a much larger one.

FIGURE 4.5 The SS *Schenectady* was only one day old when it fractured in two because a small crack had grown into a massive one, as seen in this photograph.

Amorphous Materials and Polymers

We saw in Chapter 3 that the structure of amorphous materials is more random than that of crystalline solids, and since the whole structure lacks long-range order, plastic deformation cannot occur by dislocations moving through an amorphous material. Instead, when two layers of atoms in an amorphous material are subjected to a shear stress, they move relative to one another by bonds between the atoms breaking and re-forming in the same way that a liquid flows. In contrast with the situation in crystalline solids, this movement has no preferred mechanism or direction.

In the case of polymers like thermoplastics, which show plastic deformation (see Section 4.1.1), the deformation occurs by polymer chains sliding past one another and any tangles becoming straightened out. If enough stress is applied for the chains to become almost lined up with one another, a substantial increase in stress will be required to produce further strain, as the chains will now be held together by van der Waals forces.

4.1.3 TESTING, TESTING

We have already seen that ductile materials have a relatively large extension before reaching their elastic limit, whereas brittle materials do not. In this section we will look at various ways we can scientifically test for different mechanical characteristics like this.

Bend Tests

One way of scientifically distinguishing between brittle, ductile, and elastic materials is to carry out a bend test. In this a thin sample of the material under investigation is supported at either end, then bent by a load applied to the unsupported centre of the sample. The sample is then investigated to see if it has broken or cracked.

The results of this test are qualitative rather than quantitative. This means that whilst differences in ductility between different materials are revealed relative to one another, no actual numerical values are obtained. Elastic materials such as rubber are identified by the fact that they return to their original size and shape after the load has been removed. As we would expect, brittle materials like glass break before they can be bent very much. By contrast, ductile materials can be bent much more than brittle materials before they break, but unlike elastic materials they will permanently change shape once the load has exceeded a certain threshold level (different for each ductile material) and will break if a large enough load is applied.

Having just said that this test does not yield numerical results, it can, however, be deliberately set up to do so. It is used in this way for brittle materials, such as certain ceramics, that are difficult to make into specimens the correct shape for a tensile test. (Tensile testing is discussed in detail in the next subsection.) From measurements of the specimen dimensions, positioning on the supports, and applied load, the stress as the specimen of the material breaks can be calculated. The stress at fracture obtained from a bend test is known as the "bend strength" or "modulus of rupture". Table 4.2 lists the bend strengths of some widely used ceramic materials.

Tensile Testing

Useful though bend tests are—as they are quick and simple to carry out—the tensile test is the standard way for evaluating the effect of a force (load) on a material. So

TABLE 4.2

The Bend Strengths of Some Widely Used Ceramic Materials

Material	Bend Strength (MPa)
Titanium carbide (TiC)	1100
Aluminium oxide (Al_2O_3)	200–345
Beryllium oxide (BeO)	140–275
Silicon carbide (SiC)	170
Magnesium oxide (MgO)	105
Spinel ($MgAl_2O_4$)	90
Fused silica	110
Glass	70

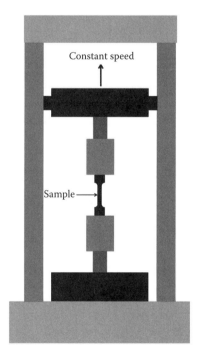

FIGURE 4.6 Schematic diagram of a tensile testing machine, that gradually increases the tensile load on the test piece until it fractures.

that the results from different tensile tests can be compared, samples of the material to be studied must be created in the standard shape and size of a tensile test piece. Different countries may have different standard sizes for test pieces.

The idea of a tensile test is to gradually increase the tensile load on the test piece until it fractures. Figure 4.6 shows a simplified schematic of a tensile testing machine. The lower horizontal bar remains stationary, while the upper crossbar is gradually moved upwards. The further the crossbar moves, the greater the force on the sample.

EXAMPLE QUESTION 4.4   TENSILE STRENGTH

Which of these materials has the greatest tensile strength, and which has the lowest?

(a) cast iron

(b) carbon nanotubes (Z. F. Ren et al., *Science* 282, 1105–1107 [1998].)

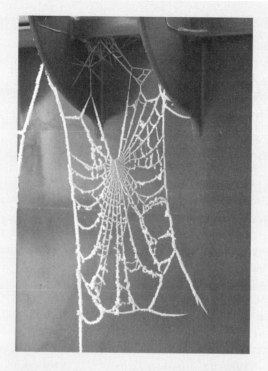

(c) main strut of a spider's web (Courtesy of Alan Piercy.)

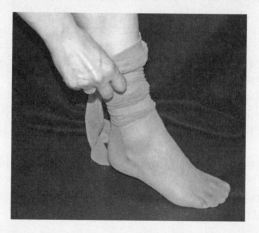

(d) nylon

Answer on page 110.

ANSWER TO QUESTION 4.4 ON TENSILE STRENGTH

The carbon nanotubes have the highest tensile strength of the materials shown here. Their tensile strengths are in the range of 50–200 GPa, and are at the time of writing this book the strongest material known.

The spider silk is the next strongest material, with a tensile strength of 870–1420 MPa, which is larger than that of cast iron, which has tensile strength values in the range of 400–1200 MPa.

Nylon therefore has the lowest value of tensile strength, which averages around 60–100 MPa.

The results of tensile tests are plotted as force-extension curves, which have the same sort of form as stress-strain curves.

If a material is going to be under compression when it is used, rather than under tension, a variation on the tensile test can be carried out in which the sample is squashed rather than pulled.

Impact Tests

Like bend tests, impact tests can also be used to tell the difference between brittle and ductile materials. They are particularly useful for testing whether a heat treatment of a material has been successful or not, as tensile testing can sometimes produce similar results for a particular material regardless of changes in heat treatment, whereas the comparative impact test results will look very different. (Heat treatments are discussed in detail in Section 4.2.2.)

There are two types of impact test (the Izod test and the Charpy test), and whilst each uses a different shape of test piece, they both involve a pendulum with a sharp edge swinging down and striking a notched test piece. For most materials the test piece will break, and whether it breaks or not the height which the pendulum swings back to after striking the test piece is proportional to the amount of energy that has been absorbed by the test piece. In the case of plastics, which do not tend to break but dent instead, the results of impact tests are given in terms of energy per unit cross-sectional area of the test piece, while for metals and alloys—which tend to break—just the energy is given. The more energy used to break (or dent) the test piece, the lower the pendulum rises on its return swing. Results for all materials are listed as the "impact strength". This is defined as the energy needed to break half of a large sample set of the same type of material.

EXAMPLE QUESTION 4.5   IMPACT TESTS

If you have carried out two impact tests on a sample of PVC and discovered that at 20°C the impact strength is 3 kJm^{-2}, while at 40°C it is 10 kJm^{-2}, at which temperature is the PVC the most ductile?

Answer on page 114.

Virtual Testing

Before any material is put to use there will be extensive testing of the material in order to determine whether it is suitable for the particular function it has been chosen for. In many cases this testing is expensive and time consuming, and can be supplemented by, and possibly in the future may be replaced by, virtual testing. This involves using complex computer simulations based on mathematical models—which have been shown to agree with experiment—that can take account of behaviour right down to the atomic level.

4.1.4 ELASTICITY

As we have already seen, elastic materials are materials that return to their original shape and size after being stretched. The elastic limit is the point up to which any stress applied to an elastic material will produce a strain that is proportional to that stress (and a strain that will disappear when the stress is removed). If stress continues to be applied to a material beyond its elastic limit, it will no longer return to its original shape. Elastic bands provide a good example of elastic behaviour when they are stretched by a moderate amount as Figure 4.7 shows. However, most of us will have discovered by playing with elastic bands as children that if they are stretched by a large amount, they become permanently stretched. All or part of them becomes much thinner, and if they continue to be stretched once they have reached this plastic stage they will eventually break.

Moduli of Elasticity

We also saw in Section 4.1.1 that the relationship between stress and strain below the elastic limit can be described by Young's modulus (equation 4.3), which is one of several different moduli of elasticity that correspond to different types of strain. While Young's modulus applies to solids under tension, the *bulk modulus*

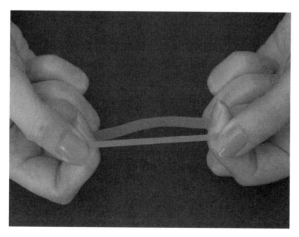

FIGURE 4.7 When an elastic band is stretched, under a certain amount of stress the band returns to its original shape, but if it is pulled really hard it becomes permanently stretched and decreases in width before finally snapping once its break point is reached.

BOX 4.1   TIGHT SQUEEZE

Lighter airliners that consume less fuel are an attractive prospect, and "you could save tons of weight by stripping out the internal copper wire communications system and replacing it with optical fibres and lasers," says Dr. David Faux from the University of Surrey in the U.K. This is currently impossible because existing semiconductor lasers—which send the signals through optical fibres—don't function properly at the extremes of temperature an aircraft is subjected to. However, "quantum dots", which Dr. Faux's team are studying, could one day form the basis of a new type of semiconductor laser that should work over a much wider range of temperatures, increasing the number of different environments optical fibre communications systems can be used in.

Quantum dots (QDs) are tiny regions of semiconductor around one billionth of a metre in size. They consist of a small group of semiconductor atoms buried inside another type of semiconductor material. To make successful QD lasers, the electronic and optical properties of quantum dots will need to be fully understood. These properties are affected by strain, and all quantum dots are under strain because there is a mismatch either in interatomic distances or in atomic sizes between the two types of semiconductor.

The Surrey researchers began their studies by investigating indium arsenide (InAs) quantum dots embedded in gallium arsenide (GaAs) (see Plate 4.1). "The indium atom is physically larger than the gallium atom, so the indium arsenide crystal has to be squeezed in order to have the same atomic spacing

PLATE 4.1   This illustration shows a pyramidal indium arsenide (InAs) quantum dot with a square base 6.7836 nm wide. A technique known as Molecular Beam Epitaxy (MBE) can be used to form InAs quantum dots. This involves directing beams of indium atoms and arsenic atoms at a heated substrate (base) of gallium arsenide inside an ultra-high vacuum chamber. Indium arsenide quantum dots—which are thought to be shaped like pyramids—form all over the substrate surface, and are finally completely surrounded and covered by a layer of gallium arsenide which forms when beams of gallium atoms and arsenic atoms are then directed at the substrate. It is hoped that quantum dots could one day form the basis of a new type of semiconductor laser that would operate over a much wider range of temperatures than existing semiconductor lasers. (Courtesy of David Faux and Stuart Ellaway, Advanced Technology Institute, University of Surrey, U.K.)

as the gallium arsenide crystal that surrounds it," explains Dr. Faux. To enable the strain this causes to be calculated as accurately as possible—and thereby improve the chances of completely understanding the optical and electronic properties of the quantum dot—Dr. Faux's team used a computer-based technique known as a Molecular Dynamics simulation.

Molecular Dynamics simulations use values of interatomic forces and distances to work out how large numbers of atoms move when a crystal at any given temperature is distorted. For the strain throughout the QD to be accurately calculated, any change in the elastic properties—that represent the extent to which it returns to its original shape and size after being stretched or squashed—caused by strain need to be taken into account. So the team used the simulation to determine the elastic stiffnesses (measures of elasticity along particular directions) of a simulated cube of indium arsenide for a range of different applied strains. "The indium arsenide crystal is subjected to about 7% strain. This is a massive amount of strain, equivalent to stretching 1 metre of wire by 7 centimetres. The question we needed to answer was will the elastic 'constants' of indium arsenide be constant over this range of strain?" says Dr. Faux.

Early in 2002 his team announced that two out of the three elastic stiffnesses measured altered significantly as the strain on the quantum dot varied (see Plate 4.2). This meant that the elastic constants—which are derived from

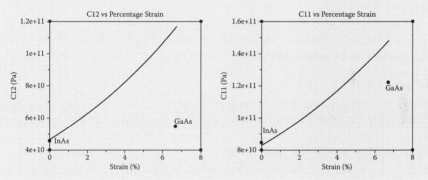

PLATE 4.2   For crystals like indium arsenide (InAs) that have cubic symmetry, the elastic properties can be described by just three elastic stiffnesses—C_{11}, C_{12}, and C_{44}—which each represent the elasticity along a different direction. These graphs show how C_{11} and C_{12} of a simulated InAs quantum dot increase significantly as the amount of strain on the dot is increased. C_{44}, however, was found to be almost constant under strain. The solid circles in each graph represent the accepted values of C_{11} and C_{12} for both indium arsenide when it is not under strain in a quantum dot, and for gallium arsenide, which surrounds the InAs dot. From the positions of the GaAs circles, it is clear that the elastic stiffnesses of the quantum dot are not the same as those of the material surrounding it. Including the strain dependence of the elastic stiffnesses in calculations should provide an accurate value of the strain in quantum dots. This would improve the understanding of their electronic and optical characteristics, and increase the chances of successful semiconductor lasers being made from them. (Courtesy of David Faux and Stuart Ellaway, Advanced Technology Institute, University of Surrey, U.K.)

the elastic stiffnesses—were not in fact constant at all. Speaking shortly after, Dr. Faux said he now planned to study other quantum dots, and hoped that including the strain dependence of their elastic properties in calculations would increase the chances of successful lasers—that could be used in aeroplanes and even homes—being made from them. "Incredibly quick communications coming into households would have an enormous impact," he stated.

ANSWER TO QUESTION 4.5 ON IMPACT TESTS

The more ductile a material is, the more energy it can absorb before breaking, so the PVC sample is more ductile at 40°C than at 20°C.

(also known as the "volume elasticity") applies to solids—and in fact fluids—under compression. The bulk modulus (K) is defined as the ratio of the pressure applied to a material to the fractional change in volume that the material undergoes as a result of this pressure. The reciprocal of the bulk modulus (1/K) is known as the "compressibility". This is therefore the fractional change in volume per unit pressure applied.

Another commonly used modulus is the "rigidity modulus" or "shear modulus", which is similar to Young's modulus but represents the ratio of shear stress to shear strain rather than tensional stress.

Poisson's Ratio

When materials are elastically stretched, compressed, or sheared they will change shape. In the case of a rod under tension, if the tensile stress is applied along the longest axis and elongates the rod, the diameter will be reduced. This is just the same situation that occurs when an elastic band (like the one shown in Figure 4.7) begins to be stretched and becomes simultaneously longer, and thinner, and less wide. Conversely, when an object is compressed, it will generally bulge outwards along both of the directions that are perpendicular to the applied compressive stress. (You can test this out by squashing a pencil eraser between your fingers.)

The amount a material bulges when it is squeezed can be measured in terms of a ratio known as Poisson's ratio, v, which is defined as

$$v = -\frac{\varepsilon_x}{\varepsilon_z} = -\frac{\varepsilon_y}{\varepsilon_z} \tag{4.4}$$

where ε_z is the strain in the z direction for a tensile stress applied along the z direction which stretches the material, and ε_x and ε_y are the compressive strains in the x direction and y directions, respectively. It is named after the French mathematician and

---

TABLE 4.3
Poisson's Ratio for Various
Metal Alloys

Metal or Metal Alloy	Poisson's Ratio
Aluminium (Al)	0.33
Brass (CuZn)	0.34
Copper (Cu)	0.34
Magnesium (Mg)	0.29
Nickel	0.31
Steel	0.30
Titanium	0.34
Tungsten	0.28

---

mathematical physicist Siméon-Denis Poisson (1781–1840), who helped establish a mathematical theory of elasticity.

Whether a compressive strain or a tensile strain is considered, it is clear from referring back to equation 4.2 that the sign of the strain along the ε_z axis will be opposite to that along either the ε_x or ε_y axes. This is why Poisson's ratio contains a negative sign—it keeps the ratio positive.

The maximum value that Poisson's ratio can have is 0.5, which indicates that there has been no change in the overall volume of the material under strain. Plastics such as nylon have values of Poisson's ratio around 0.4, and rubbers approach the maximum with a value of 0.49. However, for most metals and alloys (see Section 4.2.1), Poisson's ratio ranges from about 0.25–0.35, as Table 4.3 shows.

Atomic Movement

So far in these calculations, we have assumed that the solid is a continuum. We have not taken into account the individual atoms that the solid is made from and so have only looked at elasticity from a purely classical point of view. We will look in more detail at elasticity in terms of the movements of the atoms in the solid being stretched in Chapter 5. In the meantime, it is worth noting that if the deformation of any crystalline solid is uniform (the same) throughout it, every primitive cell within the crystal will be deformed in exactly the same way.

4.1.5 HARDNESS

Hardness is defined as the resistance of a material to being dented or scratched. Diamond is the hardest naturally occurring substance, and is used for industrial cutting as it can cut through any other material including itself.

There are several tests for measuring hardness. The oldest of these was invented in 1812 by the German mineralogist Freidrich Mohs (1773–1839). He introduced a relative measure of hardness known as Mohs scale, in which each of the ten materials on the scale can scratch either themselves or any materials with numbers below their

EXAMPLE QUESTION 4.6   POISSON'S RATIO

Given that Young's modulus for copper is 110 GPa, while Poisson's ratio is 0.34, how much force needs to be applied to a 5 mm diameter copper rod to produce a 1×10^{-3} mm decrease in its diameter?

ANSWER

To reduce the diameter of the rod, a tensile stress needs to be applied along the longest axis, which is the z-axis. As we know the initial diameter (d) and the final diameter (d_0), we can first work out the strain in the x direction as follows:

$$\varepsilon_x = \frac{\Delta d}{d_0} = -\frac{0.001}{5} = -2 \times 10^{-4}$$

(This strain is negative because the diameter of the rod is being reduced by the stress.)

We do not know how much elongation there is in the z direction, but we can work out the strain in this direction, as we have been given a value for Poisson's ratio. So

$$\varepsilon_z = -\frac{\varepsilon_x}{\nu} = -\frac{(-2 \times 10^{-4})}{0.34} = 5.88 \times 10^{-4}$$

As we have been told that the deformation is elastic, we know that Hooke's law must apply. This means we can use our knowledge of the Young's modulus of copper to work out the stress in the z direction, and from this we will finally be able to calculate the force required to produce this shape change as follows:

$$\sigma = \varepsilon_z E = 5.88 \times 10^{-4} \times 110 \times 10^3 = 64.68 \text{ MPa} = 64.68 \times 10^6 \text{ Nm}^{-2}$$

and if A_0 is the original cross-sectional area before stretching,

$$F = \sigma A_0 = \sigma \left(\frac{d_0}{2}\right)^2 \pi = 64.68 \times 10^6 \left(\frac{5 \times 10^{-3}}{2}\right)^2 \pi = 1270 \text{ N}$$

own, but none of the materials with numbers above theirs. For example, quartz can scratch feldspar but not topaz. Diamond, which can only be scratched by another diamond, is 10 on this scale, while talc, which cannot scratch anything, is material number 1 (see Table 4.4).

Whilst Mohs scale is a useful guide for quickly determining roughly how hard a material is, it is not very scientific in today's terms, so mechanical hardness tests are generally used. These involve pressing either a pyramid-shaped diamond or a steel ball into the surface of the material under study. In both the Brinell and Vickers

TABLE 4.4
Mohs Hardness Scale

Number	Material
10	Diamond
9	Corundum
8	Topaz
7	Quartz
6	Feldspar
5	Apatite
4	Fluorite
3	Calcite
2	Gypsum
1	Talc

Note: Each material can scratch those with numbers below its own, so for example quartz can scratch feldspar but not topaz. Diamond is the hardest material on this scale, and indeed is the hardest naturally occurring substance known.

hardness tests, in which a steel ball and diamond cone are used as indenters, respectively, the hardness is then calculated from the surface area of the indentation produced. By contrast for the Rockwell test, which can use either a diamond cone or a steel ball to produce the indentation, the hardness is worked out from how far the indenter penetrates the sample. As each of the three tests measures a slightly different form of deformation, there is no simple mathematical relationship between the various hardness scales, although Figure 4.8 shows graphically how the various scales relate to each other and where some common materials lie on these scales.

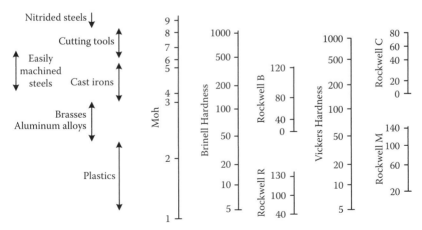

FIGURE 4.8 A comparison of the various hardness scales. On the left the hardness ranges of some common materials are indicated.

4.2   THE RIGHT MATERIAL FOR THE JOB

Of course the mechanical properties of materials are not the only considerations when choosing a material for a particular job. Cost, ease of processing, the impact on the environment, nonmechanical properties such as the electrical and thermal conductivity, and whether the material can be recycled are just some of the other factors which may need to be taken into consideration.

Sometimes no one material has the desired properties, and in the next two sections we will look at ways in which the properties of materials can be altered for particular applications and how on some occasions a mixture of materials needs to be used.

4.2.1   ALLOYS AND COMPOSITES

Alloys

Mixtures of two or more metals, that can also include nonmetallic elements, are known as *alloys*. Pure metals rarely have the properties needed for industrial applications, so alloys, which can be manufactured to have more or less exactly the required combination of properties, are often used instead. For example, chromium can be used to improve the corrosion resistance of steel, which is itself an alloy of iron and carbon. Table 4.5 lists some common alloys.

Alloys are usually used if a metal with a high strength is needed, as most pure metals are quite soft. Table 4.6 shows some of the properties of a range of nonferrous

TABLE 4.5

Some Common Alloys and Their Basic Constituents

Alloy	Constituents
Gilding brass	85% Copper and 15% Zinc
Iconel	78% Nickel, 15% Chromium, and 7% Iron
Nickel silver	≈20% Nickel, 60% Copper, and 20% Zinc
Phosphor bronze	≈95% Copper, 5% Tin, and 0.02–0.4% Phosphorus
Stainless steel (for cutlery)	≈86% Iron, 0.32% Carbon, and 13% Chromium

TABLE 4.6

The Tensile Strengths and Strength-to-Weight Ratios of a Range of Alloys

Main Constituent of Alloy	Tensile Strength (N mm^{-2})	Tensile Strength/Density (N mm^{-2}/10^3 kgm^{-3})
Aluminium	100–550	37–200
Copper	200–1300	25–160
Nickel	400–1300	47–146
Titanium	400–1600	89–365
Zinc	200–350	30–52

alloys (alloys that do not contain iron), including the variation in tensile strength. If the only alloys available were the ones listed in Table 4.6, titanium alloys would be the best choice, as they have the highest strength. However, copper alloys and nickel alloys also have high strength and, if cost was a factor, would be a better choice, as titanium alloys are by far the most expensive of these alloys. Titanium alloys are also the most corrosion resistant of the alloys considered here, although aluminium also has good corrosion resistance.

EXAMPLE QUESTION 4.7   ALLOYS

Which of the alloys listed in Table 4.6 would be the best choice for making the following items and why?

(a) a replacement hip joint
(b) an aircraft frame

Answer on page 120.

In order to make things like automobile bodies and medical implants (see Example question 4.7) the material must be able to be machined easily; that is, it must not be difficult to cut or shape into a particular part. In general, ductile materials such as pure metals are not easy to machine. They tend to drag on any tools used to cut them, so alloys are often used instead because they combine the useful properties of the metal with qualities such as machinability and castability provided by their other component materials. Although many alloys are artificially created, natural alloys do exist. For example, some types of meteorite contain iron-nickel alloys.

Composites

Alloys are not the only way of combining the properties of two or more materials to make a material suitable for a particular application. Like alloys, *composite* materials combine the properties of two or more materials when no single material is up to the job. Examples of composites include fibreglass and reinforced concrete. The latter has steel rods embedded in it so that the concrete—which is usually weak under tension—is much stronger when stretched. Reinforced concrete has become an important modern building material, although composite building materials have existed for hundreds of years. The "wattle and daub" used by medieval housebuilders in Europe is a composite material too.

4.2.2   ALTERING THE MECHANICAL PROPERTIES OF A SOLID

If we want bread to be more crispy when we eat it, we would cook it in some way. This alters its structure and produces toast. In a similar way, "cooking" solids like metals, alloys, and glasses and then cooling them at different rates alters their structure. Any process involving heating and cooling a solid to change its properties is known as a *heat treatment*.

ANSWER TO QUESTION 4.7 ON ALLOYS

(a) The high level of corrosion resistance coupled with the high strength and strength to weight ratio is why titanium alloys are used for medical implants such as replacement joints, and pins holding broken bones in place while they heal. Cost is not really a factor for this application as only relatively small quantities of titanium are involved, and the well-being of the patient is of primary concern. By contrast one of the most common types of medical implant—namely the crowns and bridges used in dentistry—are made from a gold alloy. Dental alloys in fact usually contain 60%–75% gold as well as silver, palladium, platinum, copper, and zinc.

(b) For applications like aircraft and racing cars, the strength-to-weight ratio is important, as it allows them to travel as quickly as possible with the minimum amount of fuel. Materials that are to be used for things like automobile bodies must of course be corrosion resistant as well, and this therefore narrows the choices down to titanium and aluminium alloys. Because of the much higher cost of titanium, aluminium alloys are widely used in the aviation industry and to make parts for racing cars and bicycles.

One of the most common heat treatments is *annealing*. This involves heating the solid, often to a high temperature (although this temperature is always below the melting point of the solid), keeping the solid at that temperature for a certain amount of time, then slowly cooling it. Annealing makes metals and alloys softer and more ductile, and therefore easier to work with, and strengthens glasses. In all cases, the annealing process removes stresses that have been produced during the fabrication of the materials.

For certain applications, solids need to be made harder, and this can be achieved via a process known as *quenching*. Quenching involves heating a solid and then cooling it very rapidly. The faster the cooling rate, the harder the solid produced, and plunging a hot solid into brine gives one of the fastest cooling rates. Using water as the quenching medium cools solids almost as fast as brine, while oil gives a slower cooling rate than water. Care needs to be taken when quenching to ensure the high cooling rates do not distort or crack the solids. Distortion can be minimised to some extent by inserting whichever face of an object has the smallest cross-sectional area into the quenching medium first. In addition, shaking the quenching "bath" helps prevent bubbles of steam sticking on the surface of the object being quenched and causing different parts of the solid to cool at different rates, and so cause stresses to develop. Quenched steels can also be tempered, that is, heated to a temperature lower than the original quenching temperature, then cooled in air to remove some of the stresses caused by the quenching.

Sometimes it is useful to harden just the surface layers of a metal or alloy, and there are several ways of doing this. For example, one method of surface hardening carbon steels is to heat their surface with an oxyacetylene flame, then quench them before the rest of the steel has reached the temperature of the surface.

It is also possible to harden crystalline materials without heating them. This *work hardening* process, or "cold working" as it sometimes called, involves applying a stress greater than the elastic limit to a crystal. The strain this stress causes creates dislocations, which then become tangled with the existing dislocations and so make the material harder, as it is more difficult for slip to occur when dislocations are locked together. Annealing a metal or alloy after work hardening can restore its ductility and malleability.

4.2.3   RECYCLING

Even when a material has been found (or been created) with the necessary properties for the use it is to be put to, other issues such as the cost of not only the raw material but also the creation of the end product from this material need to considered. On top of any economic concerns, the environmental impact of the materials should also influence any decisions. This may not only be how the environment will be affected by the material when it is in use in its particular application, but also how it can be disposed of when it is no longer required and also how much pollution is produced manufacturing it in the first place.

As we use up more and more of the world's natural resources, and as the amount of waste we dispose of increases—in the European Union, for example, approximately 1.3 billion tonnes of waste was thrown away in 2008—the need for recycling becomes ever greater. Companies do of course need to make a profit to remain in business, so the benefits to the environment have to weighed up against manufacturing costs and whether any increased costs for making "green" products are likely to be borne happily by the consumer. Although it often costs more overall to produce a product made from recycled materials, certain parts of the production process are however substantially cheaper. For example, around 28 times less energy is used to recycle aluminium drinks cans than is needed to refine raw aluminium ores—in other words change them from their natural state to a point where they are useable. Energy savings like this also do their bit for the environment by reducing overall energy consumption.

Some materials are easier to recycle than others. Rubber for instance is tricky to recycle when it has been vulcanised as it becomes a thermoset, and the rubber used for vehicle tyres has to be mixed with various additives—which can be tricky to remove—to prevent it from combusting at high speed. However, various objects ranging from toys and doormats to stationery and handbags have been successfully made from used rubber. It has even found its way into the art world, where it has provided the raw material for some sculptures.

By contrast, most thermoplastics can be relatively easily recycled, and many countries operate recycling schemes for PET (polyethylene terephthalate) soft drinks bottles. (In Sweden, for instance, 89% of the PET bottles used for packaging in 2006 were recovered.) The recovered bottles are first washed and their labels are removed. They are then cut up into small flakes before being melted. The molten material is

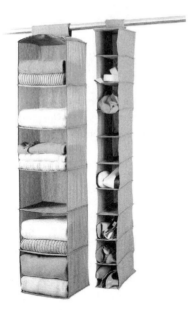

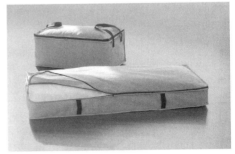

FIGURE 4.9 The clothes storage bags shown here are made from a fabric known as PETSPUN, which is manufactured from recycled PET (polyethylene terephthalate) soft drinks bottles. PET is a type of lightweight, strong, and transparent polyester that provides an effective barrier to moisture and gas and is widely used to make containers for food, drinks, and household cleaning fluids. Once these storage bags have come to the end of their useful life, the PETSPUN fabric can be recycled, although it has to be mixed with other recycled fabric products in order to be reused. Many countries run collection schemes for recyclable materials like PET. (Courtesy of Lakeland Limited.)

finally converted into either small beads which can be made into objects such as new bottles, or a thread which can be spun until it is very fine then used to weave textile products such as fleeces and carpets (see Figure 4.9). In the late 1990s, the U.K. alone was using an average of 5 million tonnes of plastic annually, and over half this amount was being disposed of as waste each year. So recycling plastic products makes a dramatic reduction in the amount of waste being buried in landfill sites.

Since landfill produces the greenhouse gas methane, any reduction in this method of waste disposal is good news for the environment. As an example, the total amount of recycling in the U.K. in 2007 reduced greenhouse gas emissions by the same figure that would be achieved if 3.5 million cars were taken off the U.K.'s roads. Part of this saving was caused by the used products not going into landfill, and the rest by not needing to use new materials to make whatever the recycled material is being turned into.

Aluminium—in the form of used drinks cans and automobile bodies—is another widely recycled material, while some alloys and most composite materials either require very complex techniques to be recycled or cannot be recycled with existing technologies. A viable alternative to recycling is to make biodegradable products. For example some plastic carrier bags are completely degradable.

FURTHER READING

Davidson, P. "A problem of perception in plastics recycling". *Materials World* 33 (March 2004).
Kahn, J. "Nano's big future", *National Geographic* 209, no. 6 (June 2006): 98–119.
(More references for further reading, as well as Web links, are available on the following Web
    page: http://www.crcpress.com/product/isbn/9780750309721.)

SELECTED QUESTIONS FROM
QUESTIONS AND ANSWERS MANUAL

Q4.2   Name the unit strain is measured in.

Q4.3

  (a) What is the "limit of proportionality" often known as?
  (b) What type of diagram are you most likely to see the answer to Q4.3(a) indicated on?
  (c) Describe the relationship between stress and strain for a solid below its limit of proportionality.
  (d) Who discovered this relationship?
  (e) Express this relationship mathematically.

Q4.5   What will the increase in length be in a 12 cm long piece of aluminium (with a Young's modulus of 69 GPa) that is subjected to a tensile stress of 240 MPa? (Assume the deformation is elastic.)

Q4.7   What is the difference between elastic deformation and plastic deformation?

Q4.9

  (a) What is the "compressibility" of a material a measure of?
  (b) What does Poisson's ratio reveal?
  (c) Fluorite has a Mohs hardness value of 4, while feldspar has a Mohs value of 6. Which material is the hardest?

5 In, Out, Shake It All About

Diffraction, Phonons, and Thermal Properties of Solids

CONTENTS

5.1 DIFFRACTION

It's 1781, and the French mineralogist René Just Haüy (1743–1822) is having a bad day. He's just dropped some of the calcite crystals he is studying on the floor. But this accident is about to turn into the best—albeit unplanned—move of his career. One of the crystals has shattered into pieces, and as Haüy looks at the fragments he discovers to his surprise that they are all trigonal in shape.

After deliberately breaking some more calcite crystals and getting an identical result, Haüy concluded that the atoms calcite is composed of always group together to form the same basic shape. Further work led him to suggest that there were only six fundamental geometrical shapes that crystals could have. Although we now know there are in fact seven crystal systems, this was still a huge breakthrough in the understanding of crystal structures, and scientists would have to wait over 130 years for the next major step forward.

This began in 1912 when the German physicist Max von Laue (1879–1960) discovered X-ray diffraction, and concluded the following year when British father and son team William H. Bragg (1862–1942) and William L. Bragg (1890–1971) discovered their famous law. There was now a way of seeing exactly where the individual atoms inside crystals were positioned, and X-ray diffraction images taken by British crystallographer Rosalind Franklin (1920–1958) played a part (the extent of which is hotly debated) in the determination of the structure of the DNA molecule by Watson and Crick in 1953. Diffraction is still used to identify unknown solids and molecules, but today both electrons and neutrons are used as well as X-rays to investigate the widest possible range of solids.

5.1.1 PROPAGATION OF ELECTROMAGNETIC RADIATION

The propagation of electromagnetic (sometimes shortened to "e-m") radiation, in other words the way in which e-m radiation travels along, affects many areas of physics. Whilst the fact that the speed of light in a vacuum is constant is the cornerstone of relativity, and the supposition that light can behave both as a wave and as a stream of particles is fundamental to quantum mechanics (see Appendix D for more on wave-particle duality), solid state physics is concerned with how e-m radiation is absorbed, generated, scattered, transmitted, or reflected by solids.

In Chapter 7 we will look at absorption and emission, at optoelectronic devices that emit light (such as LEDs and semiconductor lasers), and at devices like solar cells that detect or respond to light. In this chapter, we will consider instead the way in which e-m radiation interacts with the periodic array of atoms that make up a crystal.

Diffraction of Electromagnetic Waves

When a beam of light—or indeed any other form of electromagnetic radiation—passes through a hole or slit, or round the outside of an object, it breaks up into a number of smaller beams. This process is known as *diffraction*, and "diffraction patterns" consisting of bright and dark regions are produced when diffraction occurs and the broken-up beams interfere with one another. The light regions of the pattern are the result of constructive interference between any two (or more) diffracted beams, while the dark regions occur wherever destructive interference has taken place. (See Appendix B for a reminder of the properties and behaviour of waves, including interference effects.)

5.1.2 HOW WAVES INTERACT WITH CRYSTALLINE SOLIDS

When the term "diffraction" is used in the context of solid state physics, it is taken to mean diffraction of e-m radiation by the atoms of a solid that has resulted in constructive interference. As a consequence, a diffraction pattern will have been produced and generally recorded in some way. Crystalline solids can therefore be considered to act like huge three-dimensional (3-D) diffraction gratings. In fact the spacing between the planes of atoms in a solid can be determined in the same way that the spacing between the slits in a diffraction grating is revealed when the diffraction pattern created by light of a known wavelength is analysed (see Appendix B).

Diffraction patterns are created by crystals as a result of the way in which e-m radiation interacts with the component atoms. Although light can produce diffraction patterns when shone through an optical diffraction grating, it is not used for diffraction experiments. This is because compared with an atom, the wavelength of visible light is large and it washes over each atom like a large ocean wave washes over a small pebble on a beach without its path being changed in any way by the obstacle. If however e-m radiation of a smaller wavelength (such as X-rays) is shone at a crystal, the atoms are large enough in comparison with these waves to interact with them—in the same way that a high concrete wall would affect a large ocean wave that crashed into it.

The Bragg Law

The Braggs were the first to show that X-ray diffraction patterns could reveal the arrangement of atoms inside crystals. They stated that the atoms lay in planes that could act like mirrors and "reflect" the X-rays that entered the crystal. Figure 5.1 shows a set of parallel planes with a beam of X-rays shining on them. A "diffraction spot" will be produced whenever there is constructive interference between the X-rays

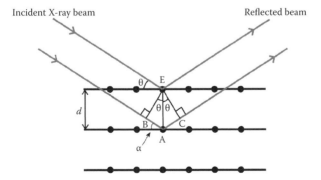

FIGURE 5.1 When a beam of X-rays hits a crystal, it behaves as if it has been reflected by some of the atomic planes. This diagram shows what happens when two rays from a beam hit a pair of atomic planes separated by a distance, d. (The atoms marked A, B, C, and E are used in conjunction with the main text to help derive the Bragg law.)

FIGURE 5.2   An experimentally measured X-ray diffraction pattern obtained by the powder method, which will be described in Section 5.1.3. Each line represents the diffraction from a single set of planes. (Courtesy of Alan Piercy.)

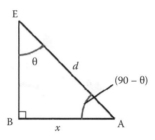

FIGURE 5.3   The triangle EBA from Figure 5.1.

reflected from neighbouring planes. Figure 5.2 shows an experimentally measured diffraction pattern consisting of an array of diffraction lines.

The conditions under which constructive interference of reflected X-rays occurs are summarized in the Bragg law, which can be derived by looking closely at Figure 5.1. If we just consider the top two atomic planes in the diagram, which are separated by the distance d (usually referred to as the inter-plane spacing), it is clear that the lower of the two X-rays has to travel further than the upper ray before leaving the crystal. In fact its path length is BAC longer. If the angle of incidence of the X-rays with respect to the crystal planes is taken to be θ, the angle marked α will also be θ, as it is a complementary angle. If perpendiculars are drawn from the atom at point E to the lower ray (hitting the ray at points B and C), and another line is drawn connecting the atoms at point E and point A, two little right-angled triangles are formed. The angles at the top of these triangles will be θ because there is already a right angle—since $\alpha = \theta$—plus the angle $\beta = (90 - \theta)°$ at the bottom of each triangle. To make this easier to visualise, Figure 5.3 shows one of these little triangles separately.

From Figure 5.3 we can now find the distance, x, from B to A, as $\sin \theta = x/d$, and therefore $x = d \sin \theta$. The other little triangle will produce the same result for the distance AC, so that the total path length BAC will be equal to $2d \sin \theta$.

It is only possible to get constructive interference from two parallel rays if the difference in their path length is an integral number of wavelengths. So BAC must equal $n\lambda$, where n is an integer which indicates the order of the bright image (see Appendix B) and, as BAC = $2d \sin \theta$,

$$2d \sin \theta = n\lambda \qquad (5.1)$$

which is known as the Bragg law.

When constructive interference is obtained from a crystal in real life, it is not just made up from reflections from the top two planes. X-ray beams penetrate quite deeply into crystals, and anything from 10^3 to 10^5 planes of atoms can contribute to the reflected beam. There are also many different sets of lattice planes within a crystal that can produce reflections.

For a thick crystal, if the incident radiation is not quite at the Bragg angle, a situation can arise that results in no diffraction pattern being produced. In this case the reflections from successively deeper planes are increasingly out of phase with the reflection from the uppermost plane. In a thin crystal, this would just mean that the diffraction spots or lines would not be as intense as those produced from X-rays incident at the Bragg angle, as the contributions from the different planes would only partially reinforce. However, with a thick crystal, there are so many atomic layers that eventually a point is reached where the light "reflected" from one of the layers is exactly out of phase with that from the first layer. Destructive interference of these two diffracted waves then occurs, so the contributions from these two planes cancel each other out.

The same process can occur for other pairs of planes, and in some cases every contribution is cancelled out so no diffraction pattern is seen. For thick samples, therefore, it is extremely important that the angle of incidence of the X-rays to the crystal planes is as close to the Bragg angle θ as possible so that the contributions from all planes in a set of crystal planes are in phase.

We saw earlier that the Braggs suggested planes of atoms were "reflecting" the X-rays, and this term is often used in discussions on X-ray diffraction because the X-rays under consideration are behaving as if they have been reflected. In reality, however, the X-rays are scattered in all directions by the orbiting electrons in the atoms that make up each plane, but the only scattered X-rays that can be detected are the ones which constructively interfere—and so obey the Bragg law.

Armed with this law, we can determine the shape and size of the unit cell of an unknown solid, as the angle θ (sometimes referred to as the Bragg angle) of the maxima in a diffraction pattern reveals the distance between the planes of atoms making up the solid. However, in order to work out the arrangement and type of atoms within the unit cell—and so have enough information to determine the crystal structure—the intensities of the diffracted beams also need to be taken into account, as the next two subsections explain.

Atomic Scattering Factor

As we have just seen, X-rays are scattered by the electrons in the atoms that make up solids. The intensity of a diffracted beam will depend not only on the amount of scattering a single atom produces, but also on the scattering from the unit cell. We will look at how the arrangement of the atoms in a unit cell affects the intensity in the next subsection, after first considering the scattering for a single atom.

In general, the more electrons in an atom, the stronger its interaction with the X-rays is, and the stronger the diffracted beam will be. The intensity of the diffracted beam for any given type of atom depends, in fact, on its "atomic scattering factor", f_s.

If θ_s is the angle through which the incident radiation is scattered, the atomic scattering factor, f_s is defined as

$$f_s(\theta_s) = \frac{\text{scattering amplitude at } \theta_s \text{ from the actual electron distribution in an atom}}{\text{scattering amplitude at } \theta_s \text{ from a free electron}} \quad (5.2)$$

Atoms with a large atomic number (Z) have a large value of f_s, as f_s is approximately proportional to the number of electrons per atom. (See Appendix C for a reminder of atomic structure.)

This dependence of the scattering intensity on the atomic number of the atoms under investigation means that X-ray diffraction from light atoms can be very weak. As a rough guide, atoms heavier than carbon tend to give diffracted beams strong enough for analysis, whereas those lighter than carbon do not. As a consequence, some structures—including those of organic substances—are better investigated using either electron or neutron diffraction (which will be discussed in Section 5.1.4).

Structure Factor

We now know that for diffraction to occur, it is not enough for the Bragg condition (equation 5.1) to be satisfied. The atomic scattering factor must also be near to the maximum value for the particular atoms under consideration (in other words as close to Z as possible) if the diffracted beam is to be intense enough to be observed. Of course real solids do not consist of just one atom. So in addition, a quantity known as the "geometrical structure factor" (F_{hkl}), which takes into account the positions of the atoms within the unit cell of the solid, needs to be considered when determining whether diffraction will occur from a particular set of atomic planes.

The geometrical structure factor can be zero, small, or large for a given (hkl) plane. It tells us how much the interference from atoms in the unit cell reduces the scattering intensity for that plane: the greater its value the greater the intensity of the diffraction observed. (There is no need to consider the contributions from every single atom of a crystalline solid because, as we saw in Chapter 2, if a unit cell is repeated again and again, the entire structure of the crystal is built up.)

As equation (5.3) reveals, the geometrical structure factor does not just depend on where the atoms of a unit cell actually are. It also depends on the atomic scattering factor for the atoms as shown:

$$F_{hkl} = \sum_{x=1}^{N} f_{s_x} \exp\left[2\pi i \left(hS_x + kS_y + lS_z \right) \right] \tag{5.3}$$

where N is the number of atoms in the unit cell, f_{sx} the atomic scattering factor of the xth atom of this unit cell, and the exponential term represents the phase difference between an X-ray scattered by an atom on the (hkl) plane with atomic coordinates S_x, S_y, and S_z and an X-ray scattered from an atom situated at the origin of the same unit cell. (Atomic coordinates were discussed in Section 2.2.7.)

However, in cases like that shown in Figure 5.4(b), which represents the diffraction from the (001) planes of a body-centred orthorhombic lattice, it is irrelevant what the atomic scattering factors of the atoms in the unit cell are. This is because there is no diffraction at all from this set of planes. To understand how this happens, we first need to look at the situation that arises when X-rays hit the (001) planes in a base-centred orthorhombic lattice.

As Figure 5.4(a) reveals, if the Bragg law is assumed to be satisfied for this particular combination of wavelength and incident angle, beams 1 and 2 must have a path

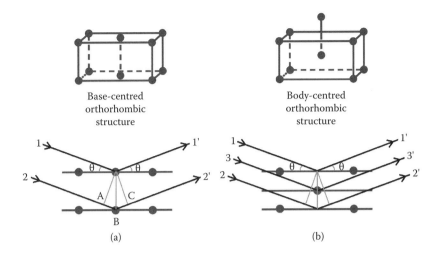

FIGURE 5.4   Diffraction from (a) the (001) planes of the base-centred orthorhombic lattice, and (b) the (001) planes of the body-centred orthorhombic lattice.

difference, ABC, of an exact integral number of wavelengths and therefore interfere constructively and add together to produce diffraction. With the addition of an extra plane of atoms halfway between these two planes however—which is the position the atom from the body-centre of the body-centred orthorhombic lattice shown in (b) occupies—destructive interference occurs between beams 1 and 3, and they cancel each other out as they are exactly half a wavelength out of phase. (A half-wavelength phase difference means the peaks of one wave line up with the troughs of the other [see Figure B6 in Appendix B], and so the waves interfere destructively with one another.)

This process continues throughout the crystal, with, for example, the plane of atoms directly below those illustrated producing a beam that cancels out beam 2.

It is now becoming clear how diffraction patterns can reveal information about solids. These patterns will have "missing" spots or lines, and the spots or lines that are present will have particular intensities. As mentioned previously, measuring the angle, θ, of the maxima in a diffraction pattern will reveal the distance between the adjacent pairs of atomic planes that created those maxima. The combination of these results should then be enough to enable identification of a crystal by comparing it with the data collated by the International Centre for Diffraction Data.

As we will see in the next section, diffraction patterns look different, depending on the method used to obtain them. Whether they are in the form of spots or lines, however, the purpose of the diffraction pattern is to obtain information on how the atoms are arranged inside the crystal being investigated. Samples can either be single crystals or polycrystalline. In fact, art conservationists and historians can identify the pigments used in paintings by obtaining X-ray diffraction patterns from polycrystalline paint samples. In this case a surgical scalpel is used to detach a tiny fragment of paint either from the edge of the painting being investigated—which is normally covered by the frame—or from a damaged area. The diffraction pattern produced by

EXAMPLE QUESTION 5.1   GEOMETRICAL STRUCTURE FACTOR

An fcc crystal can be described by the following four sets of atomic coordinates: 000, 0½½, ½0½, and ½½0. Is there a Bragg reflection (in other words a reflection that satisfies the Bragg law) from the (100) plane of an fcc crystal?

ANSWER

Inserting both these coordinates and the Miller indices of the plane into equation 5.3 gives us

$$F_{100} = \sum_{x=1}^{N} f_{s_x} \exp\left[2\pi i \left(hS_x + kS_y + lS_z \right) \right]$$

$$= f_s e^{2\pi i \left(1\times 0 + 0\times 0 + 0\times 0 \right)} + f_s e^{2\pi i \left(1\times 0 + 0\times \frac{1}{2} + 0\times \frac{1}{2} \right)} + f_s e^{2\pi i \left(1\times \frac{1}{2} + 0\times 0 + 0\times \frac{1}{2} \right)} + f_s e^{2\pi i \left(1\times \frac{1}{2} + 0\times \frac{1}{2} + 0\times 0 \right)}$$

$$= f_s e^{2\pi i 0} + f_s e^{2\pi i 0} + f_s e^{\pi i} + f_s e^{\pi i}$$

$$= f_s \left(1 + 1 - 1 - 1 \right)$$

$$\therefore F_{100} = 0$$

as $e^0 = 1$ and $e^{\pi i} = -1$. Therefore, for an fcc crystal there will be no Bragg reflection from the (100) planes.

the sample is then compared with patterns previously obtained from known pigments so that it can be identified. Knowledge of the pigments used gives historians a better understanding of the impact old paintings would have had before their colours faded. It also allows conservationists to carry out more accurate restoration work.

Other solids including amorphous solids and even human hair can be studied using X-ray diffraction. For example, cosmetics companies carry out X-ray diffraction experiments on hair before and after treatments such as perming and blow-drying. This allows them to see what effects the treatments are having on the structure of the hair, and whether any of their products can in fact repair or prevent damage to our hair, and thus improve its appearance.

5.1.3 Obtaining X-Ray Diffraction Patterns

There are several different experimental set-ups that can be used to obtain X-ray diffraction pictures, and the choice of technique will often be determined by the type of sample available for analysis. In this section we will look at three of the most widely used experimental arrangements.

We saw in Section 5.1.2 that one of the fundamental conditions for constructive interference to take place is that the Bragg law (equation 5.1) is satisfied. This will only occur when θ, d, and λ have particular combinations of values. Therefore, when single crystals are being investigated, the methods used must either involve directing X-rays of many different wavelengths at the crystal, or making sure the sample is moved about so that a monochromatic (single wavelength) X-ray beam can enter the crystal at a variety of different angles. It is pointless to just direct a monochromatic beam of X-rays at a single crystal from a fixed arbitrary angle, as it is highly unlikely to produce diffraction.

Laue Method

In the Laue method, a polychromatic beam of X-rays (in other words a beam containing a broad spectrum of wavelengths) is shone at a single crystal. In this case, whatever the angle of incidence of the X-ray beam to the crystal, there will be a wavelength in the beam that will satisfy the Bragg condition and give rise to diffraction.

As Figure 5.5—which shows the experimental set-up for the Laue technique—reveals, the diffraction pattern is recorded on a photographic plate positioned on the opposite side of the sample to the X-ray source. Although the diffraction pattern of spots is characteristic of both the crystal structure and orientation, it is normally only used to discover the orientation of a known crystal. This is because each spot can in fact be made up of several different reflections which occur when the various wavelengths in the X-ray beam produce different orders of reflection from the same set of atomic planes. Not surprisingly this makes it tricky to determine new structures.

It is worth noting that whatever method is used, analysing X-ray diffraction patterns is not straightforward, and there are entire books devoted to the subject. Some of this complexity is caused by defects. For simplicity we are only considering "perfect"

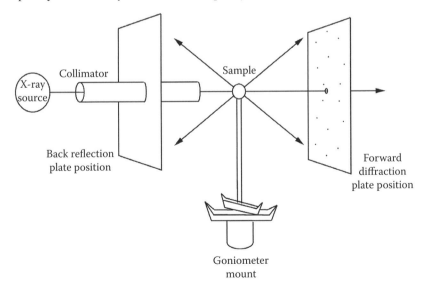

FIGURE 5.5   Experimental set-up for the Laue method of X-ray diffraction. (Copied from J. S. Blakemore, *Solid State Physics*, 2nd edition (1985), © Cambridge University Press. With permission.)

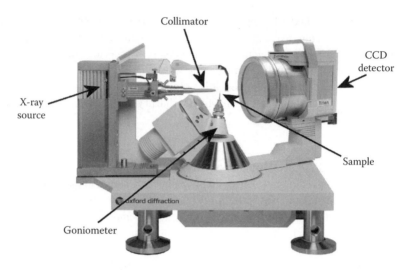

FIGURE 5.6 This annotated photograph shows Oxford Diffraction's Xcalibur Nova system for protein crystallography. The sample is mounted on a goniometer, which moves through many different angles in order to bring different planes within the sample into the Bragg diffraction condition. The monochromatic X-ray beam, which is made parallel by the collimator, typically has a wavelength of 1.54 Å when the structure of proteins is being investigated. The resulting diffraction is collected by the CCD detector, and hundreds or even thousands of diffraction patterns are recorded in order to determine the 3-D structure of a protein. Figure 5.7 shows an example of a diffraction pattern produced by this system. (Photo courtesy of Oxford Diffraction.)

crystals in this section, but as we saw in Chapter 3 there is no such thing as a perfect crystal in reality, and defects will affect the appearance of diffraction patterns. For example, the more dislocations in a crystal the less sharp the diffraction pattern is.

Rotating Crystal Method

By contrast, both the rotating crystal and powder methods (see next subsection) of diffraction are routinely used to determine unknown crystal structures. The rotating crystal method uses a monochromatic X-ray beam and, as its name suggests, involves moving the sample in order to bring planes into the Bragg angle.

Figure 5.6 shows a commercial system for protein crystallography in which the sample is positioned by the goniometer, allowing many different crystal planes to be brought into the Bragg diffraction condition. The diffraction patterns produced are recorded by a CCD detector (see Section 7.2.5) on the opposite side of the sample to the X-ray source.

Powder Method

In the powder (alternatively known as the Debye-Scherrer) method the sample is in powdered form, so there will always be some crystallites within the polycrystalline sample that will be sitting at the right orientation for diffraction to be produced from the monochromatic X-ray beam shining at them. (See Section 2.2.6 for a reminder of the structure of polycrystals.)

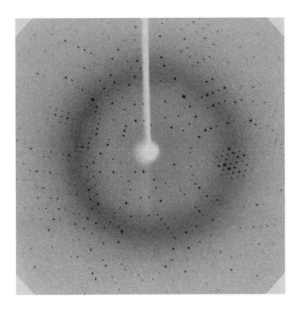

FIGURE 5.7   Diffraction pattern for insulin, produced by the system shown in Figure 5.6. (Courtesy of Oxford Diffraction.)

Unlike the rotating crystal method, which produces a diffraction pattern in the form of spots, the powder technique produces a series of circles or arcs, depending on the size of the photographic plate (see Figure 5.2).

The basic experimental arrangements used in both the rotating crystal and powder methods can also be used for neutron diffraction experiments (see Section 5.1.4), while the layout used for the powder method is also used in electron diffraction. It is also worth noting, as Box 5.1 reveals, that samples do not have to be at room temperatures or pressures when an X-ray diffraction experiment is taking place.

5.1.4   Electron and Neutron Diffraction

Most of our discussions so far have revolved around X-ray diffraction, but in some cases, X-rays are not the most suitable type of radiation to use to obtain a diffraction pattern. As we saw in Section 5.1.2, X-rays interact primarily with the electrons in atoms. This means that, for crystals containing atoms of elements lighter than carbon which contain few electrons, and for looking at surfaces, electrons—which interact strongly with every type of atom—are more suitable.

As electrons are scattered in air, electron diffraction experiments have to be carried out in a vacuum, and in fact nowadays are performed inside electron microscopes. Being charged, electrons can only penetrate short distances into crystals before their energy is completely absorbed by their interactions with the atoms. This means electrons are generally used to study either surfaces or thin films. In fact one important application of electron diffraction is in the electronics industry, where it is used to assess the surface structure of semiconductors. Two different set-ups are used for this, as we will see later.

BOX 5.1   UNDER PRESSURE

When we use a pencil to write something, it is sometimes difficult—and rather depressing—to accept that the carbon atoms making up the graphite pencil lead would form into diamond under the right conditions. But that is exactly what happens when carbon is subjected to extremely high temperatures and pressures, and it is by no means the only element to show a similarly dramatic alteration in structure and properties. "We have found that the elements tellurium and selenium have high pressure structures that are unlike anything else seen in an element," says Dr. Malcolm McMahon from the Centre for Science at Extreme Conditions at the University of Edinburgh in Scotland.

In 2003, Dr. McMahon's team announced the results of a series of experiments designed to reveal how materials behave at pressures over 100,000 times greater than everyday atmospheric pressures. Their hope was that information obtained from these experiments could eventually allow materials with unusual characteristics to be developed, and also help scientists understand more about the cores of planets including the Earth.

Since journeys to the centre of the Earth are only possible in the movies, scientists who want to learn more about the Earth's core need to find ways of re-creating the huge pressures found there. One way of doing so is to use a device known as a Diamond Anvil Cell (DAC) in which a solid sample is squeezed between two gem-quality diamonds with flattened tips (see Plate 5.1). "An experiment done in a small DAC in a laboratory can tell you important information on the state of iron at the earth's core, or the behaviour of gases and ice at the centre of other planets," explains Dr. McMahon.

The Scottish researchers used this method to squeeze the very small samples of the different elements they were investigating, and whilst each element was under pressure in the DAC an X-ray diffraction picture was taken. X-ray diffraction involves shining a beam of very high intensity X-rays at the sample, and recording the pattern produced on a detector as these X-rays are reflected by the planes of atoms in the sample. Analysis of these diffraction patterns reveals the positions of the individual atoms inside the material, so different structures show up as different patterns.

Under normal conditions, both tellurium and selenium have a simple structure consisting of helical chains of atoms that repeat regularly throughout the crystal. However, high pressures distort this structure, and the diffraction patterns revealed that at pressures above 45,000 atmospheres the sample adopted a crystal structure that was "incommensurate". In an incommensurate structure, the positions of the atoms are displaced in such a way that the pattern of atoms never repeats itself anywhere else in the structure (see Plate 5.2).

Previous studies by Dr. McMahon's team had revealed that the structures of various other elements, including barium, bismuth, and caesium, were also very unusual when the elements were under extremely high pressures and

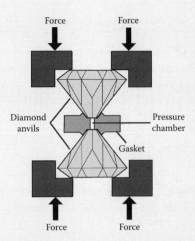

PLATE 5.1 This diagram shows the basic components of a Diamond Anvil Cell (DAC). In a DAC, two gem-quality diamonds with flat tips are squeezed together in order to subject any materials placed between them to enormous pressures. The particular design used by Dr. McMahon's team has a thin piece of metal known as the "gasket" between the diamond tips. The gasket is around 0.1 mm thick, and has a small hole approximately 0.2 mm in diameter drilled in it. This hole acts as a small sample chamber, and once the sample is in place, a drop of a methanol/ethanol mixture is added into the chamber, so when the diamonds are pushed together, they push on the liquid, which in turn pushes with fairly even pressure on the sample. Because diamonds are optically transparent, it is possible to view the sample when it is in place, and as diamonds are fairly transparent to X-rays, it is also easy to shine X-rays at the sample and get beams diffracted by the sample back out.

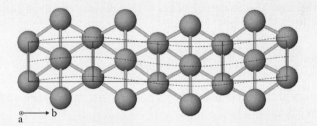

PLATE 5.2 This illustration shows the incommensurate Te-III structure at 8.5 GPa, as viewed down its a-axis (which is the axis coming out of the page at right angles to the plane of the paper). A normal crystal structure comprises atoms arranged within a unit cell which is repeated throughout the crystal. The arrangement of atoms within each unit cell is identical. In an incommensurate structure, the atoms are distorted away from these regularly repeating positions, as shown by the dashed line. The wavelength of this distortion wave is such that the arrangement of atoms within every unit cell is now different, and the structure therefore never repeats itself. (Courtesy of Malcolm McMahon, University of Edinburgh, Scotland.)

much more complex than the structures those elements have at standard atmospheric pressures.

Speaking in 2003, Dr. McMahon said that in the future the Edinburgh team intended to measure other physical properties that might also be very different at high pressures. "We can then use pressure to 'tune' the properties of materials to improve their characteristics," he explained, adding that he hoped such high-pressure properties could be reproduced at atmospheric pressures by methods routinely used to change the properties of materials such as adding atoms of another element to the material.

Whatever type of radiation is used to investigate a crystal, it must have a wavelength roughly the same size as the interatomic spacings if diffraction is to be observed. This will require beams of very different energies, depending on whether X-rays, electrons, or neutrons are being used to probe the crystal.

The wavelengths of electrons and neutrons are given by the following equation:

$$\lambda = \frac{h}{p} \tag{5.4}$$

where h is Planck's constant, and p is the momentum. This equation, which is known as the de Broglie equation, is named after the French physicist Louis de Broglie (1892–1987), who in 1924 proposed in his doctoral thesis that the dual wave-particle nature of radiation was also a feature of matter. This meant particles—like waves—had both wavelike and particlelike properties. (See Appendix D for a revision of quantum mechanics.)

So as well as having wavelike properties, as predicted by de Broglie, electrons and neutrons also behave as particles, and so their kinetic energy is that of a particle and is given by

$$K.E. = \frac{1}{2}mv^2 \tag{5.5}$$

The momentum of a particle is given by $p = mv$, so rearranging this and inserting it into equation 5.5 gives

$$K.E. = \frac{1}{2}m\left(\frac{p}{m}\right)^2 = \frac{p^2}{2m} \tag{5.6}$$

If we now rearrange the de Broglie equation so that $p = h/\lambda$ and insert this into the expression for the kinetic energy of the particles (equation 5.5), we have

$$K.E. = \frac{h^2}{2m\lambda^2} \tag{5.7}$$

In general, planes of atoms in crystals are about 0.1 nm apart, so to provide diffraction, the wavelength of the incident radiation must also be approximately 0.1 nm; otherwise the radiation will just "wash over" the atoms without interacting with them. Equation 5.7 reveals the energy that electrons or neutrons must have in order for diffraction to occur, as shown in Example question 5.2.

Neutron diffraction experiments are even more complex than electron diffraction, as the neutrons have to be supplied by a nuclear reactor. In fact it is via a small hole in the reactor wall that a beam of neutrons—produced in the nuclear fission process—is allowed to escape. Neutrons from thermal reactors are the best ones to use, as they have wavelengths of around 0.1 nm, which as we have already seen is the correct size to produce diffraction from crystals.

Despite the experimental difficulties, neutron diffraction is extremely useful for studying organic crystals that contain hydrogen, as neutrons are scattered strongly by hydrogen atoms, which only scatter X-rays weakly. This is because neutrons interact with the nuclei of atoms rather than their electrons.

LEED and RHEED

Low-energy electron diffraction (LEED) is used to study the positions of atoms in the surface layers of solids. These positions are usually different to those within the bulk of the solid. LEED involves directing a beam of electrons with energies between ≈20 and ≈250 eV at the sample, and recording the resulting diffraction pattern on a fluorescent screen placed on the same side of the sample as the "electron gun" that produces the beam of electrons. It is only electrons diffracted from the first three or four layers of atoms that make it back out of the sample, so the pattern will only represent the atomic structure of the surface. This experimental arrangement enabled Davisson and Germer to discover the wavelike nature of electrons back in 1927. (See Section D1 of Appendix D for more about their experiment.)

By contrast, reflection high-energy electron diffraction (RHEED) patterns tend to be made up from electrons of ≈50 keV. These are fired at the sample at such a large angle of incidence that they just graze the surface and cannot penetrate far into the solid. One of the main uses of RHEED is assessing how smooth the surface of a sample is. For example, RHEED can be used to monitor the growth of new atomic layers on crystal substrates, which as we will see in Chapter 7 is one of the main processes used in the production of electronic devices.

5.2 LATTICE VIBRATIONS AND PHONONS

So far, when we have considered the atoms in a crystal lattice, we have assumed that they are stationary. But like students in a lecture, atoms never sit completely still. They always have some motion about their lattice point. In fact if they did not, they would violate Heisenberg's uncertainty principle, which states that if the position of a particle is known exactly, its momentum cannot be known as well, and vice versa. So although diffraction reveals the positions of the atoms in a crystal, it only in fact reveals the positions of the atoms at the instant they reflect the radiation being used in the diffraction experiment.

EXAMPLE QUESTION 5.2   THE LOWEST PARTICLE ENERGY THAT WILL ALLOW BRAGG DIFFRACTION

Given that the spacing between adjacent planes of atoms ($d_{(hkl)}$) for cubic crystals is

$$d_{(hkl)} = \frac{a}{\sqrt{h^2 + k^2 + l^2}}$$

what is the lowest particle energy that will allow Bragg diffraction to occur during an electron diffraction experiment from the {111} planes of a silicon crystal which has a lattice parameter, a, equal to 0.542 nm?

ANSWER

We have been told the value of a, and we are dealing with the {111} plane in this case, so putting in the numbers into the equation we have been given for the spacing between planes we get

$$d_{(hkl)} = \frac{0.542}{\sqrt{3}} = 0.313 \text{ nm}$$

For diffraction to occur, the Bragg law $2d \sin \theta = n\lambda$ must be satisfied. The maximum value θ can have is 90°, so as $\sin 90° = 1$,

$$n\lambda/2d \le 1$$

This means the upper limit on the wavelength λ_{max} is equal to $2d$, and so in this case

$$\lambda_{max} = 0.313 \times 2 = 0.626 \text{ nm}$$

We can see from the equation for the kinetic energy of particles (equation 5.7) that the larger the wavelength of the particles, the lower their energy will be. So if we put λ_{max} into this equation it will yield the lowest possible energy the particles can have for Bragg diffraction to occur. (Both the mass of the electron, m_e, and Planck's constant, h, will be on most scientific calculators, but if not their values can be looked up in tables—if you have not yet memorised them!) So,

$$\text{K.E.} = \frac{h^2}{2m_e\lambda^2}$$

$$= \frac{\left(6.626 \times 10^{-34}\right)^2}{2 \times 9.11 \times 10^{-31} \times \left(0.626 \times 10^{-9}\right)^2}$$

$$= \frac{4.390 \times 10^{-67}}{7.139 \times 10^{-49}}$$

$$= 6.149 \times 10^{-19} \text{ J}$$

$$= 3.838 \text{ eV}$$

(Remember that, to convert joules into electron volts, you divide the quantity in joules by the charge on the electron, $e = 1.602 \times 10^{-19}$ C. Also, if you use the physical constant button on your calculator you will get a slightly different answer to the calculation above, as the constants will consist of far more significant figures than I have written here.)

Apart from the fact the particle mass will be different, the method will be identical if you are asked to work this out for neutrons.

An instant later, their positions will have changed. However, this change in position is incredibly small, so it makes no difference to identifying a structure by diffraction.

Although the structure of a particular crystal may look the same when different diffraction patterns from it are analysed, there are other properties of crystals that are affected much more by the motion of the atoms about their lattice positions. In particular the thermal properties of solids (see Section 5.3) are entirely due to atomic motion, whilst—as we will see in Chapter 6—the electrical properties of solids are influenced by the vibrations of their atoms.

5.2.1 ATOMIC VIBRATIONS IN CRYSTALLINE SOLIDS

Since atoms in solids do not remain still at their lattice positions, they need some form of energy in order to be able to move. This energy is either heat energy or mechanical energy. The hotter a solid is, the more the atoms vibrate about their positions in the lattice. (This can actually be revealed by diffraction experiments because there is a decrease in intensity of the diffracted beams as the temperature of a sample is raised. This is because the higher the temperature, the more the atoms are moving, so at the instant an X-ray diffraction image is taken there are likely to be fewer atoms sitting at their respective lattice sites and producing in-phase reflections of the beam.) As we have just seen, it would violate Heisenberg's uncertainty principle for the atoms in crystals to be totally still, so although atoms move much less at low temperatures than at high temperatures, they still move even at absolute zero. This movement is known as *zero-point motion*, and the kinetic energy possessed by atoms with zero-point motion is termed "zero-point energy".

If an atom is considered individually, its motion about its own lattice point is approximately the same as that of simple harmonic motion (SHM). However, atoms cannot vibrate independently of each other because they are connected together by bonds. So any vibration in one atom gets passed on to its neighbours, which in turn pass the vibration on to their neighbours, and so on, producing a vibratory wave that passes through the solid.

Acoustic and Optical Waves in Solids

One way of thinking about the atoms in a crystal is as a collection of balls linked by springs, where the balls represent the atoms and the springs represent the bonds as illustrated in Figure 5.8.

There are two ways in which an elastic wave—in other words, a wave that travels along because of the vibration of the particles of the material it is in—can be set up in a one-dimensional layer of atoms like the one shown in Figure 5.8. One is for two balls to be moved further apart and then allowed to bounce back together again in a horizontal direction (see Figure 5.9). This is what occurs when a particular area of a crystal is heated. The expansion in the heated region causes the atoms to move further apart, then they move back together again as they cool. As a result of this motion, a longitudinal wave is created that travels through the solid. The movement is passed from one neighbour to another in a knock-on effect because they are all linked by bonds.

These longitudinal waves are actually sound waves, although that does not mean sound waves in solids are audible. They are sound waves simply because sound waves are longitudinal vibrational waves; that is, vibrational waves moving along the same direction as the vibration making up the wave (see Appendix B for a reminder of the main properties of vibrations and waves). So audible sound waves in air are produced by the same mechanism as sound waves in solids, but to avoid confusion, sound waves in solids tend to be referred to as "acoustic" waves.

The other way in which a wave can be set up in a solid is for the atoms to start oscillating in a vertical direction, as shown in Figure 5.10. This sets up a transverse wave, in which the wave travels in a direction perpendicular to the direction of

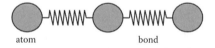

FIGURE 5.8   A representation of a one-dimensional chain of atoms linked by atomic bonds.

FIGURE 5.9   Representation of an acoustic (longitudinal) wave travelling along a one-dimensional chain of atoms linked by atomic bonds.

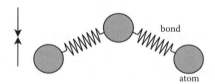

FIGURE 5.10   Representation of an optical (transverse) wave travelling along a one-dimensional chain of atoms linked by atomic bonds.

motion of the particles making it up, and since light is a transverse wave, these waves are known as "optical" waves.

5.2.2 PHONONS

In the same way that light can be regarded as either a wave or a stream of particles known as photons, vibrations in solids can be thought of either as waves or particles known as *phonons*. This means that, as with other quantum mechanical particles, phonons can only have certain values of energy. In fact a phonon vibrating at a frequency ω can only have values of energy that are $\hbar\omega$ apart from one another.

If a phonon is thought of as a quantised packet of vibratory energy, a region of a solid containing larger amplitude atomic vibrations than another area can be said to contain more phonons. Equally, if a solid is considered as a whole, the greater the atomic vibrations are throughout the solid, the more phonons are said to be in the solid.

The particle nature of phonons means they can be considered to "move" through solids, and many properties of solids that are affected by temperature can be explained in terms of phonons being scattered. This scattering can either be by point defects or dislocations, by one another, or by the surface of the solid. In each case there will be a *mean free path*, l, which represents how far a phonon travels on average before being scattered. (The concept of the mean free path is used in many different areas of physics, and it reveals the average distance that a particle travels in a medium before either decaying or undergoing some sort of interaction.)

Like all moving objects, phonons have a wavelength associated with them (see Section D1 of Appendix D for more on wave-particle duality), and the extent to which phonons are scattered by lattice defects depends on the temperature of the solid they are in because phonons have different wavelengths at different temperatures. There is actually a range of phonon wavelengths at all temperatures, but on average the phonon wavelength is longer at low temperatures and shorter at high temperatures. In fact as a very rough guide, the dominant phonon wavelength can be considered to be in the order of a few hundred atomic spacings just above absolute zero, while at room temperature and above it is around twice the value of the lattice parameter, a. This means that at low temperatures, the phonon wavelength is much larger than any point defect, so the defect is unable to scatter the wave (just as a pebble on a beach will not alter the path of a large ocean wave). The amount of scattering is therefore much less at low temperatures than that at higher temperatures, where the phonon wavelength is of a comparable size to that of the point defects.

As higher temperatures mean a larger amount of movement of the atoms, and the larger the amplitude of any two waves the greater the chance of them interacting, the scattering of phonons by one another is also greater at higher temperatures. Lower amplitude vibrations are like ripples on a pond that do not scatter each other when they meet, so phonon-phonon interactions are negligible at very low temperatures. By contrast, the strain field around dislocations is large enough to scatter phonons at any temperature.

Even when the effect from scattering by defects is at a minimum, there is still a limit to how far a phonon can travel without being scattered. This is because even if it does not encounter anything else on its path, a phonon will be scattered when it reaches the boundary of the solid it is travelling in. Somewhat surprisingly, this does not mean that the mean free path of phonons affected by boundary scattering is limited to a maximum value of the width of the specimen. If the surface of the sample is highly polished, any impinging phonons can be reflected back by the boundary, increasing the length of the mean free path.

5.3   THERMAL PROPERTIES

When a solid is heated it absorbs energy from the heat and its temperature rises. In some cases the heat is transferred from a hot region of the solid to a cooler area. The solid will melt if it is heated to its melting point (or above), while below this temperature it will expand when heated. The thermal properties of materials are an important consideration when designing objects ranging from buildings and bridges to computer chips, and in this section we will look at how solids react to heat on both the atomic and macroscopic scale.

5.3.1   SPECIFIC HEAT

Definition

If it often useful to know exactly how much heat needs to be supplied to a substance in order to raise its temperature, and the ratio of heat supplied to temperature rise is known as the heat capacity. The *specific heat capacity*, C—or "specific heat" for short—is more widely used, however, as it has a more precise definition, namely the amount of heat required to raise the temperature of one kilogram of a substance by one kelvin. In other words, the ratio of heat supplied to 1 kg of a substance to the temperature rise. It has units therefore of J K^{-1} kg^{-1}. (Alternatively, the "molar heat capacity" can be used, which is the heat energy that has to be added to a mole of a substance to raise its temperature by one degree kelvin. We will concentrate on the specific heat capacity here, as this is the quantity generally used when discussing solids.)

The specific heat can be measured in two ways: either with the specimen under consideration kept at constant pressure, or with it kept at constant volume. The symbols for these two different types of specific heat are C_p and C_v, respectively. Both types of specific heat are important when studying gases, and can have very different values. For solids at room and lower temperatures, however, C_p and C_v are quite similar in value. Having said that, C_p is always a few percent higher than C_v for a solid because it includes a component related to the work done in changing the volume of the solid as well as the change in internal energy produced by the introduction of heat.

Table 5.1 gives the specific heat capacity for several solids and for water. The higher the value of the specific heat capacity for a substance, the greater the amount of thermal energy it needs to receive in order for its temperature to be raised.

Specific Heat at Different Temperatures

In 1912, Dutch-American physicist Peter Debye (1884–1966) came up with the currently accepted theory that all the internal energy in a solid is assumed to be contained within phonons (see Section 5.2.2). So as the internal energy is increased in

TABLE 5.1

The Specific Heat Capacity and Debye Temperature for a Range of Substances

Substance	Specific Heat Capacity, C_p (J/g · K)	Debye Temperature, Θ_D (K)
Aluminium (Al)	0.897	375
Copper (Cu)	0.385	340
Lead (Pb)	0.129	95
Silver (Ag)	0.235	230
Water (H_2O)	4.184	N/A

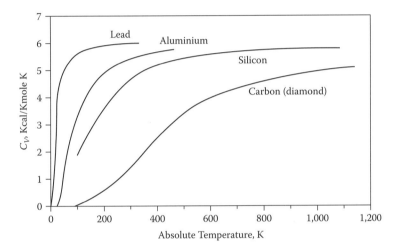

FIGURE 5.11 This graph shows the specific heat at constant volume as a function of temperature for several elements. Dulong and Petit's law holds fairly well for lead (Pb), aluminium (Al), and silicon (Si) at temperatures above about 300K, but fails for all the elements shown at lower temperatures. (From Perspectives of Modern Physics, A. Beiser, 1981, © The McGraw-Hill Companies, Inc. Reproduced with permission.)

solids by an increase in heat, this must also mean there is an increase in the number of phonons. In other words, as an object gets hotter, its atoms vibrate more, and so the amplitudes and number of vibrational waves passing through them gets larger.

This theory contrasted with that of the French physicists Alexis-Thérèse Petit (1791–1820) and Pierre Dulong (1785–1836), who had previously discovered a relationship known as the Dulong-Petit law. This law states that the specific heat of any solid multiplied by its atomic weight is a constant with an approximate value of 25 J mol^{-1} K^{-1}. As Figure 5.11, which shows the specific heat at constant volume for a range of solids, indicates, the Dulong-Petit law provides a good approximation of the experimental results for many solids at high temperatures. However, some of the lighter elements—such as boron, and carbon in the form of diamond—do not obey the law, which also fails for all solids at low temperatures, as the specific heat drops

BOX 5.2   MAKING WAVES

An imaging technique developed by physicists in Japan could help the micro-electronics industry produce more efficient components for communications networks and mobile phones. The system, designed by Prof. Oliver Wright and his team from Hokkaido University, is able to record movies of tiny sound waves rippling across the surfaces of crystals.

The researchers generate these sound waves—which are in fact the collective vibrations of some of the crystal's atoms—by focusing extremely short bursts of laser light onto the surface of the crystal. Each laser pulse rapidly heats up the region it falls on, causing it to expand. The atoms in this region move from their normal positions in the crystal lattice during the expansion. Because each atom is linked to its neighbours by bonds, such movement causes the surrounding atoms to vibrate. This knock-on effect continues forming a vibrational wave known as a "surface acoustic wave" (SAW) that travels across the crystal.

To measure the movement of the surface as a SAW ripples across it, the team reflects two short "probe" laser pulses off the surface in quick succession. The moving sound wave affects the two pulses differently, and the interference pattern produced when the reflected pulses meet represents the change in surface height that occurs as the wave passes the spot being probed. By scanning this laser spot across the crystal, a complete image of the surface is built up, and repeated scanning of the whole surface allows a movie of the surface acoustic waves to be obtained (see Plate 5.3). "Apart from being noncontact and nondestructive, our technique is also very sensitive, allowing surface displacements of 0.1 atomic diameters to be measured easily," says Prof. Wright.

In the same way that light can be regarded as either a wave or a stream of particles known as photons, lattice vibrations in solids can be thought of as either sound waves or packages of vibratory energy called phonons. Different crystal structures bring phonons to a focus in different directions, creating a range of patterns that can be predicted theoretically. Since phonons also become scattered when they reach defects, deviations from the predicted patterns highlight impurities and structural faults (see Plate 5.4). "The ability to directly visualize the propagation of the surface acoustic waves can reveal information about the shape of a defective region that may not be possible to obtain by other techniques," explains Prof. Wright.

Since the technique allows real-time imaging of thin films on opaque substrates (which form the basis of electronic devices), it should prove useful to the microelectronics industry not only for detecting defects, but also for evaluating SAW devices. There is currently a lot of interest in improving the performance of SAW filters—that use surface acoustic waves to remove unwanted frequencies from electrical signals—because of their widespread use in mobile phones and cable-based communications networks.

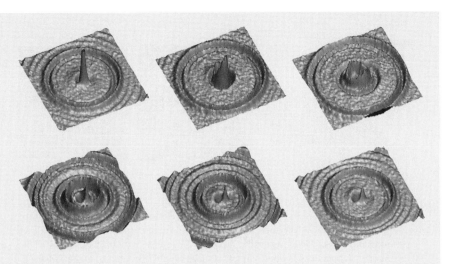

PLATE 5.3 This series of movie stills shows surface acoustic waves travelling across a 90 μm^2 square region of glass that has been coated with a thin polycrystalline film of gold. The symmetry of the pattern is determined by the glass substrate, which the waves penetrate about 10 μm into. The speed with which they travel in any given direction depends on the strength of the atomic bonds in that direction, and since the bond strength is the same in all directions for this glass substrate, these waves are circular. The waves can be seen broadening as they move further away from the centre. This dispersion occurs because their shorter wavelength components are mainly confined to the gold film, and since sound waves travel more slowly in gold than in glass, they get left behind the longer wavelength components that penetrate into the substrate. Analysis of the dispersion reveals the film thickness.

Speaking shortly after announcing the system in 2002, Prof. Wright said he felt it would take until at least 2005 to develop the technique enough for it to be routinely used by the microelectronics industry. However, he joked that it had already found a somewhat less serious application. "We projected our phonon focusing movies at the local discotheque and it seemed to be appreciated by the dancers!" he explained.

and tends towards zero as the temperature is decreased. This partial failure is due to the fact that deriving the Dulong-Petit law mathematically requires the use of a classical model that considers the specific heat to be due to the vibrations of individual atoms that are not coupled together, and which we now know is incorrect.

Looking again at the low-temperature region in Figure 5.11, you can see the specific heat tending to zero at absolute zero and increasing in value between this point and the flattened-off high-temperature region of the graph. Above absolute zero, the lattice waves have a greater and greater amount of energy as the temperature rises, and below the region in which the classical model dominates, the specific heat is found to be proportional to T^3.

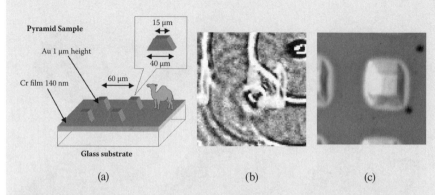

(a)   (b)   (c)

PLATE 5.4   Image (b) shows how tiny gold pyramids that sit on top of a chromium-coated glass substrate refract surface acoustic waves as they pass through. (With no defects present, these waves would be circular, like those shown in Plate 5.3.) This refraction occurs because sound waves travel more slowly in gold than in the surrounding chromium. The waves also scatter from the pyramid edge as they leave it. Diagram (a) shows the dimensions of this sample, while (c) is an optical reflectivity image showing the position of the gold pyramid responsible for the SAW refraction and scattering seen in (b). As well as revealing the presence of defects, the ability to image surface acoustic waves at this resolution should help physicists learn more about their propagation. (Plates 5.3 and 5.4 courtesy of Oliver Wright, Hokkaido University, Japan.)

TABLE 5.2
The Debye Temperature
for a Range of Substances

Substance	Θ_D (K)
Iron (Fe)	360
Carbon (diamond) (C)	1850
Gallium arsenide (GaAs)	204
Calcium fluoride (CaF_2)	474

The Debye temperature (Θ_D) provides a useful but approximate division between the high-temperature behaviour—which can be described by the classical model—and the lower-temperature region, which is best explained in terms of phonons. Tables 5.1 and 5.2 list the Debye temperatures of a selection of solids.

This does not mean that Debye's theory fails at high temperatures. In fact, although phonons are considered to be a "phonon gas" obeying Bose-Einstein statistics (see Appendix E for a reminder of statistical mechanics), their behaviour above the Debye temperature is in fact the same as that predicted by the classical model, and matches experimental data well. So quantum statistics and the classical statistics used to derive the Dulong-Petit law give the same result at high temperatures. However, unlike classical statistics, quantum statistics explain why the specific heat approaches zero as T approaches zero.

As with all models, the Debye model is a simplification of reality, and for some solids at certain temperatures, modifications need to be made. For example, we will see in Chapter 6 that although the specific heat capacity is mainly due to phonons, there is also a contribution from some of the free electrons in metals which is most noticeable at extremely high and extremely low temperatures.

5.3.2 Thermal Conductivity

The transfer of heat from hot regions of a substance to cooler regions is known as *thermal conduction*. We are all familiar with this process from everyday life, as the heat of a fire or an oven will cook—that is, heat up—any cold food that is put in contact with it. Some materials are better conductors of heat than others, and the *thermal conductivity* of a material, κ, represents its ability to conduct heat. The higher the value of κ the better that material is at conducting heat. As Figure 6.1 of Chapter 6 shows, solid materials possess a wide range of values of electrical conductivity. This contrasts quite dramatically with the much smaller range of thermal conductivities. The reason for the big difference is that solids can be excellent electrical conductors at one end of the scale and almost totally insulating at the other end, whereas all solids can conduct heat to some extent. The thermal conductivities of a selection of solids are given in Table 5.3.

Both good and bad thermal conductors have many different uses. For example, good conductors are needed to conduct heat away from the processors in electronic devices, which would otherwise overheat and become unreliable. By contrast, poor conductors—in other words, thermal insulators—such as foam are used as insulation in lofts and around domestic hot water tanks. Animal fat is another example

TABLE 5.3
The Thermal Conductivities of a Range of Materials at 293K

Material	κ (W m^{-1} K^{-1})
Silver	429
Copper	401
Aluminium	237
Sodium	141
Zinc	116
Cadmium	97
Iron	80
Lead	35
Fused silica	1.4
Carrot	0.53
Banana	0.475
Nylon 6,6	0.24
Polystyrene	0.13

of a good insulator, which is why long-distance swimmers cover themselves in fat before embarking on record-breaking attempts and one of the reasons why polar bears can survive in extremely cold temperatures. Not only does the bears' fur trap heat, but a thick layer of fat underneath their skin prevents them from losing too much body heat.

Heat is conducted through solids by the vibrations of the atoms that make up the solids (in other words by phonons) and by any free electrons that the solid may contain. Since hotter atoms vibrate more about their normal positions in the lattice than colder atoms and bonds between atoms link them together, the movements get passed on, and so the heat travels through the solid. As we saw in Section 5.2.1, any expansion caused by heating also causes atoms to move—in this case further apart as the region of the solid they are in expands, then back together as their surroundings cool. This sets up acoustic waves that travel through the solid, providing another way for phonons to contribute to thermal conduction.

Free electrons can transport heat through a solid because any in a hot part of a solid gain kinetic energy from the heat. When they move to a colder part of the solid they then give this kinetic energy up as they collide with lattice defects or phonons, and so this additional kinetic energy changes into lattice vibrational energy—that is, more phonons. So the total thermal conductivity, κ, is given by

$$\kappa = \kappa_l + \kappa_e \tag{5.8}$$

where κ_l is the contribution to the thermal conductivity from the lattice vibrations, and κ_e is the portion of the thermal conductivity that is due to free electrons moving through the solid.

κ_e increases relative to κ_l as the number of free electrons increases because there are more electrons available to transfer heat. This means κ_e will be greater than κ_l in pure metals, which contain large numbers of free electrons. In metals containing impurities, the electron mean free path is reduced by collisions with the impurities to such an extent that κ_e becomes similar in value to κ_l. κ_l dominates in nonmetals due to the lack or greatly reduced number of free electrons, and for amorphous materials and polymers both κ_l and κ_e are very low. This is because the lack of regularity in the structure of amorphous materials and polymers (see Chapter 3) gives them a very small mean free path, so neither phonons nor free electrons can travel far through them.

5.3.3 THERMAL EXPANSION

The fact that solids expand and contract as their temperature changes is something that architects and designers have to take into account when designing any construction or object that will be subjected to a range of temperatures during its working life. For example, engineers stress the continuous welded rails (widely used on mainline routes) to prevent expansion on hot days. Despite such precautions, rails can still buckle in exceptionally hot weather as Figure 5.12 shows.

FIGURE 5.12 Despite the precautions of engineers who stress continuous welded rails to prevent expansion on hot days, rails can still buckle in exceptionally hot weather as shown in this picture of track at Moses Gate in Manchester, U.K., on 18 July 2006. The rail temperature often rises much higher than the surrounding air temperature, and even in the U.K., rail temperatures above 50°C have been recorded on hot summer days. (Courtesy of Network Rail.)

Even though the amount of expansion of a structure may not seem a lot compared with its overall size—for example, the arch of Sydney Harbour Bridge can become 18 cm higher on a really hot day—it is enough to induce significant stresses.

Coefficients of Expansion

Over a fairly wide range of temperatures, the increase of the length of a solid rod when it is heated is proportional to the increase in temperature. This can be expressed by the following equation:

$$\frac{\Delta l}{l_0} = \alpha_l \Delta T \tag{5.9}$$

where l_0 is the initial length of the rod, Δl the change in length, ΔT the change in temperature, and α_l the "linear coefficient of thermal expansion".

The value of the linear expansion coefficient is generally an order of magnitude larger for polymers than for other solids as Table 5.4 shows. α_l is also temperature dependent—increasing with increasing temperature—and the values in Table 5.4 were measured at room temperature. If an object is to be made from two or more

TABLE 5.4

The Linear Thermal Expansion Coefficient, α_l, at Room Temperature for a Range of Solids

Material	Linear Coefficient of Thermal Expansion, $\alpha_l \times 10^{-5}$ (K^{-1})
Polyethylene (low density)	18–40
PTFE	12–21
Aluminium	2.39
Brass	1.8
Glass (Pyrex)	0.32
Quartz	0.05

Note: The value for polymers depends on their thermal and mechanical history as well as how crystalline they are.

materials, it is better to choose materials with similar thermal expansion properties, otherwise failure can occur prematurely due to the stresses caused by different rates and amounts of thermal expansion. (Thermal stresses are discussed in the next subsection.)

A similar expression can also be written for the volume changes when a solid expands. In this case the lengths in equation 5.9 are replaced by volumes, and α_l by α_v, the "volume coefficient of thermal expansion". For many materials α_v is anisotropic; in other words, it is different along different directions.

It is possible to see how thermal expansion occurs on an atomic level by referring back to Figure 2.4, which shows the potential energy of a pair of ions in a crystal as a function of their separation. This diagram also represents the situation for two atoms in a solid. As a solid is heated, the atoms will gain energy from the heat. This is illustrated in Figure 5.13, in which E_1 is the vibratory energy at temperature T_1, E_2 is the energy at the higher temperature T_2, and so on. The average amplitude of the vibrations an atom has at each temperature corresponds to the width of the trough in the potential energy curve, and the mean position of a vibration (given by the halfway points labelled r_1 through r_3 on the graph) represents the average interatomic distance. So as Figure 5.13(a) reveals, the asymmetry of the potential energy curve causes the atoms to move further apart with an increase in temperature. If it were symmetrical, as shown in Figure 5.13(b), there would be no increase in distance between the atoms with a rise in temperature.

Thermal Stresses

If a solid metal rod is heated up in a laboratory to measure its thermal conductivity, it does not matter that the rod expands as it is heated. It is free to expand and contract at will. However, if that same rod was being used as part of an engineering structure

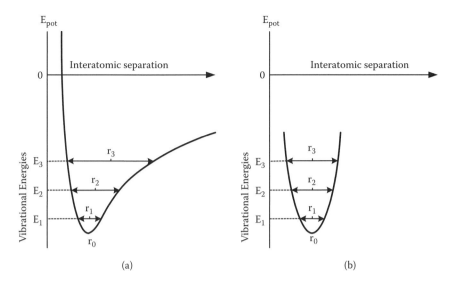

FIGURE 5.13 (a) shows part of the potential energy curve first introduced in Figure 2.4. As the temperature increases, the two neighbouring atoms represented by the graph gain energy. Because of the asymmetric shape of the curve, this has the effect of increasing their distance apart from one another. If the potential energy curve was symmetrical as shown in (b), there would be no increase in separation distance with increasing temperature and therefore no expansion.

and was restrained at both ends by a material that did not expand and contract at the same rate as the rod when it is heated, mechanical stress will build up in the rod.

Stresses can also be set up when solids are heated or cooled quickly. If, for example, a solid object gets hot extremely rapidly, the outside of the object will increase in temperature faster than the inside. The outside will therefore expand more than the centre, introducing stress. The amount of stress induced in a particular solid will depend on its thermal conductivity, the rate of the temperature change, and the size and shape of the object.

It is possible to reduce thermally induced stresses in ductile metals and polymers by plastic deformation. Nonductile materials like ceramics have a tendency to fracture under thermal stresses. For example, surface cracks tend to appear when brittle materials are cooled quickly. Brittle fracture caused by thermal stresses is known as thermal shock.

There are various ways of reducing the likelihood of thermal shock occurring. These include reducing the heating or cooling rates and introducing large pores (holes) into a material to prevent cracks from growing too large. Glasses are particularly likely to suffer from thermal shock. The introduction of additives, or carrying out annealing, can reduce the chances of glasses breaking under thermal stresses. The type of glass known as Pyrex has an additive of B_2O_3, which reduces its coefficient of expansion to a figure low enough that the glass will not break if it is used for cooking in a conventional oven.

EXAMPLE QUESTION 5.3 THERMAL EXPANSION

You are asked to lay a railway track across a quarry. The rails you have available are made from a low-carbon steel with a linear coefficient of thermal expansion of 1.2×10^{-5} K^{-1}, and the railway must remain operational in temperatures ranging from 10°C to 40°C. If each section of rail is 20 m long, how much of a gap do you need to leave between each section to allow for thermal expansion on a hot day if you lay the rails when the temperature is 10°C?

ANSWER

Since you are given the linear coefficient of thermal expansion for the material to be used, you can use the following equation:

$$\frac{\Delta l}{l_0} = \alpha_l \Delta T$$

Rearranging gives

$$\Delta l = \alpha_l \Delta T l_0$$

$$= 1.2 \times 10^{-5} \times 30 \times 20$$

$$= 7.2 \times 10^{-3} \, \text{m} = 7.2 \, \text{mm}$$

So you would need to leave a gap of 7.2 mm between each section of rail.

FURTHER READING

Cullity, B. D., and Stock, S. R. *Elements of X-ray Diffraction*. 3rd ed. Englewood Cliffs, NJ: Prentice Hall, 2001.

Leroy, F. "Because you're worth it". *Physics World* May 22, 2001.

Liang, X. G., Zhang, Y., and Ge, X. "The measurement of thermal conductivities of solid fruits and vegetables". *Measurement Science and Technology* 10, no. 7 (1999): N82–86.

(More references for further reading, as well as Web links, are available on the following Web page: http://www.crcpress.com/product/isbn/9780750309721.)

SELECTED QUESTIONS FROM QUESTIONS AND ANSWERS MANUAL

Q5.1

(a) State the Bragg law.

(b) What do each of the terms represent?

EXAMPLE QUESTION 5.4   PHYSICS IN THE HOME

If you had to choose one of the following to help you open a glass food jar with a metal lid that won't budge, which one would you go for and why? What precautions, if any, would you need to take in order to use this method safely?

(a) a can opener
(b) a knife
(c) a damp cloth
(d) hot water

Answer on page 156.

Q5.4

(a) What types of radiation other than X-rays are commonly used to obtain diffraction patterns?
(b) What do the initials LEED stand for?
(c) Calculate the lowest neutron energy that will allow Bragg diffraction from the {110} planes of a silicon crystal with a lattice parameter, a, of 0.542 nm.

Q5.7   Which statement—(a), (b), or (c)—best describes a phonon?

(a) A large amplitude atomic vibration
(b) A quantized packet of vibratory energy
(c) The sound equivalent of an electron

Q5.9   Why is C_p (the specific heat at constant pressure) of a given solid slightly higher than the value of C_v (the specific heat at constant volume)?

Q5.14   If a 10 cm long brass rod is clamped at one end and heated from 20°C to 40°C, how much would it expand lengthways?

ANSWER TO QUESTION 5.4 ON PHYSICS IN THE HOME

Pouring hot water onto the metal lid of the jar would give you the best chance of removing it, as the heat from the water would cause the metal to expand and so loosen its grip slightly on the threads holding it in place. You would need to take care not to crack the glass, particularly if it was very cold when you started to pour the hot water on the lid. Although the glass will not expand as much as the metal, heat from the water—and even from the lid as it warms up—could set up thermal stresses in the glass that might be large enough to make it break if the heating is rapid enough.

6 Unable to Resist
Metals, Semiconductors, and Superconductors

CONTENTS

6.1   FREE ELECTRON MODELS OF ELECTRICAL CONDUCTION

The ancient Greek philosopher Thales of Miletus is reputed to have discovered in around 600 b.c. that if a piece of amber is rubbed with wool or fur it becomes electrically charged. Whether this story is true or not, it is probably safe to say that Thales would not have had much inkling as to the origin of this charge or indeed that much of the world would one day rely heavily on electricity.

We now understand much more than Thales about electricity, but as this chapter will reveal, although most aspects of conduction have been adequately explained, physicists are still investigating the finer points. To give an idea of how research has been progressing, we will consider the different models of electrical conduction that have been developed over the last 100 years or so as well as discuss the currently accepted models.

When studying solids, it is often useful to divide them up into different groups that contain materials with similar properties. One of the most common ways of classifying solids is according to how well they conduct electricity—that is, by grouping them into conductors, insulators, and semiconductors. In the second half of this chapter we will discuss some of the basics of semiconductor physics, before looking briefly at a type of conductor known as a superconductor that, below a critical temperature, exhibits no resistance to the flow of an electric current. Insulators will be covered in detail in Chapter 7, along with the applications of semiconductors.

6.1.1   OVERVIEW OF ELECTRICAL CONDUCTION

Under normal circumstances most solids—including conductors—are electrically neutral. Conductors contain electrons that are free to move about inside them, and generally the electrons move randomly in every direction so that no net electric polarisation is produced in any particular region of the material.

However, when a conductor is placed in an electric field, for example, by connecting it to the terminals of a battery, these "free" electrons start to flow in the same direction and therefore create an electric current. This is because the negatively charged electrons are attracted to the positive terminal of the battery. Insulators (or dielectrics, as they are often called) can be thought of as the opposite of conductors, as they have very few—if any—free electrons to take part in conduction. This means that barely any electric current will flow if an insulator is connected to a battery, so insulators are not only used to shield us—and delicate electronic components—from harmful currents, they are one of the main constituents of capacitors, which store electric charge.

In Section 2.1 we saw how the type of bonding influences the macroscopic properties of a solid. As far as conduction is concerned, ionic solids are good electrical and thermal insulators because almost all of the electrons are bound closely to particular ions, leaving very few "free" electrons available for conduction. The van der Waals solids are also poor conductors of heat and electricity because they do not contain any free electrons either. By contrast, solids held together by metallic bonding are good conductors because the valence electrons of their atoms are free to move around rather than being bound to the positive ions, as they are in ionic solids. As far as covalent solids are concerned, some are insulators, while others are semiconductors (which can conduct under certain circumstances). This is because the valence electrons in some covalent solids are not as tightly bound as those in other covalent solids. In a diamond, for example, there is particularly strong covalent bonding, so this is an insulator, but in semiconductors like silicon the valence electrons are less tightly bound. This means they can move to a certain extent as they are shared between atoms and can therefore contribute to current flow.

Conductivity and Resistivity

Figure 6.1 shows the conductivities of some well-known materials. The *electrical conductivity*, σ, of a material is a measure of its ability to conduct electricity and, as equation 6.1 shows, σ is the reciprocal of the *resistivity*, ρ.

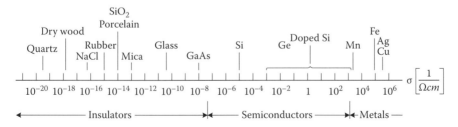

FIGURE 6.1 Room temperature conductivities of several well-known materials. (This diagram has been copied from *Understanding Materials Science*, R. E. Hummel, corrected second printing 1999, Fig. 11.1, p. 181, © 1998 Springer-Verlag New York, Inc. With kind permission of Springer Science and Business Media.)

$$\sigma = \frac{1}{\rho} \tag{6.1}$$

The resistivity is related to the *resistance*, R, of a particular device or sample of material as follows:

$$R = \frac{L\rho}{A} \tag{6.2}$$

where L is the length of the device or sample, and A its cross-sectional area.

Although the value of the resistance is related to the dimensions of whatever sample of the material is being considered, the resistivity is independent of sample size. Electrical conductivity has units of reciprocal ohm-metres $((\Omega\text{-m})^{-1})$, although it is not unusual for conductivities to be quoted in reciprocal ohm-centimetres instead, as in Figure 6.1. Resistance is measured in ohms (Ω), while the unit of resistivity is the ohm-meter $(\Omega\text{-m})$.

Both the resistance and resistivity represent how much a material resists the flow of an electric current, and the greater the values of these quantities for any material, the less current that can pass through it. This means insulators will have much higher values of these two quantities than semiconductors and conductors.

In a similar way, the better a material is at conducting an electric current, the greater the value of the conductivity. Of all the physical properties of solids, the electrical conductivity at room temperature has the widest range of values, as the best conductors have conductivities over 25 orders of magnitude larger than those of the best insulators.

From Figure 6.1 we can see that, as we might expect, metals are conductors and silicon (which almost all of our electronic devices are made from) is a semiconductor, while dry wood, glass, and rubber are insulators. In fact, rubber and various other polymers are such good electrical and thermal insulators that they are used to encase electrical equipment and cover electrical wires to prevent electric shock or burns to the user. However, although most polymers are insulators, there are a handful which have conductivities either in the conducting or semiconducting range. At the time of writing this book, research on these solids is still at an early stage, although some are already in use; for example, the cathodes for the batteries in implantable pacemakers are made from a lightweight conducting polymer.

Origins of the Resistivity

Both the electrical conductivity and resistivity are dependent on temperature. Figure 6.2 shows the increase of resistivity with increase in temperature for the metal copper.

If we imagine free electrons moving through a metal under the influence of an electric field, they will sometimes collide with the atoms as they travel forwards. Each time they strike an atom they will lose energy. The atoms are therefore providing resistance to the flow of the current. The reason this resistance (and hence the resistivity) increases with temperature is that the higher the temperature, the more the atoms of the metal will vibrate about their lattice sites, and the greater the

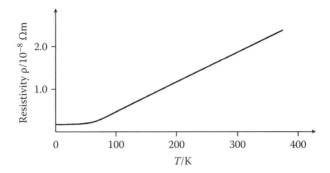

FIGURE 6.2   Temperature dependence of resistivity for copper.

opportunity is for electrons to collide with them. An increase in temperature is not the only thing that will increase the resistivity, though.

Lattice defects including dislocations, impurities, and vacancies will also contribute to the resistivity, as the free electrons can hit into them. Figure 6.3 shows the various types of resistance that a free electron might encounter as it makes its way through a piece of metal.

It is actually an oversimplification to say that electrons "collide" with either lattice atoms or defects. In fact they interact electrostatically with them, but the net result is the same: The electrons lose energy and therefore are encountering resistance to their flow.

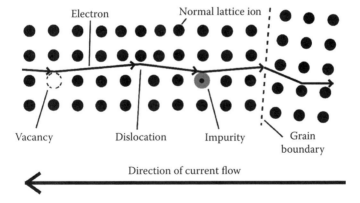

FIGURE 6.3   A free electron trying to make its way through a piece of metal under the influence of an electric field can encounter obstacles in the form of lattice atoms, impurities, vacancies, dislocations, and grain boundaries. Although it makes life simpler to say that the electrons "collide" with these obstacles, they do not actually hit into them, but interact electrostatically with them as they try to pass. This interaction causes the electrons to lose energy, so the metal is providing resistance to the flow of an electric current. This diagram should not be taken too literally, however, as in reality collisions are less frequent than this suggests. The mean free path (distance between collisions) is generally in the order of several hundred atomic spacings.

The resistivity caused by defects is known as the "residual resistivity", as unlike the thermal component of the resistivity, it still exists even if the temperature is near to absolute zero. The residual resistivity only increases with an increase in the number of defects in the material.

The total resistivity for a conductor is given by the following:

$$\rho = \rho_{th} + \rho_{res} \tag{6.3}$$

where ρ is the total resistivity, ρ_{res} is the resistivity due to impurities and defects, and ρ_{th} is the resistivity due to the thermal motion of the atoms, which is often described in terms of the free electrons being scattered by phonons. (See Example question 6.1.)

Electron Mean Free Path

The average distance an electron under the influence of an electric field travels through a solid before it encounters an obstacle is known as the "electron mean free path". This is analogous to the phonon mean free path discussed in Section 5.2.2. Once the electron interacts with an obstacle—whether that obstacle is a grain boundary, dislocation, vacancy, impurity, or lattice atom—the direction in which it is travelling (in other words its path) changes. This situation is illustrated in Figure 6.4.

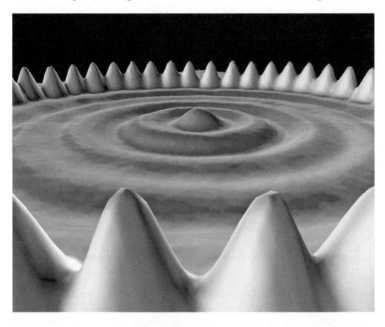

FIGURE 6.4 This scanning tunnelling microscope image shows part of a circle of 48 iron atoms that have been deliberately placed in these positions. When free electrons travelling along the surface of the crystal come up against these atoms, they are scattered back in the direction they have come. However, in this picture it is not the particlelike behaviour of the electrons we are seeing, but their wavelike behaviour. The ripples inside this circle are the interference patterns produced from the various electron waves that are reflected off the atoms. (Photo courtesy IBM.)

EXAMPLE QUESTION 6.1   RESISTIVITY

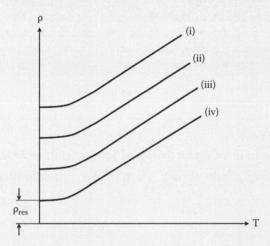

The curves in this graph represent the resistivity of pure copper and three copper nickel alloys. Alloy A contains 1 atomic percent Ni, alloy B contains 2 atomic percent Ni, while alloy C has 3 atomic percent Ni. Which of the curves represents which material?

Anwer on page 164.

Of course free electrons can also interact electrostatically with one another, but somewhat surprisingly, given how close together the free electrons are within a metal ($\approx$2 Å apart), any electron can travel around 10 cm at 1K and 1 µm at room temperature before colliding with another electron. These long mean free paths are mainly due to the Pauli exclusion principle. Because of the change in its direction of travel, each electron will have a different wave vector after a collision with another electron, and the exclusion principle only allows the postcollision electrons to have certain values of wave vector, which effectively limits the number of collisions that can take place.

By contrast, collisions between electrons and phonons are more likely than electron-electron collisions in metals at room temperature, and the mean free path for electron-phonon interaction is usually at least one order of magnitude shorter than the electron-electron mean free path. As we saw in the last subsection, it is the collisions between electrons and phonons that help explain the relationship between resistivity and temperature: The higher the temperature, the more phonons there are, and so the more electron-phonon collisions take place. In fact electron-electron collisions are so rare in comparison that they only give a noticeable contribution to the resistivity at liquid helium temperatures.

Wiedemann–Franz Law

We have already seen that the range of values for the electrical conductivity of different materials is very large. By contrast, thermal conductivities (see Section 5.3.2) do

TABLE 6.1
Comparative Conductivities of Some Metals at 293K

Metal	σ (10^7 Ω^{-1} m^{-1})	κ (W m^{-1} K^{-1})	(κ/σ) (10^{-6} W Ω K^{-1})
Silver	6.21	429	6.9
Copper	5.88	401	6.8
Aluminium	3.65	237	6.5
Sodium	2.11	141	6.9
Zinc	1.69	116	6.9
Cadmium	1.38	97	7.0
Iron	1.02	80	7.8
Lead	0.48	35	7.3

not vary by as many orders of magnitude, as all solids conduct heat to some extent. However, the two types of conductivity do have something in common.

If just metals and alloys are considered, the thermal conductivity (κ) and electrical conductivity (σ) vary in a similar pattern, so that values of κ/σ at a given T are almost constant over a fairly wide range of temperatures (see Table 6.1). This relationship was first observed by German physicists Gustave Heinrich Wiedemann (1826–1899) and Rudolf Franz (1827–1902) in the middle of the 19th century, and it is known as the Wiedemann–Franz law.

6.1.2 DRUDE'S CLASSICAL FREE ELECTRON MODEL

In 1900, the German physicist Paul Drude (1863–1906) proposed the first model of conduction. His model stated that each of the atoms a metal is composed of has lost one or more of its valence electrons and so has become a positively charged ion. This part of the model is the same idea that was used in Section 2.1.4 to help visualise metallic bonding. However, Drude's model is a classical model that goes on to assume that the liberated electrons in a metal behave like the atoms or molecules of a classical ideal gas. This means they are free to move throughout the metal in a random direction but may collide either with each other or with the fixed positive ions in the crystal lattice. (This movement is usually at speeds of hundreds of kilometres per hour in the absence of a voltage, but at much slower speeds of around a few centimetres per second when flowing as part of a current.) The model does not, however, take into account any of the details of the electrostatic interactions between either the electrons or the electrons and ions.

Despite its limitations, Drude's model predicts correctly that metals which can conduct heat well are also good electrical conductors, and so explains the Wiedemann–Franz law that was described in the last section. It also predicts that the current in a metal obeys Ohm's law:

$$V = IR \qquad (6.4)$$

where I is the current flowing in a circuit containing a conductor, V is the voltage across the conductor being measured, and R is the resistance of the conductor. (It is named after its discoverer, the German physicist Georg Simon Ohm [1787–1854].)

By contrast, Drude's model produces a value for the heat capacity that is much larger than the true value. This is because the electrons are assumed to be an ideal gas, so the distribution of their kinetic energies will be given by the Maxwell-Boltzmann distribution. (See Appendix E for a reminder of the classical statistics of an ideal gas.) According to the classical gas model, the average thermal energy of each electron is $\langle E \rangle = 3k_BT/2$, and if this is multiplied by the number of free electrons per unit volume of a mole of the substance under consideration, a molar heat capacity is produced. The value of this is, however, a factor of 100 higher than experimentally measured values for molar specific heat. In reality, as we saw in Section 5.3.1, the heat capacity is due to the thermal vibrations of the lattice ions, and the free electrons make almost no contribution.

So to get nearer to the actual situation that occurs in solids, quantum theory needs to be taken into account, and that is what Pauli's quantum free electron model does.

6.1.3   Pauli's Quantum Free Electron Model

This model is quite similar to Drude's model in that it still treats the conduction electrons as if they were the molecules of a gas that are free to move through the crystal. However, it takes quantum theory into account and so considers the electrons to be fermions. In this case the electrons are then subject to Pauli's exclusion principle (see Appendix D). This prevents any two fermions (particles with a spin quantum number of ½, such as electrons) from having exactly the same quantum state. So that the free electron gas in this model is not confused with the free electron gas in Drude's model, it is known as the free electron Fermi gas, which is often shortened to "Fermi gas".

States of the Free Electron Model

As the electrons in the Fermi gas obey the Pauli exclusion principle, they are only rarely scattered by each other. This is in direct contrast to Drude's model in which the valence electrons can collide both with each other and the fixed lattice ions. In another contrast to Drude's model, the electrons in a Fermi gas do not collide with the positive ions in the lattice when they are in their correct lattice positions. This is because matter waves propagate without any deflection within a periodic lattice.

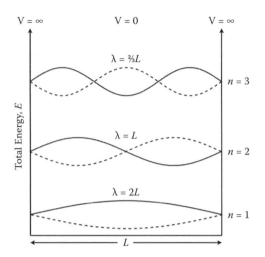

FIGURE 6.5   The first few allowed energies for electrons confined in a 1-D metal of length L. The waves representing the electrons are also shown, and look like the standing waves produced when a guitar string is plucked.

To understand more about how the electrons behave in this model, it is useful to imagine the electrons moving in just one dimension (1-D). The metal can then be represented by a square well potential (see Appendix D) where the distance 0 to L is the entire length of the crystal. We know from quantum mechanics that these electrons cannot just have any energy within this well. They can only have energies given by

$$E_n = \frac{\hbar^2}{2m}\left(\frac{n\pi}{L}\right)^2 \tag{6.5}$$

where m is the mass of the electron, and n the quantum number of the energy level.

The first few allowed energies for electrons confined in a 1-D metal of length L are shown in Figure 6.5.

The Pauli exclusion principle does not allow more than one electron to have the same set of quantum numbers, so a maximum of two electrons can occupy each level (or state) shown in Figure 6.5. Two can be accommodated because one can have a spin quantum number of $+\frac{1}{2}$ (spin up), while the other can have a spin quantum number of $-\frac{1}{2}$ (spin down).

At absolute zero, the electrons in a metal will fill up the possible states two at a time, starting with the state that has the lowest energy, so that the total energy of the system is kept as low as possible.

Fermi Energy

So if we imagine there are N electrons in our 1-D solid, the energy states will be filled from the bottom upwards, with two electrons going into the $n = 1$ level, then two electrons filling the $n = 2$ level, and so on until all N electrons are accounted for.

If n_F represents the highest filled energy level, then the *Fermi energy*, E_F, which is the energy of the highest filled state, is given by

$$E_F = \frac{\hbar^2}{2m} \left(\frac{N\pi}{2L} \right)^2 \tag{6.6}$$

where m is the mass of the electron, L the length of the hypothetical one-dimensional solid, and N the number of electrons. (As you will see if you refer back, this equation has the same form as equation 6.5. The 2 in the denominator of the term in parentheses in equation 6.6 comes from the fact there are two electrons in each level.)

The relevance of this energy for the Pauli quantum free electron model is that when electrons with energies below the Fermi energy acquire extra energy (for example, when the solid they are in is heated) and move above E_F, they are then able to take part in conduction. So only electrons with energies fairly close to the Fermi energy can become so-called conduction electrons. As we will see in Section 6.4, the Fermi energy is also of great importance for semiconductors, but it is worth noting immediately that for semiconductors E_F is not generally representative of the highest filled energy state, as it is for metals.

As it is impossible for more than one quantum state in any quantum mechanical system to have the same value of energy, the electrons in a Fermi gas are divided up between all the possible energy states available. The average number of these free electrons in a given energy state, E, is described by the Fermi–Dirac distribution as follows:

$$F_F(E) = \frac{1}{e^{(E-E_F)/k_B T} + 1} \tag{6.7}$$

where $F_F(E)$ is the Fermi distribution function, E_F is the Fermi energy, k_B is Boltzmann's constant, and T is the temperature in degrees kelvin. (See Appendix E for a more detailed discussion on the Fermi–Dirac distribution function.)

Figure 6.6 shows the Fermi distribution function both at 0K and at a range of nonzero temperatures. At absolute zero, it is clear that $F_F(E)$ is zero above E_F and 1 below E_F. $F_F(E) = 0$ means that no electrons are occupying any of the states above E_F. By contrast, because $F_F(E) = 1$ below the Fermi energy, all of the states in the region up to E_F must be occupied by an electron.

This is just another way of looking at the gradual filling up of energy states in a metal from the lowest state upwards. However, since it is impossible for any metal (or indeed anything at all) to become as cold as absolute zero, the graph shown in Figure 6.6(b) is in fact more representative of the real situation. Here it can be seen that the boundary at E_F is less abrupt. This is because above 0K, electrons in levels just below E_F can gain enough thermal energy to be excited into quantum states above E_F, and so take part in conduction. The higher the temperature, the more electrons can gain enough thermal energy to be excited to levels above the Fermi energy, and so the more shallow the slope of the curve around E_F, and the greater the current produced.

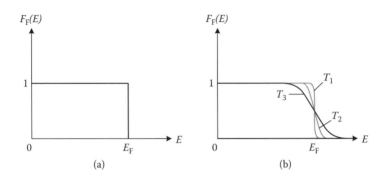

FIGURE 6.6 The Fermi distribution function at (a) 0K, and (b) a range of temperatures where $T_3 > T_2 > T_1$. The greater the value of $F_F(E)$ for any particular value of energy state E, the greater the number of electrons in that energy state. Any states for which $F_F(E) = 0$ are empty. (The Fermi distribution function is discussed further in Section E4 of Appendix E.)

E-k **Relationship**

In free electron theory, the conduction electrons are described by free electron wave functions, and these have certain allowed values of wave vector k that are analogous to the quantum numbers n, l, and m that describe the permitted states of an electron in an atom. In fact the energy of a free electron in a metal is related to its wave vector, k, by

$$E = \frac{\hbar^2 k^2}{2m}$$

(6.8)

Since the energy of a free electron is entirely energy due to its motion—in other words, due to its kinetic energy—E in equation 6.8 can be replaced by $\frac{1}{2} mv^2$ as follows:

$$\frac{1}{2} mv^2 = \frac{\hbar^2 k^2}{2m}$$

(6.9)

which cancels into

$$mv = \hbar k$$

(6.10)

So you can see from equation 6.10 that the wave vector, k, is a measure of the momentum of a totally free electron, as momentum, p, is equal to mv. (In fact **k** is often referred to as the "quasi-momentum". This is because although it gives an indication of the momentum—and hence the velocity—of a totally free conduction electron, as we will see in the next subsection, it does not always do so for conduction electrons influenced by the lattice ions.)

Returning to equation 6.8, the form of the relationship between E and k is parabolic because in Cartesian coordinates any parabola has a standard equation of the following form:

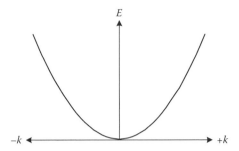

FIGURE 6.7 *E-k* relationship for a totally free electron, which does not interact with the crystal lattice but behaves as though it is a particle in a quantum gas.

$$y = ax^2 \qquad (6.11)$$

This means that if we were to plot the *E-k* relationship, the resulting graph would look like Figure 6.7.

The way an electron behaves in response to an electric field can be determined by the *E-k* plot. This is because, as we have just seen, *k* is a measure of the momentum, so *E-k* diagrams are nothing more mysterious than graphs of energy vs. momentum.

EXAMPLE QUESTION 6.2   *E-k* RELATIONSHIP

The curve in Figure 6.7 is shown for negative values of *k* as well as positive values. What do negative values of *k* represent?
Answer on page 170.

Effective Mass

If an electron is travelling through a vacuum and being accelerated by an electric field, its motion can be described by Newton's laws of motion. Somewhat surprisingly for a system governed by quantum mechanics, Newton's laws can also describe the motion of a free electron that is being accelerated within a crystal by the electric field produced by an applied voltage. However, as the electrons are interacting with different potentials as they make their way through the lattice, in order for the simple Newtonian physics to work, the normal mass of the electron, m_e, must be replaced by the *effective mass*, m^*. The effective mass takes all of the complex interactions between the electrons and the lattice into account, and so adjusts the classical equations enough for them to still be adequate.

In terms of an equation,

$$m^* = \frac{\hbar^2}{\dfrac{d^2 E}{dk^2}} \qquad (6.12)$$

ANSWER TO QUESTION 6.2 ON *E-k* RELATIONSHIP

Negative values of the wave vector *k* represent electrons travelling in an opposite direction to those with a positive value of *k*.

where m^* is the effective mass of the electron. So if the *E-k* curve for an electron is drawn, its mass is inversely proportional to the curvature (in other words, the rate of change of the slope) of the *E-k* curve, as $\dfrac{dE}{dk}$ = the slope of the curved line, and $\dfrac{d^2E}{dk^2}$ = the rate of change of the slope.

Equation 6.12 also holds for a totally free electron (or indeed other types of particle). So in all cases the value of the mass of a particle can be obtained by taking the second differential of its *E-k* curve.

Fermi Surfaces

In general, all of these sorts of diagrams are a plot of energy against wave vector in just one dimension. In other words, they show E against k, rather than E against $\boldsymbol{k}$, which is made up from k_x, k_y, and k_z as follows:

$$\boldsymbol{k}^2 = k_x^2 + k_y^2 + k_z^2 \tag{6.13}$$

Solids are of course three dimensional (3-D) in real life, and so a more accurate way of representing which electron energy levels are occupied and which are empty is to construct a Fermi surface.

The Fermi surface is the surface representing the Fermi energy E_F when the energy levels are plotted as points in 3-D $\boldsymbol{k}$-space. As we have already seen, we can build up an N–electron ground state for an electron gas by making use of the Pauli exclusion principle. In this case, with each allowed wave vector $\boldsymbol{k}$ there are two electronic levels: one for "spin up" and one for "spin down". So to build up the ground state, two electrons are placed in the level $\boldsymbol{k} = 0$; then the levels are successively filled with electrons in a similar way, starting with the level of least energy each time.

If there is a large number of electrons, N, the occupied region in $\boldsymbol{k}$-space will be indistinguishable from a sphere. As each of the occupied levels is represented by a point within this "Fermi sphere", the sphere surface—in other words, the Fermi surface—separates the occupied from the unoccupied levels. Figure 6.8 shows the Fermi surface of a hypothetical metal described by Pauli's quantum free electron model. The radius of this Fermi sphere is k_F, so the volume is $4/3\pi k_F^3$. (Fermi surfaces can be constructed for both free electrons and nearly free electrons—which will be discussed in the next section.)

In reality, most Fermi surfaces are not spherical. In fact they can have extraordinary shapes. The shape of the Fermi surface determines the electrical properties of a metal because the conductivity is due to changes in the occupancy of states

EXAMPLE QUESTION 6.3   EFFECTIVE MASS

For points A, B, and C in the curve below, is the effective mass, m^*, positive, negative, or infinite? What do your results mean in terms of the physical properties of a particle with that effective mass? (Hint: It may help to sketch each mathematical step.)

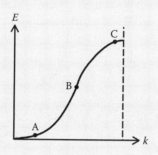

ANSWER

Since we will need to work out the rate of change of the slope in the diagram, in other words $\dfrac{d^2E}{dk^2}$, we can start by drawing a diagram of $\dfrac{dE}{dk}$, in other words the slope of the curved line, as follows:

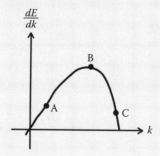

The curve is curving slightly upwards at point A, so the slope at A is small but positive. The curve moves upwards more steeply at B, so the slope has a larger positive value than A. Finally, the slope is much less at point C.

If we then take the slope of the curve in the diagram above, we will have a sketch of $\dfrac{d^2E}{dk^2}$ as follows:

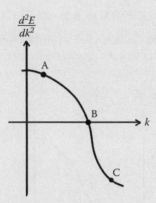

The slope at point B in the diagram above is zero because it is the maximum of the curve. The slope is less at C, so C must now be below the zero of point B. By contrast, point A had a positive slope, and so will appear above B on this diagram.

The final sketch takes equation 6.12 into account, and so shows the reciprocal of the diagram above.

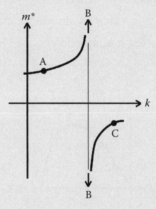

The reciprocal of zero is infinity, so point B represents a particle with an infinite effective mass. The reciprocal of a negative number is also negative, so point C represents a particle with a negative effective mass, while the reciprocal of point A is positive so point A represents a particle with a positive effective mass.

Physically, a particle like the one represented by point B that has an infinite mass would be so heavy it could not move at all. Meanwhile, a particle with a negative effective mass would accelerate in the opposite direction to a particle with a positive effective mass in the presence of applied electric field. So while point A in this question represents an electron (with positive mass and negative charge), point C represents a particle known as a "hole" which can be thought of as having a positive mass and positive charge (equal in magnitude to the charge on the electron).

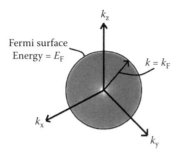

FIGURE 6.8 The Fermi surface of a hypothetical solid containing N free electrons. The occupied electron energy levels fill a sphere of radius k_F, which is known as the Fermi wave number.

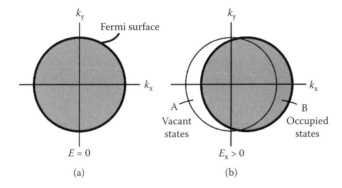

FIGURE 6.9 A Fermi surface for a hypothetical 2-D metal (a) with no electric field present and (b) under the influence of an electric field.

near the Fermi surface. The surface shape is influenced strongly by the numbers of valence electrons of each particular metal.

If we look again at the picture of a Fermi surface in Figure 6.8, it can be seen that with the values of wave vector plotted in three dimensions, the result is a sphere centred on the origin of the different k axes, in other words, centred at $k = 0$. As k is a measure of the momentum, this must mean there are equal numbers of electrons travelling in every possible direction. If we were to think about this idea in just two dimensions (2-D) for a moment (just to help visualise the situation), it means that as many electrons would be travelling to the right as there would be to the left. So adding together the contributions from all of the free electrons would give an average electron velocity of zero, and no net electric current. This situation is depicted in Figure 6.9(a).

By contrast, Figure 6.9(b) shows what happens to the surface of a 2-D Fermi sphere when there is an electric field, E, applied to our hypothetical metal. Every electron in the Fermi gas will be accelerated in a direction opposite to the field, so there will now be more electrons travelling in one direction than the other. This will shift the Fermi sphere, and of course produce a flow of current.

When a current flows in a conductor it does not just keep increasing in size with the same electric field applied. Instead it quickly rises up to a certain value that depends on the size of the field, then remains steady at that value. This can be explained in terms of the Fermi surface.

Free electrons in the energy states near to the Fermi surface (which has the energy E_F) suffer collisions with other electrons, lattice ions, and defects (as we saw in Section 6.1.1), which causes them to be scattered. This in turn will mean that their wave vector will change, or in other words, the change in direction and speed that the scattering produces means they have to go to a new k state. It is only those electrons near the Fermi surface (labelled as "occupied states" in Figure 6.9[b]) that can find vacant states into which they can be scattered. Another look at Figure 6.9(b) shows the vacant k states (new values of wave vector) that the scattered electrons take are those with a lower value of energy situated to the left side of the Fermi sphere.

The current does not therefore increase indefinitely in size, because to do so the electrons would have to keep increasing in energy (so they could travel along faster). This would mean they would have to move to k states with higher and higher energies rather than moving into states with lower energies. For example, an electron in state B in Figure 6.9(b) will move into state A rather than a higher energy state, limiting the current.

Electronic Contribution to the Specific Heat

A glance back at Table 5.2 will reveal that it lists Debye temperatures for both metals and nonmetals. This must mean that the free electrons present in metals make no contribution to the specific heat.

In fact they make a very small contribution at very low and very high temperatures, but for most temperatures only a very small number of the free electrons can contribute. These are the free electrons we have just discussed that have energies near to the Fermi energy, and can therefore move into a higher energy state when they acquire extra energy from an increase in temperature. So applying quantum mechanics to the free electron theory, and treating the electrons as a Fermi gas, explains the electronic specific heat that Drude's model was unable to provide a sensible explanation—and indeed a value—for.

6.2 ENERGY BAND FORMATION

As we have just seen, Pauli's free electron theory gives a good explanation of many of the properties of metals, including their thermal and electrical conductivities. But it fails to explain why some materials are conductors, some semiconductors, and some insulators, and cannot account for the electron mean free path being around a hundred or more atomic spacings. A better way to understand these topics is to view solids in terms of energy bands instead, and as we will see in the next two subsections, there are two totally different ways of considering how energy bands come about.

6.2.1 NEARLY FREE ELECTRON MODEL

This model is an extension of Pauli's quantum free electron model that takes into account the effect that the ions in the lattice have on the sea of electrons. In Pauli's model, the conduction electrons are assumed to have a uniform potential energy because the positive charge of the lattice ions is effectively spread out across the whole of the specimen of metal being considered. As we saw in Section 6.1.3, the potential energy of electrons within a metal is actually taken to be zero for simplicity, while the potential energy outside the metal is assumed to be infinite so that the electrons are confined within the metal.

However, the potential energy of the electrons is like that shown in Figure 6.10 once the interaction of the conduction electrons with the positively charged lattice ions is considered. As the diagram reveals, the potential energy of an electron is lower than its average value when that electron is near a positive ion, and is at its highest when the electron is exactly in between two ions. This is because any object will have a lower potential energy when it is near to something it is attracted to than when it is far apart. In fact the situation is similar to an object being pulled by gravity down to Earth. For example, when an apple is ripe it will fall from the tree it has been growing on, and the potential energy of the apple will be higher when it is still attached to the tree than when it has fallen off and is lying on the ground.

If electrons are then considered in terms of their wavelike properties, and the energies that quantum mechanical calculations (that are well beyond the scope of this book) show the electrons can have are plotted as a function of the wave vector k, a broken curve like that shown in Figure 6.11 appears.

The dotted line in Figure 6.11(a) shows the E-k relationship from the Pauli model. By contrast, the E-k curve for the nearly free electron model, with its missing portions, shows there are values of energy that electrons in solids cannot occupy. These can be represented by a "band diagram", which is shown in Figure 6.11(b), and would normally be displayed without the E-k curve overlaid. (See figure caption for more details.) The "forbidden" bands of levels, known as *bandgaps* or "energy gaps", are indicated by the plain white regions. These correspond to the areas in the crystal where the electron wavelength satisfies the Bragg condition, and so the Bragg reflections from the lattice ions are preventing electrons with energies within the band

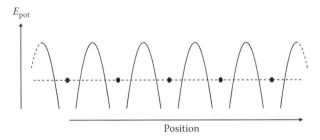

FIGURE 6.10   The potential energy for conduction electrons in a hypothetical 1-D solid in which the dots represent the positive lattice ions.

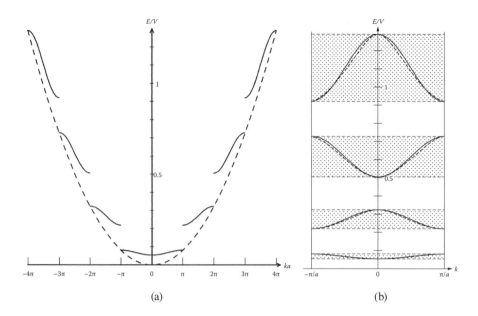

(a)   (b)

FIGURE 6.11   (a) The dashed line shows the $E\text{-}k$ curve that the Pauli model predicts (and we first saw in Figure 6.7). In this case conduction electrons can have any value of energy, whereas the nearly free electron model predicts the existence of ranges of energy that electrons cannot have. These are indicated by the gaps in the solid-line, nearly free electron $E\text{-}k$ curve for a hypothetical 1-D solid of lattice constant, a. (In fact the lowest part of the $E\text{-}k$ curve for nearly free electrons should pass through zero just like the free electron $E\text{-}k$ curve, but the curves are often separated in diagrams of this sort so it is easier to compare the two shapes.) Part (b) shows another form of the same diagram. If we first look at the curved lines, it can be seen by comparing it with (a) that this is just a more compact way of plotting the $E\text{-}k$ curve for nearly free electrons. A third way of representing this information is provided by the shaded sections underlying these lines, which form what is known as a band diagram. In general, band diagrams will not show the $E\text{-}k$ curve as well, and will indicate ranges of allowed energies known as "energy bands" simply by shaded regions. In this case the energy bands are shown in dotted shading, while the forbidden ranges of energy known as "energy gaps" are white. We will see more band diagrams in the next few subsections. (Courtesy of Gary Swanson.)

gaps from travelling along. Electrons with all other values of energy in conjunction with any of the values of wave vector, k, that fit on the broken $E\text{-}k$ curve can travel through the crystal, and the regions representing these allowed energy states in band diagrams are known as *energy bands*—here represented by dotted shading.

This means that, for electrons with wave vectors that do not satisfy the Bragg condition, a perfectly periodic lattice has no effect at all on their passage through a metal. This is because if a lattice is identical in every direction, no matter where any given electron is sitting at a particular time, it will be just as strongly attracted to the ions in one direction as to those in another direction. The net effect is that it will not be scattered by the lattice unless there are defects in the lattice. Although the presence of defects is one of the major factors giving the electron mean free path a finite length, the lack of scattering by the bulk of the lattice allows the electrons to travel along hundreds of atomic spacings without having their paths altered.

However, it should not be forgotten that although there is only electron scattering from the lattice under certain conditions, the whole basis of the nearly free electron model is that the periodic potential that the positive ions in the lattice generate does change how the possible electron states are distributed in comparison with the free electron model.

Although the nearly free electron model leads us to a picture of solids in terms of energy bands—which, as we will see later, explain the existence of conductors, insulators, and semiconductors—it does not say much about the wave functions of the electrons. These can be very important for some calculations, so the tight-binding model described in the next section can be used instead as an alternative route to forming energy bands. In fact, most of the research that is carried out on band theory involves tight-binding calculations rather than the nearly free electron method. This is not least because the nearly free electron model only applies to crystalline metals. Semiconductors and insulators cannot be described by it because they do not have a structure like that shown in Figure 6.10. Amorphous materials cannot be accounted for, either, because the model relies on electrons being Bragg-scattered by a periodic lattice, and there is no long-range order in amorphous materials (see Section 3.2.1), so no periodic lattice for the free electrons to be scattered from.

6.2.2   Tight-Binding Model

Diagrams of electron shells (orbits) surrounding a nucleus are a common sight when learning about atomic physics. However, in solid state physics it is more usual to depict these shells as straight lines. Each electron shell has a different value of energy, as does each corresponding line—which is known as an energy level. Figure 6.12 shows how these two ways of depicting electron energy levels relate to one another.

Of course the diagram on the right of Figure 6.12 shows how the energy levels would look if a single silicon atom existed on its own, and is only likely to be a reasonably accurate representation of the situation if a gas of silicon atoms was being studied.

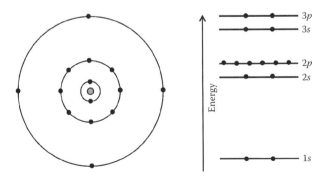

FIGURE 6.12   The electron energy levels of a silicon atom depicted on the left as electron orbitals and on the right as energy levels.

The tight-binding model reveals what happens when a very large number of atoms an infinite distance apart come closer together to form a solid. This occurs, for example, during the chemical vapour deposition technique (which will be described in Chapter 7), as the gas in the deposition chamber condenses onto a substrate (base) to form a thin layer of solid material. As the atoms approach one another to form the solid, the wave functions of their electrons start to overlap. The interactions between the electrons cause the discrete energy levels of each of the isolated atoms to split into a huge number of energy levels forming a band of levels. This prevents the Pauli exclusion principle from being violated, as while two atoms can have their electrons in exactly the same levels as one another if they are isolated, this cannot occur with the atoms side by side in a solid. If it did, this would mean electrons with exactly the same set of quantum numbers would be in the same energy level, which is not a situation allowed by the Pauli principle. So the split levels created when a solid is formed get round this problem by providing enough separate states to accommodate all the electrons in the solid with the same quantum numbers. These energy levels are in fact so close together within the bands that they can be considered to be a continuum of levels rather than a collection of discrete energy levels. Figure 6.13 shows the changes in the electron energy levels as a huge number of the same type of atoms approach one another.

This idea is just an extension of the situation that occurs when two atoms come together to form a molecule. Taking hydrogen as a simple example, the single electron

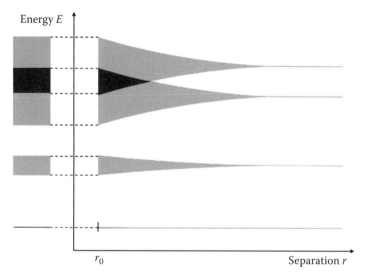

FIGURE 6.13   When atoms are a long distance apart from each other, their electron energy levels are separate, as the right of the diagram shows. However, as a very large number of similar atoms approach one another, their energy levels start to widen into energy bands. When the atoms are a distance r_0 apart, a solid forms, and the band structure of that solid is shown on the far left of the picture. (This diagram originally appeared as Figure 2.23, p. 82, of *Quantum Physics of Matter*, edited by Alan Durrant, © 2000, The Open University. Reproduced with permission.)

of each hydrogen atom is in the $1s$ state, and as a pair of hydrogen atoms come towards each other, their wave functions begin to overlap. This causes the $1s$ level to split into two separate energy levels. If a large number of hydrogen atoms were brought together, the $1s$ energy level would split into the same number of separate levels as there were atoms.

Of course other solids are more complicated than hydrogen, as their atoms contain more than one electron. In fact as the separate atoms are brought together it is, not surprisingly, the wave functions of the outermost electron shells that overlap more than those of the electrons closest to the nucleus. This results in the energy bands formed from the outermost electron shells being broader—and so containing more energy levels—than those closer to the nucleus. In fact the outermost electrons can take part in conduction because the overlap of the outermost electron wave functions is so great it actually extends throughout the whole solid. By contrast, the electrons nearest to the nucleus remain bound to the nucleus. So this model gets its name because most of the electrons in metals are "tightly bound" to their respective atoms. Only a small number actually take part in conduction, which we know is true from experimental measurements of the electronic specific heat (as we discussed in Section 6.1.3).

As in the nearly free-electron approach to energy bands, the regions in between the bands created by the tight-binding method are known as "band gaps" or "energy gaps" (although, as we will see later, the term "energy gap" tends to be used for just one particular forbidden band). These forbidden bands are effectively another version of the "forbidden" regions that exist in isolated atoms between the electron shells, but they are narrower. Quantum mechanics dictates that electrons cannot possess any of the values of energy in these forbidden bands, so electrons cannot move about in the crystal within the band gaps. However, electrons can have any of the values of energy contained within the energy bands. This means they can move about within the energy bands because moving means changing their kinetic energy, which in turn means changing their energy level, and in a band there are empty levels nearby that electrons can move into. This, as we will see in the next section, together with the fact that each different type of solid has a different band structure, explains the difference between conductors, insulators, and semiconductors.

Both the tight-binding model and the nearly free electron model produce energy bands that are exactly the same, but although the tight-binding method has the advantage of covering metals, insulators, and semiconductors (as we will see in the next section), it is still just a model—and therefore an approximation. One thing missing from it is that it fails to take account of any electrostatic interactions between electrons. Having said that, tight-binding calculations (quantum mechanical calculations that use the tight-binding model as their basis) are important tools for the researcher, as Figure 6.14 helps illustrate. Other than the conductivity in complex systems such as these nanotubes, quantities that can be calculated using these methods include energies of localised electron states caused by defects in semiconductors, band structures, the density of states (see Section 6.3.2), and Fermi surfaces of metals.

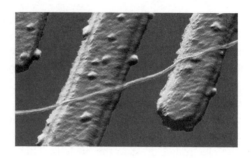

FIGURE 6.14 This atomic force microscope (AFM) image shows a single-wall carbon nanotube approximately 1 nm in diameter lying on two 15-nm-thick platinum electrodes. By applying a bias voltage to the electrodes and measuring the current flowing through the tube, Prof. Cees Dekker and his team at Delft University of Technology in the Netherlands were able to investigate the electrical transport through the nanotube and determine that it behaved like a tiny metal wire. Further experiments using a scanning tunnelling microscope (STM) to measure the electronic density of states (see Section 6.3.2) in the nanotube have revealed that the conduction electrons travel throughout its length, and that there are discrete energy levels that they move through. As the nanotubes are so narrow, electron movement is confined to just 1-D, and so this conduction can be described by the "particle in a box" model shown in Figure 6.5, with the length L equal to the length of the nanotube. The current-voltage characteristics recorded for the sample above consisted of a series of steps like the steps on a staircase. Each of these "step" increases in current results from an extra electron energy level becoming available for carrying current, so the width of each measured current step is proportional to the energy difference between the quantized energy levels.

The Delft team has also measured the electron wave functions corresponding to these discrete energy levels in a separate STM experiment on short metallic carbon nanotubes. In this case the regions with a high density of states revealed the positions of the standing wave peaks. The structure of a single-wall nanotube can be thought of as a sheet of carbon atoms—arranged in the same hexagonal formation as the carbon making up the layers of a graphite crystal—rolled up into a tube which has each of its ends capped with half of a bucky-ball. Additional STM experiments together with tight-binding calculations have both shown the conductivities of nanotubes to depend on their diameter and the orientation angle of the carbon hexagons to the tube axis. As about one third of nanotubes behave like metals, while the rest are semiconducting, this opens up the possibility of "molecular electronics" based on devices made from these single molecules. (Courtesy of Cees Dekker.)

6.3 SIMPLE BAND THEORY

6.3.1 APPLICATION OF BAND THEORY TO REAL SOLIDS

At the start of this chapter we saw that a common way of classifying solids was as conductors, semiconductors, and insulators. This distinction is widely used because not only do metals, semiconductors, and insulators have different electrical and thermal characteristics, they also have different optical properties—some of which are exploited in important devices like solar cells and light-emitting diodes, as we will see in Chapter 7.

Energy Band Formation for a Conductor

Figure 6.15 shows what happens when the conductor sodium forms into a solid. The electron configuration of separate sodium atoms is $1s^2 2s^2 3p^6 3s^1$ (see Appendix C for

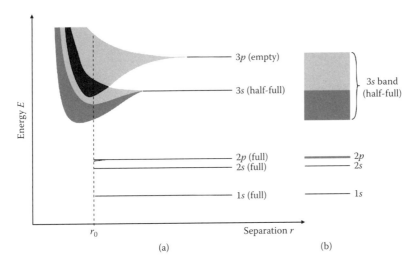

FIGURE 6.15 Energy band formation as a very large number of sodium atoms come together to form a solid. (This diagram originally appeared as Figure 2.24, p. 83, of *Quantum Physics of Matter*, edited by Alan Durrant, © 2000, The Open University. Reproduced with permission.)

a revision of electron shell notation). As we saw in the previous section, when atoms approach each other to form a solid, these discrete energy levels broaden out into energy bands.

These bands can be very different sizes, as Figure 6.15 illustrates, and the number of states within each band will equal the number of atoms forming the solid multiplied by the number of energy levels in the corresponding original subshell of an isolated atom. For example, any p subshell has three possible energy levels, so in a solid an energy band formed from a p subshell will contain $3N$ levels, where N is the number of atoms making up the solid. (When the solid is first formed, the electrons from the various subshells will sit in their corresponding energy band. It is only if the solid is then subjected to some form of change in its environment that the electrons will move from these positions.)

As we have already seen, the energy bands formed from the outermost electron shells are broader—and so contain more energy levels—than those closer to the nucleus. So the electrons in the lowest energy shells remain bound fairly closely to the nuclei of their parent atoms. In the case of sodium the $1s$ and $2s$ levels are closest to the nucleus and so do not broaden at all, while $2p$ only broadens by a tiny amount.

We also saw in the previous subsection that the wave functions of electrons in outer subshells of atoms tend to overlap a lot as the atoms approach one another, and so these energy levels split into broad bands like the $3s$ band shown in Figure 6.15. Any electrons within this broad band are no longer bound to their atoms but are free to wander about the solid and hence can take part in conduction. In fact for any solid, not just sodium, it is only the so-called valence electrons in the uppermost levels (outermost electron shells) that are free to take part in conduction. (The energy levels of all the higher subshells also broaden out into bands, but for a solid like sodium that

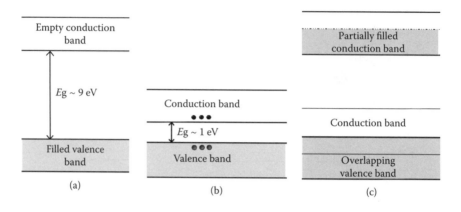

FIGURE 6.16   Energy band diagrams for (a) an insulator, and (b) a semiconductor. Part (c) shows two possible band structures for a conductor.

has a partially filled outer shell, these make no difference to the conductivity of the solid, so Figure 6.15(b) does not include an empty $3p$ band.)

So just what is it about the electrons in the $3s$ band of sodium that allows them to form an electric current when the solid is under the influence of an electric field?

To answer this we need to recall from atomic physics (see Appendix C) that the $3s$ energy level can contain a maximum of two electrons. Since the $3s$ level only contains one electron for sodium, the $3s$ band in solid sodium has only half of the available energy levels within it occupied by electrons. This means that the electrons can move about with ease into neighbouring empty energy levels and so can form an electric current by travelling through the solid. It is therefore this half-full band that makes sodium—along with all the other metals in Group 1A of the periodic table that also have a single valence electron in their outermost s subshell—a conductor.

The energy band in which electrons move when they are taking part in conduction is known as the *conduction band*. In conductors the conduction band is the highest band that contains electrons (the highest occupied band), whereas in semiconductors and insulators the conduction band is the empty band directly above the highest occupied band, which is known as the *valence band* because it contains the valence electrons for these solids. (As we saw in Chapter 2, it is the valence electrons that also take part in bonding between atoms.) So for sodium, the $3s$ band is the conduction band.

The upper diagram in Figure 6.16(c) shows a simplified version of the energy band structure for conductors such as sodium. (Since it is only the uppermost bands that decide whether a solid can conduct electricity or not, only these bands are shown in these diagrams.) Below this is another type of band structure that can be found in conductors. In this case the highest filled band and lowest empty band overlap in energy.

Energy Bands for Insulators

Although, as we saw in Figure 6.11, there are several forbidden "energy gaps" in a complete band diagram, only one of these gaps is relevant to discussions about conduction. This is known as the *bandgap*, which is denoted by E_g, and is the energy difference between the top of the valence band and the bottom of the conduction band. Figure 6.16 gives an idea of the values of E_g for insulators and semiconductors (which are discussed in Section 6.4). Silicon, for example, has a bandgap of 1.1 eV at room temperature. (Semiconductor bandgaps vary slightly with temperature.) By contrast, as Figure 6.16(c) shows, conductors have no bandgap.

If we compare Figure 6.16(a) with Figure 6.16(c) we can see that an insulator differs from a conductor by having a totally full uppermost band. For insulators the next empty band up—the conduction band—is also so much higher in energy that electrons cannot gain enough energy, either from an applied electric field or from the solid being heated up, to move into it. It is therefore impossible for an electric current to flow through such a solid because there are no empty states available for electrons to move into.

As useful as this model is, like all models, even the latest version is only an approximation of the true behaviour. (In fact the theories behind almost every aspect of solid state physics are still being worked on, so the most recent models presented in this chapter could easily look as out of date in the next century as Drude's classical free electron model appears to us today.) Ionic crystals like NaCl, for example, are well known insulators (see Figure 6.1) and have much too large a bandgap for electrons to cross and therefore contribute to an electric current. However, their ions can actually conduct a very small amount of electricity by moving into the spaces created by Schottky defects, otherwise known as vacancies. (Vacancy diffusion is discussed in Section 3.1.1.)

The overall electrostatic charge on an ionic crystal must be neutral. So if impurities with a higher value of valency are added to a monovalent ionic crystal (in other words a crystal in which each of the component ions has a valency of +1), vacancies must be created to cancel out the excess charge. For example, if Ca^{2+} impurities are added to a crystal of NaCl as shown in Figure 6.17, not only does the Ca^{2+} ion substitute for one Na^+ ion, but an Na^+ vacancy must be created in order to balance out the charges. So in this case two Na^+ ions are lost for every Ca^{2+} ion added. As the only way for an electric current to flow in solid NaCl is via the host ions moving through the crystal from one vacancy to the next, this means the greater the number of Ca^{2+} impurities, the higher the current will be.

These ideas are certainly not just theory, however. It has been experimentally shown that the conduction in alkali and silver halide crystals is ionic and not electronic by measuring the amount of charge flowing in an electric current through a crystal and comparing this with the mass of material deposited on an electrode in contact with the crystal. (Ions will be deposited on the electrode, while electrons will remain in the solid.) Ionic conductivity in these solids is in fact regularly studied by researchers interested in defects in these crystals, as at certain temperatures the ionic conductivity is proportional to the amount of divalent impurities (impurity atoms with a valency of +2) added to the crystal.

Conduction in Polymers

As we saw in Section 6.1.1, polymers are widely known as insulators, and have been used for many years as insulating casings for electrical products, and to isolate conducting parts of electronic circuits. It is their lack of free electrons that makes them

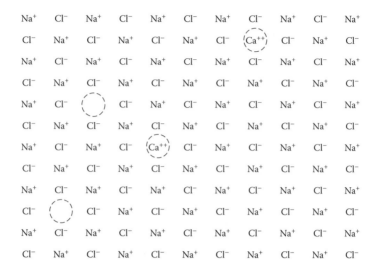

FIGURE 6.17 Ionic conduction can occur in ionic crystals, which would normally be insulators, if vacancies are present. The ions of the crystal can then migrate through it by moving from one vacancy to the next and so make up an electric current. In this picture, part of an NaCl crystal is shown which has had Ca^{2+} ions deliberately added to it. Two Na^+ ions are lost for every Ca^{2+} ion gained, so the crystal remains neutrally charged overall. (This diagram was originally published in *Solid State Physics*, International Edition, Neil W. Ashcroft and N. David Mermin, p. 623, © W. B. Saunders Company [Elsevier], 1976. Reproduced with permission.)

insulators, but in recent years some conducting polymers have been manufactured with conductivities to rival that of copper.

In order to achieve values of conductivity this high, impurity atoms have to be added to the polymers. The way in which conduction is produced by these impurity atoms is not yet well understood, but it is thought that when the dopant atoms are added, new energy bands are created. These bands overlap the valence and conduction bands that the polymer has in its original form. The partially filled conduction band this creates looks like the band in the upper part of Figure 6.16(c) and contains free electrons that can take part in conduction.

Conducting polymers could become very important in the future because they are so flexible and lightweight. At the time of writing this book, however, many are unstable or even toxic, and one of the only major uses of conducting polymers is (as mentioned earlier in this chapter) for the cathodes of the rechargeable batteries used in pacemakers. The polymer used for these lasts about ten years, and has the advantage of reducing the overall weight of the battery. Other polymers can be made that behave like semiconductors (see Section 6.4) and even superconductors (see Section 6.5).

6.3.2 DENSITY OF STATES IN ENERGY BANDS

The pictures of band structures we saw in the last section are actually simplified views of the actual situation. They fail to account for the fact that there are often

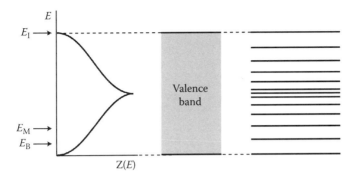

FIGURE 6.18 Energy bands often contain regions in which there are a greater concentration of energy states than in other areas. For this hypothetical valence band, the density of energy states is highest near the middle of the band, as illustrated by the curve on the left. (This diagram first appeared as Figure 11.7, p. 187, of *Understanding Materials Science*, by R. E. Hummel, corrected second printing 1999, © 1998, Springer-Verlag New York, Inc. Reproduced with kind permission of Springer Science and Business Media.)

more energy states in one region of an energy band than another. This idea is illustrated in Figure 6.18.

One of the major ideas of band theory (and quantum free electron theory) is that only electrons near to the Fermi energy can take part in conduction rather than all the free electrons—as classical free electron theory states. This was illustrated pictorially in Figure 6.9, and as we saw then, the reason only electrons near the Fermi surface can take part in conduction is that they are the only electrons that have vacant states nearby into which they can be scattered.

The number of electrons near to E_F will depend on the density of the available electron states in that region, so a knowledge of the density of states is vital to understanding the conduction of individual metals. Copper, silver, and gold, for example, have a large density of states near E_F, along with a large number of electrons in them, and as a consequence have high conductivities.

6.4 ELEMENTAL AND COMPOUND SEMICONDUCTORS

6.4.1 INTRINSIC AND EXTRINSIC SEMICONDUCTORS

As we saw in Section 6.1.1, semiconductors have conductivities in between those of insulators and metals. There is nothing inherently helpful about having a value of conductivity in this range, however. The reason semiconductors are so useful is that their conductivity is sensitive to light, temperature, magnetic fields, and even the tiniest quantity of impurity atoms. In many ways the greatest difference in conductivity is caused by impurities. When semiconductors are in their pure state and have not had any impurities deliberately added, they are known as *intrinsic semiconductors*. By contrast, any semiconductor that has had the atoms of another material added to it to alter its properties is known as an *extrinsic semiconductor*.

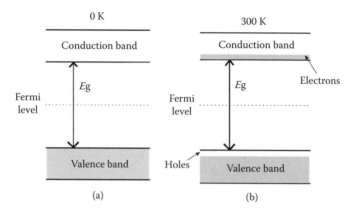

FIGURE 6.19   Band diagrams for an intrinsic semiconductor at (a) 0K and (b) 300K.

Conductivity of Intrinsic Semiconductors

At absolute zero, the valence band of an intrinsic semiconductor is completely full, while its conduction band is completely empty (see Figure 6.19[a]). As there are no electrons in the conduction band to take part in conduction, and no empty states in the valence band that electrons could potentially be excited into, the semiconductor is unable to conduct electricity. However, as the temperature is raised, electrons from the valence band can gain enough thermal energy to move across the relatively small bandgap ($E_g \approx 1$ eV or less) into the conduction band and therefore take part in conduction. As the temperature is increased, more electrons will be promoted into the conduction band, so the conductivity increases with an increase in temperature.

One point that is important to remember when considering band diagrams is that they are energy diagrams and so show what values of energy electrons and holes have rather than their location. So although the word "move" of course indicates motion in an everyday context, when electrons are said to "move" between energy levels, they are not physically moving anywhere within the crystal. Instead they are "moving" up or down in energy. For example, when an electric field is applied to a metal it will cause conduction electrons to start moving along as part of a current. The extra energy each electron gains from that electric field not only allows it to move through the solid but also causes it to "move" into an energy state higher up in the conduction band than it would occupy if there was no current flowing. (This is why a conduction band can never be completely full; if it was full there would be no higher energy states available for electrons taking part in conduction to move into.)

The bandgap is much larger in insulating materials than it is in semiconductors, as a glance back at Figure 6.16 will confirm. It is also larger in compound semiconductors (semiconductors that are composed of more than one element) than in elemental semiconductors, as Table 6.2 shows. The notation used to describe compound semiconductors comes from the groups of the periodic table that the constituent elements come from. (See Appendix C for a copy of the periodic table.) For example, Zn is found in Group II of the periodic table, while sulphur appears in Group VI, so zinc sulphide (ZnS) is known as a II-VI semiconductor. As well as II-VI and III-V

TABLE 6.2
Bandgap Values of Several
Common Semiconducting Materials

Semiconducting Material	Type	E_g (eV)
Si	Elemental	1.12
Ge	Elemental	0.67
GaP	III-V compound	2.25
GaAs	III-V compound	1.42
CdS	II-VI compound	2.40
ZnTe	II-VI compound	2.26

compounds (which include gallium phosphide, GaP, and gallium arsenide, GaAs), there are also some IV-VI semiconducting compounds and one IV-IV semiconductor, namely silicon carbide (SiC). Sometimes these compounds contain more than two elements. For example, AlGaAs is a III-V semiconductor, as Al and Ga are both Group III and As is Group V.

As we saw in Section 2.1 of Chapter 2, both silicon and germanium are covalently bonded, while gallium arsenide has mixed bonding and is partially ionic and partially covalent. Like GaAs, the other III-V semiconductors, including indium phosphide (InP) and aluminium gallium arsenide (AlGaAs), have mixed bonding that is mainly covalent but also slightly ionic in character. By contrast, II-VI semiconductors such as zinc selenide (ZnSe), while still having a mixture of covalent and ionic bonding, are more ionic than the III-V semiconducting compounds. (Referring back to Figure 2.9 should help you to visualise the situation.)

In terms of the bonds between atoms in a covalent solid, conduction only occurs in an intrinsic semiconductor when the temperature is so high that the thermal vibrations of the atoms are large enough to break the covalent bonds holding the atoms together. Electrons that have broken free from the covalent bonds they were part of can participate in electrical conduction, which is why the conductivity is dependent on the temperature. Conductivity is also dependent on the size of the bandgap, as the larger E_g is, the greater the temperature needed to excite an electron into the conduction band.

The movement of each thermally excited electron into the conduction band leaves behind an empty energy level known as a *hole* in the valence band. So whenever thermal energy is sufficient to promote an electron into the conduction band, an "electron-hole pair" is in fact produced, as shown in Figure 6.20(a). Holes can be thought of as positively charged particles that move, under the influence of an applied electric field, in an opposite direction to electrons (see Figure 6.20[b]). This means they contribute to conduction in semiconductors. (We first encountered this idea earlier in this chapter when the answer to the question on effective mass revealed a particle that had a negative effective mass behaved as though it had a positive mass and positive electric charge.) In reality, holes—in the sense of positively charged particles—do not exist. The conduction that occurs in the valence band is still electron

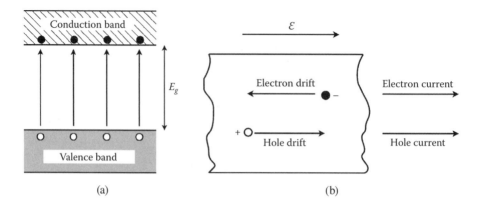

FIGURE 6.20 (a) Thermal energy creates electron-hole pairs in an intrinsic semiconductor. If this semiconductor has an electric field ε applied to it, electrons drift in the opposite direction to the field, while holes drift in the same direction as shown in (b). This means both electrons and holes contribute to the overall current flow.

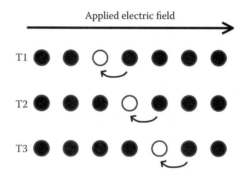

FIGURE 6.21 Hole conduction in a semiconductor.

conduction, but because the valence band contains so many electrons and so few empty energy levels for them to move into, it is easier to analyse the situation in terms of these make-believe holes.

Figure 6.21 shows the mechanism of hole conduction. As this diagram reveals, the holes do not actually move anywhere in the true sense of the word "move". What actually happens is that electrons that are under the influence of an electric field move into the empty hole states, so that the net result is that the hole has moved in the opposite direction. One way of visualising this situation is to imagine a busy car park with only one space that remains empty throughout the day. As people come and go and cars are parked in different spots, this space will appear in different locations around the car park as if it has moved there itself.

The total energy (kinetic plus potential) of electrons in the conduction band is measured upwards from E_C, which is the energy at the bottom of the conduction band. In other words, the greater the value of energy that an electron in the conduction band has, the further up in the band it will be. In the case of holes the reverse is true, so the further they are below E_v—the energy at the top of the valence band—the higher their energy is. This is because band diagrams are potential energy diagrams for negative electrons, so the potential energy of a positive hole is increasing in the opposite direction.

Intrinsic Carrier Concentrations

In an intrinsic semiconductor, at temperatures above absolute zero, the number of electrons in the conduction band is equal to the number of holes in the valence band. This is hardly surprising, as it is the movement of an electron into the conduction band that creates a corresponding hole in the valence band. Another way of looking at this is to say that the number of occupied energy states in the conduction band is equal to the number of unoccupied states in the valence band.

Since the number of electrons per unit volume of the conduction band, n, is equal to the number of holes per unit volume of the valence band, p, for an intrinsic semi-conductor the intrinsic carrier concentration, n_i, is given by the following expression:

$$n_i = n = p \qquad \qquad (6.14[a])$$

which can be rewritten as follows:

$$n_i^2 = np \qquad \qquad (6.14[b])$$

As the temperature rises, more electron-hole pairs are generated and so n_i increases. In fact n_i increases exponentially with temperature.

Doping to Produce n-type and p-type Semiconductors

To increase the conductivity, and in so doing make the semiconductor more suitable for use in electronic devices by making one type of charge carrier dominant, atoms of another material—known as "dopant" atoms—are often added to intrinsic semi-conductors. When the dopant atoms have more valence electrons than are needed to bond with the atoms of the host semiconductor, they are known as *donor* impurities. They make the semiconductor material *n-type*, which means that negatively charged electrons are the charge carriers. Conversely, when the dopant atoms have too few electrons, they are known as *acceptor* impurities and a *p-type* semiconductor is pro-duced in which positively charged holes are the charge carriers.

Figure 6.22(a) shows how an n-type semiconductor is produced from intrinsic silicon, while Figure 6.22(b) illustrates the creation of a p-type semiconductor. In (a) one of the silicon atoms has been replaced by an atom of arsenic. As arsenic has five valence electrons and silicon has only four, the arsenic atom forms covalent bonds with each of its four neighbouring silicon atoms, leaving its fifth electron free to take part in conduction when the temperature is high enough for the donor atom to have been ionised. An n-type semiconductor is therefore produced.

The ionisation temperature is much lower than it would be for an atom with the same number of electrons as its neighbours, as the fifth "donor" electron is only weakly bound to its parent atom. In fact all donors will generally be ionised at room temperature.

By contrast, if a boron atom is substituted for a silicon atom, it can only form three of the four covalent bonds that it needs to with its neighbours from its own

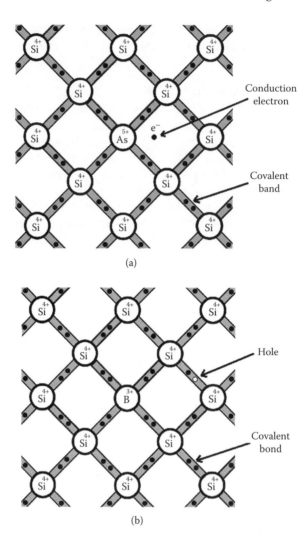

FIGURE 6.22 Schematic diagram showing silicon being doped by (a) arsenic to produce an n-type semiconductor, and (b) boron to produce a p-type semiconductor.

three valence electrons. To form the fourth bond it has to accept an electron from a silicon atom elsewhere in the lattice, which is freed when the host atom is ionised. This leaves a hole behind in place of this electron, and since there has been an addition of positively charged holes, a p-type semiconductor has been produced. As with an n-type semiconductor, the ionisation takes place at relatively low temperatures, so hole conduction can occur at room temperature.

All semiconductors can be doped, not just silicon. Gallium arsenide (GaAs), for example, can be doped with tellurium (Te) to make it n-type. The tellurium with its six valence electrons has one more electron than arsenic (As) and so can donate one free electron.

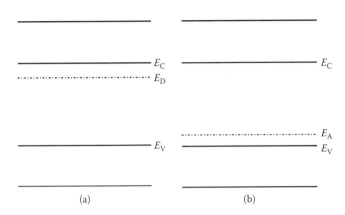

FIGURE 6.23 Impurities create energy levels within the bandgap in semiconductors. (a) shows a donor energy level, while (b) is a level caused by acceptors.

In terms of band structure, the dopant atoms introduce energy levels into the forbidden gap, as shown in Figure 6.23. In (a) the donor atoms have created a donor level E_D just below the conduction band. When the temperature is large enough, all the donor atoms will be ionised and their freed electrons will be released from the level E_D into the conduction band. From there they can then take part in conduction. $E_C - E_D$ therefore represents the energy needed to ionise the donor atoms, and because this donor energy level is only just below E_C—the bottom of the conduction band—in most cases room temperature is a high enough temperature for all the donor atoms to be ionised.

A similar situation occurs for acceptor impurities, which create an energy level E_A just above the top of the valence band, as shown in Figure 6.23(b). In this case electrons can easily be released from the valence band into vacant states in E_A, leaving holes behind in the valence band that can take part in conduction. As with movement from E_D, room temperature is high enough to cause electrons to transfer into E_A. Different impurities create donor levels at different energies, as Figure 6.24 reveals.

As this diagram indicates, the bandgap is only forbidden in intrinsic—in other words, "pure"—semiconductors. The lattice disruption caused by the addition of dopants changes the potential energies of the conduction electrons (as discussed in Section 6.2.1) and so allows energy levels to exist within the forbidden gap.

Although both donor and acceptor atoms are ionised within the semiconductor— each donor atom becoming a positive ion when it loses an electron to the conduction band, and each acceptor atom becoming a negative ion when it acquires an electron from the valence band—the semiconductor is still neutral overall. This is because an intrinsic semiconductor is electrically neutral, and all the dopant atoms in an extrinsic semiconductor are electrically neutral too. They only become ionised once inside the semiconductor. There are, however, localised changes to electrical neutrality in the regions where dopant atoms are ionised. These charged regions can affect the behaviour of semiconductor devices, as we will see in Chapter 7.

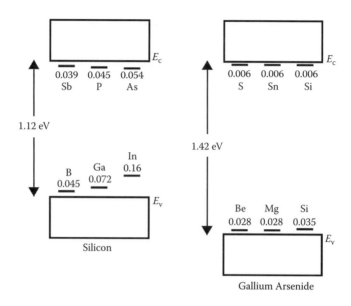

FIGURE 6.24 Different impurities create energy levels at different heights within the band-gap, as these diagrams indicate. The levels near the bottom of the conduction band are donor levels, while those just above the valence band are acceptor levels. In GaAs, silicon can act as a donor by replacing a Ga atom, or as an acceptor if it replaces As instead.

Majority and Minority Carriers

Sometimes both donor and acceptor impurities are present in a semiconductor, and in this case the type of conductivity (n-type or p-type) is determined by which-ever type of impurity has the greatest concentration. In an n-type semiconductor where the charge carriers are predominantly electrons, the electrons are referred to as the "majority carriers", while any holes that take part in conduction are termed "minority carriers". Conversely, for a p-type semiconductor, holes are the majority carriers and electrons the minority carriers.

Even when there is just the one type of dopant, if an extrinsic semiconductor is heated to a very high temperature, then it behaves as though it was an intrinsic semi-conductor. This is because once the temperature is high enough to excite electrons right across the bandgap from the valence band to the conduction band, charge will be carried through the semiconductor by these electrons and holes rather than by electrons or holes from the dopant atoms—which will all be ionised by then. Before reaching the totally intrinsic region, there will be a mixture of intrinsic and extrinsic behaviour.

The temperature at which intrinsic behaviour starts to occur depends on the width of the bandgap: the greater E_g, the higher the temperature needed for intrinsic conduction to occur. For applications where a semiconductor has to operate at high temperatures without any intrinsic behaviour, a material with the widest possible bandgap needs to be chosen.

No matter what temperature the semiconductor is at, the intrinsic carrier concentration equations 6.14(a) and (b) are valid for both intrinsic and extrinsic semiconductors. (Proof of this can be found in Section A3 of Appendix A.)

EXAMPLE QUESTION 6.4   MINORITY CARRIERS

A piece of silicon at 300K is doped with 6×10^{18} boron atoms per cm^3. What type of semiconductor does this produce, and what are the concentrations of both electrons and holes in the sample?

Assume the intrinsic carrier concentration is 1.4×10^{10} per cm^3.

ANSWER

Referring back to equation 6.14(b) we have

$$n_i^2 = np$$

Doping silicon with boron atoms makes it a p-type semiconductor, so in this case if we assume that room temperature is enough to ionise all the acceptors, then $p = 6 \times 10^{18}$ cm^{-3}.

So rearranging equation 6.14(b) will give us n as follows:

$$n = \frac{n_i^2}{p} = \frac{\left(1.4 \times 10^{10}\right)^2}{6 \times 10^{18}} = 32.67 \text{ cm}^{-3}$$

It is immediately obvious that the number of electrons is tiny compared with the number of holes, and so the electrons contribute much less to the flow of electric current than the holes, easily earning their title of "minority" carriers.

The Fermi Level in Intrinsic Semiconductors

Since electrons are fermions (see Appendix D), the probability that an electron will have a particular value of energy E is given by the Fermi distribution function $F(E)$ described by equation 6.7 in Section 6.1.3. (See Appendix E for more on Fermi–Dirac statistics.)

At 0K, the Fermi energy—which is also known as the Fermi level—represents the highest filled level in the semiconductor. In this case, a semiconductor is like a metal in the respect that at 0K all levels below E_F in a semiconductor are filled, while those above it are empty. By contrast, at higher temperatures in semiconductors, the Fermi level is the energy level at which $F(E)$ equals ½. So the probability of an electron in a semiconductor having energy E_F is exactly one half. You can see that this is the case by looking at equation 6.7, as with $E = E_F$ the exponential term will be equal to 1 (as $e^0 = 1$) and so $F(E) = 0.5$. Referring back to Figure 6.6(b) shows the Fermi distribution function (which is sometimes referred to as the "Fermi occupation factor") plotted as a function of energy for a range of different temperatures. From this diagram you can see that some of the levels above E_F are full, while some below E_F are empty. It is also possible to see from Figure 6.6(b) that the Fermi–Dirac function is symmetrical about E_F.

Before learning any more, it is important to be clear that the Fermi level is a "reference" energy level rather than an actual level created by atoms. As we will see shortly, it appears at different places on a semiconductor band diagram, depending on the doping conditions. Firstly, however, we need to consider where the Fermi energy lies in an intrinsic semiconductor.

In both extrinsic and intrinsic semiconductors, the carrier concentrations for electrons (n) and holes (p) are, respectively,

$$n = N_C \exp\left[-\frac{(E_C - E_F)}{k_B T}\right] \tag{6.15}$$

$$p = N_V \exp\left[-\frac{(E_F - E_V)}{k_B T}\right] \tag{6.16}$$

where E_F is the Fermi energy, E_C is the bottom of the conduction band, E_V is the top of the valence band, k_B is Boltzmann's constant, and N_C and N_V are the effective density of states (see Section 6.3.2) in the conduction band and valence band, respectively. In silicon at 300K (room temperature), for example, $N_C = 2.8 \times 10^{19}$ cm^{-3} and $N_V = 1.04 \times 10^{19}$ cm^{-3}. (The reason N_C and N_V are known as the "effective" density of states is that they give the density of states available calculated from classical Boltzmann statistics for simplicity rather than from Fermi–Dirac statistics.)

If we first consider an intrinsic semiconductor, combining equations 6.15 and 6.16 (see Appendix A for a complete derivation) gives the following expression for the Fermi energy:

$$E_F = \frac{E_C + E_V}{2} + \frac{k_B T}{2} \ln\left[\frac{N_V}{N_C}\right] \tag{6.17}$$

As E_C is the energy at the bottom of the conduction band and E_V is the energy at the top of the valence band, the first term describes the energy at the middle of the bandgap. At $T = 0$, the second term on the right of equation 6.17 disappears, leaving E_F exactly in the middle of the bandgap.

At room temperature the second term in equation 6.17 is so much smaller than the bandgap that the Fermi level is very close to the middle of the bandgap. The Fermi level will, however, be exactly in the middle of the bandgap no matter what the temperature of the intrinsic semiconductor is if $N_C = N_V$. (This is because ln 1 = 0.) If the density of states in one of the bands is much larger than that of the other, for example, as we saw in the last subsection N_C for silicon is larger than N_V, then the Fermi level will no longer be found in the middle of the gap. In the case of silicon, for example, it will be below the middle. In all cases, the Fermi energy of an intrinsic semiconductor can be described as the "intrinsic Fermi level" and denoted by E_i.

The Fermi Level in Extrinsic Semiconductors

When the temperature is high enough in an n-type semiconductor for complete ionisation to have taken place, the electron density, n, is equal to the donor concentration N_D. So equation 6.15 will therefore describe the position of the Fermi energy in an

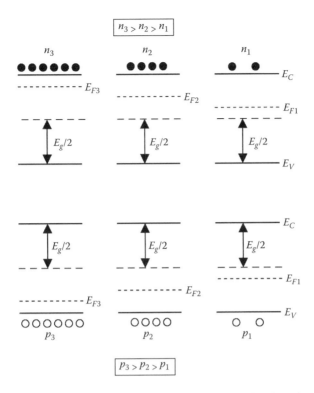

FIGURE 6.25 The Fermi level moves from the middle of the bandgap depending on the amount and type of dopant present. (This diagram first appeared as Figure 3.11, p. 41, of *Introductory Semiconductor Device Physics*, by Greg Parker, © IOP Publishing Ltd., 2004. Reproduced with kind permission of Greg Parker.)

n-type semiconductor in terms of the donor concentration. As N_D increases, $E_C - E_F$ must decrease, meaning that the Fermi level is moving nearer to the bottom of the conduction band as the number of donor atoms increases.

Similarly for a p-type semiconductor with the hole density p equal to the acceptor concentration N_A under complete ionisation, a glance at equation 6.16 reveals that the Fermi level moves towards the top of the valence band as the acceptor concentration increases. Figure 6.25 shows how the Fermi level moves as the dopant concentration changes.

More on Carrier Concentrations

Combining equations 6.15 and 6.16 reveals the total intrinsic carrier concentration, and is valid for both intrinsic and extrinsic semiconductors. The result is the following equation, and if you do not want to have a go at obtaining this yourself you can find out how it is produced in Appendix A.

$$n_i = \sqrt{N_C N_V} \exp\left[-\frac{E_g}{2k_B T}\right] \qquad (6.18)$$

Equation 6.18 is applicable to both intrinsic and extrinsic semiconductors because it does not contain E_F—which, as we saw in the previous subsection, moves as the dopant concentration changes—and so is independent of whether a semiconductor has been doped or not.

EXAMPLE QUESTION 6.5   EXTRINSIC SEMICONDUCTORS

What are the carrier concentrations, and where does the Fermi level lie at room temperature in a piece of silicon doped with 2×10^{16} arsenic atoms per cm^3?

ANSWER

Doping silicon with arsenic makes it n-type, so assuming room temperature is enough to ionise all the donor atoms, then $n = N_D = 2 \times 10^{16}\,\text{cm}^{-3}$.

If the intrinsic carrier concentration is again assumed to be $1.4 \times 10^{10}\,\text{cm}^{-3}$, then

$$p = {n_i^2}\Big/{N_D} = \frac{\left(1.4 \times 10^{10}\right)^2}{2 \times 10^{16}} = 9.8 \times 10^3 \text{ cm}^{-3}$$

We can obtain the position of the Fermi level—as measured from the bottom of the conduction band—by rearranging equation 6.15 as follows:

$$n = N_C \exp\left[-\frac{\left(E_C - E_F\right)}{k_B T}\right]$$

$$\Rightarrow \ln N_D = \ln N_C \left[-\frac{\left(E_C - E_F\right)}{k_B T}\right]$$

$$\Rightarrow E_C - E_F = k_B T \ln\left[\frac{N_C}{N_D}\right]$$

as in this case $n = N_D$.

In the previous subsection we saw that $N_C = 2.8 \times 10^{19}$ cm^{-3} in silicon at 300K, so

$$E_C - E_F = k_B \times 300 \ln\left[\frac{2.8 \times 10^{19}}{2 \times 10^{16}}\right] = 0.188 \text{ eV}$$

So in this case, the Fermi level is 0.188 eV below the bottom of the conduction band.

6.4.2 MOTION OF CHARGE CARRIERS IN SEMICONDUCTORS

Drift Velocity

As we have just learned, unlike metals in which only electrons carry charge, both electrons and holes carry charge in semiconductors. The movement of charge carriers under the influence of an electric field is known as the *drift current*, and the average velocity of these carriers is called the *drift velocity*. The drift velocity, v_{dr}, is related to the strength of the electric field, ε, in the following way

$$v_{dr} = \mu \varepsilon \tag{6.19}$$

where μ is the drift mobility, which has different values for different materials, and expresses how strongly the motion of a charge carrier is affected by an applied electric field.

The mobility of charge carriers decreases with increasing temperature because, as we saw in Section 6.1.1 when we were looking at the origins of resistivity, the lattice atoms vibrate more at higher temperatures than lower temperatures, and therefore charge carriers are more likely to collide with them. By contrast, impurity scattering, which is caused by the charged carriers being electrostatically attracted or repelled by ionized impurity atoms, tends to be less important at higher temperatures. This is because the carrier is moving past the ion at a faster speed and so has less time to interact with it.

The amount of impurity scattering does, however, depend on the concentration of impurities present. This is why the semiconductors used to make devices have to be so pure. If they contain too many impurities, the mobility of the charge carriers is reduced to such an extent that semiconductor devices could not work adequately. We will look at the various methods for producing extremely pure semiconductor crystals in Section 7.3.1.

Thermal Velocity

Drift mobility is not, however, the only motion charge carriers have. Whether or not an applied field is present, charge carriers have a thermal motion simply because they are at a temperature, T, above absolute zero. This thermal motion is random, as shown in Figure 6.26(a), because each charge carrier is continually being affected by collisions with either impurity atoms or lattice atoms, just like it is when it is under the influence of an electric field. In fact with the proviso that they have an effective mass that accounts for their interaction with the crystal lattice, the charge carriers can be considered to move like classical particles in an ideal gas and so are governed by classical statistics. (See Section E3 of Appendix E for a brief reminder of classical statistical mechanics.)

When a system like this is in thermal equilibrium, its total energy is divided up equally among its "three degrees of freedom", which account for the translational motion of the charge carriers within the 3-D space inside a piece of semiconductor. (There would only be more degrees of freedom if the charge carriers had some sort of additional motion; for example, if they were molecules they would have rotational energy as well, which would increase the number of degrees of freedom.) This division

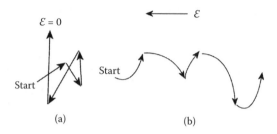

FIGURE 6.26 Path of an electron in a semiconductor in (a) random thermal motion and (b) under the influence of an applied electric field (motion due to the field combined with the random motion).

of energy is known as the equipartition of energy theorem, and for an ideal monatomic gas—whose behaviour the charge carriers in semiconductors are mimicking—the mean energy of each carrier per degree of freedom is $k_B T/2$. So the average energy of each carrier is $3k_B T/2$, where k_B is Boltzmann's constant and T is the temperature.

Since kinetic energy (in other words, the energy due to movement) is equal to $\frac{1}{2}mv^2$, we can derive an expression for the thermal velocity of a charge carrier in a semiconductor as follows:

$$\frac{1}{2}m^* v_{th}^2 = \frac{3}{2}k_B T \qquad (6.20)$$

where m^* is the effective mass of the charge carrier (electron or hole), and v_{th} is the thermal velocity of a charge carrier. Rearranging gives us

$$v_{th}^2 = \frac{3k_B T}{m^*} \Rightarrow v_{th} \neq \sqrt{\frac{3k_B T}{m^*}} \qquad (6.21)$$

Overall Motion

Whenever an electric field is applied to a semiconductor, the motion of the electrons (and holes) is a combination of the drift velocity and the random thermal motion. Figure 6.26(b) shows the motion of a charge carrier under the influence of an electric field.

The average distance that a charge carrier travels before its path is altered by a collision is known as the "mean free path", and the average time between collisions, τ_c, is known as the *mean free time*. (Please refer back to Section 5.2.2 for an introduction to the concept of a mean free path.)

Use of the mean free time gives us another way to express the drift velocity as velocity = acceleration × time. Given that acceleration = force/mass, the drift velocity can be given by

$$v_{dr} = \frac{F\tau_c}{m} \qquad (6.22)$$

EXAMPLE QUESTION 6.6   DRIFT VELOCITY
VS. THERMAL VELOCITY

Calculate the thermal velocity of an electron moving in a piece of n-type silicon at 300K, and also its drift velocity when an electric field of 2000 Vm^{-1} is applied. What do your results reveal?

Assume $m^* = 1.18\,m_0$, with $m_0 =$ the mass of a free electron $= 9.11 \times 10^{-31}$ kg; and μ for the electron $= 0.15$ m^2 V^{-1} s^{-1}.

ANSWER

The thermal velocity is given by

$$v_{th}\sqrt{\frac{3k_BT}{m^*}} = \sqrt{\frac{3 \times k_B \times 300}{1.0749 \times 10^{-30}}} = 107517.9 \text{ ms}^{-1} = 1.075 \times 10^5 \text{ ms}^{-1}$$

The drift velocity is given by

$$v_{dr} = \mu\varepsilon = 0.15 \times 2000 = 300 \text{ ms}^{-1}$$

It can immediately be seen that the drift velocity is several orders of magnitude smaller than the thermal velocity. Since the thermal velocity is barely changed by the application of an electric field, the time between collisions can be assumed to be the same whether or not an electric field is acting upon the piece of semiconductor under consideration.

where **F** is the force exerted on the carriers by the applied electric field and τ_C is the mean free time.

In fact, the force exerted on the carriers is related to the electric field producing it via the following equation from electrostatics:

$$\mathbf{F} = q\varepsilon \tag{6.23}$$

where q is the charge on the carriers and ε is the applied electric field. Substituting for **F** in equation 6.22 gives us

$$v_{dr} = \frac{q\varepsilon\tau_C}{m} \tag{6.24}$$

Combining this result with equation 6.19 gives the following expression for the mobility:

$$\mu = \frac{q\tau_C}{m^*} \tag{6.25}$$

EXAMPLE QUESTION 6.7   MEAN FREE PATH

What is the mean free path for an electron like the one considered in the previous question that is moving in a piece of n-type silicon?

As before, $m^* = 1.18\,m_0$, with $m_0 = 9.11 \times 10^{-31}$ kg, $\mu = 0.15$ m^2 V^{-1} s^{-1}, and as we have already calculated, $v_{th} = 1.075 \times 10^5$ ms^{-1}.

ANSWER

We first need to find out the mean free time. Rearranging equation 6.25 gives

$$\tau_C = \frac{\mu m^*}{q} = \frac{0.15 \times 1.0749 \times 10^{-30}}{1.602 \times 10^{-19}} = 1 \times 10^{-12}\,\text{s}$$

As distance = velocity × time, the mean free path, l, is equal to $\tau_C \times v_{th}$ so

$$\tau_C \times v_{th} = 1 \times 10^{-12} \times 1.075 \times 10^5 = 1.07 \times 10^{-7}\,\text{m} = 107\,\text{nm}$$

Conductivity

As the current is carried in semiconductors by both electrons and holes, the total conductivity is a combination of the conductivity produced by the free electrons and that produced by the holes. This can be expressed as follows:

$$\sigma = qn\mu_n + qp\mu_p \tag{6.26}$$

where q is the charge, n is the number of free electrons, p the number of holes, and μ_n and μ_p the electron and hole mobilities, respectively.

Diffusion

In fact, the last few subsections were misleading when they suggested that thermal motion and drift velocity produced the overall motion of charge carriers in a semiconductor, as there is another way charge carriers can move in a specific direction. This is known as *diffusion* and occurs when there is a concentration gradient of carriers within a semiconductor. In the same way that the molecules of a small gas leak will soon move throughout a room, charge carriers will move away from a region where their concentration is high. (In both cases, having a more uniform concentration is a lower energy situation than having a concentration gradient, so diffusion will continue until the concentration is uniform.) This movement causes a "diffusion current" to flow, so in semiconductors with a concentration gradient of charge carriers, the total current flowing will be due to both drift and diffusion. In this case the overall carrier motion will be a combination of the random thermal motion, the drift velocity, and the diffusion.

As we will see in Chapter 7, diffusion plays an important role in the operation of many semiconductor devices, and is also one of the methods by which dopants are

**EXAMPLE QUESTION 6.8   CONDUCTIVITY
IN SEMICONDUCTORS**

Calculate the conductivity in a piece of silicon at 300K doped with (a) 6×10^{16} phosphorus atoms per cm³ and (b) 5×10^{17} arsenic atoms per cm³.

ANSWER

Although the conductivity in a semiconductor is a combination of the contribution from both electrons and holes, in most cases there is such a large difference between the donor and acceptor concentrations that the contribution from the minority carrier can be ignored. In this case, the dopants are phosphorus and arsenic, which both make silicon n-type, and as there is no mention of an acceptor concentration, we can assume that it is fine for equation 6.26 to be reduced to

$$\sigma = qN_D\mu_n$$

where N_D is the donor concentration (which is equal to n, the number of free electrons). We have a value of mobility for electrons in n-type silicon at room temperature from our previous two calculations, namely 0.15 m² V⁻¹ s⁻¹. So for (a)

$$\sigma = 1.602 \times 10^{-19} \times 6 \times 10^{16} \times 0.15 = 0.0014 = 1.4 \times 10^{-3} \left(\Omega\text{-cm}\right)^{-1}$$

and for (b) we have

$$\sigma = 1.602 \times 10^{-19} \times 5 \times 10^{17} \times 0.15 = 0.012 = 1.2 \times 10^{-2} (\Omega\text{-cm})^{-1}$$

Referring back to Figure 6.1, you can see that both of these values, as expected, fall in the range of conductivities shown for doped silicon.

added into semiconductors. Both the drift and diffusion of electrons is faster than that of holes in silicon and gallium arsenide, so devices that need to operate at high speed need to be based on the movement of electrons rather than the drift and/or diffusion of holes.

6.5   SUPERCONDUCTIVITY

6.5.1   INTRODUCTION TO SUPERCONDUCTIVITY

No matter how good a conductor a metal is, it still provides some resistance to the flow of an electric current. This is because, as we saw in Section 6.1.1, the resistivity is partly due to the thermal motion of the lattice atoms and partly due to the presence of defects. Whilst the first of these contributions decreases as the temperature decreases, the latter contribution is present even at temperatures just a fraction

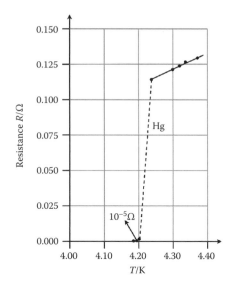

FIGURE 6.27 This graph shows the experimental data of resistance and temperature obtained by Onnes for solid mercury. He found the mercury became superconducting below 4.2K. (This diagram originally appeared as Figure 2.46(a), p. 103, of *Quantum Physics of Matter*, edited by Alan Durrant, © 2000, The Open University. Reproduced with permission.)

above absolute zero. However, as the old expression says, there are exceptions to every rule, and in 1911 the Dutch physicist Heike Kamerlingh Onnes (1853–1926) found that solid mercury lost all its resistance to electricity, in other words became a *superconductor*, when it was cooled below −269°C (4.2K).

Figure 6.27 shows the resistance of mercury disappearing at 4.2K. Above this temperature the mercury behaves like an ordinary metal, but below 4.2K it is a superconductor. The temperature at which a solid becomes superconducting—and therefore the temperature which separates the superconducting state from the normal state of a solid—is known as the *superconducting transition temperature*, T_c.

Since Onnes made his discovery, thousands of materials—including metallic alloys, many compounds, and a handful of metals—have been seen to show super-conductivity below a transition temperature which varies widely between substances, as Table 6.3 shows. In certain circumstances, such as under high pressure or for a very thin film of a sample, even some semiconductors can become superconducting.

Above T_c, the resistivity of a superconducting metal is just the same as that of a normal metal (see Section 6.1.1), with both defect and phonon scattering contributing to the resistivity of the material. However, in a superconducting metal, neither scattering mechanism has any effect below T_c, so the resistance becomes zero.

This does not mean, however, that materials capable of becoming superconductors will always be superconducting at temperatures below their respective T_c. In fact the superconducting state is destroyed by the presence of a high magnetic field, although in some unusual cases magnetism and superconductivity can coexist, as the caption to Figure 6.28 explains.

TABLE 6.3

Superconducting Transition Temperatures for a Variety of Solids

Material	Superconducting Transition Temperature, T_c (K)
Titanium (Ti)	0.4
Lead (Pb)	7.2
Magnesium diboride (MgB_2)	39
Yttrium barium copper oxide ($YBa_2Cu_3O_7$)	92

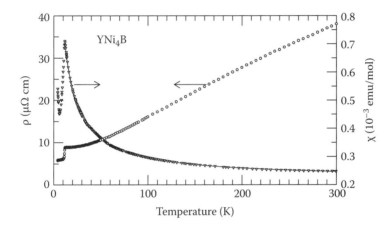

FIGURE 6.28 Since Heike Kamerlingh Onnes discovered superconductivity in solid mercury in 1911, a wide variety of superconducting materials have been found. In 1992, for example, scientists at the Tata Institute of Fundamental Research in India discovered superconductivity in a class of metal compounds known as "quaternary borocarbides", which contain nickel, boron, carbon, and one of the rare earth elements. This graph shows both the resistivity (marked by circles) and magnetic susceptibility (marked by inverted triangles) measured by the researchers for a sample of yttrium nickel boride with a nominal composition of YNi_4B. The sudden drop in resistivity at around 12K reveals that part of the sample becomes superconducting below this temperature. (If the entire sample became superconducting, the resistivity would have fallen to zero at 12K rather than tailing off to a nonzero value as seen here.) Further work showed that the superconductivity originated from a quaternary borocarbide phase containing yttrium, nickel, boron, and carbon. The graph also reveals a sharp drop in magnetic susceptibility, which is due to the diamagnetic behaviour of this superconducting phase below 12K, and therefore confirms the superconducting transition temperature of approximately 12K indicated by the resistivity. (Magnetic properties and diamagnetism are discussed in detail in Chapter 8.) (The experimental data shown in this graph is reprinted from Solid State Communications, Vol 87, Mazumdar, Nagarajan, et al, Superconductivity at 12K in Y-Ni-B system, Pages 413–416, Copyright 1993, with permission from Elsevier.)

If a continuous ring is made out of a superconducting material, the lack of resistance means that, once it is set up, a direct current flowing around the loop would theoretically circulate forever. In reality, it is likely that the resistance in a superconductor is not quite zero, and that a current in a superconducting loop would not keep flowing indefinitely. However, experiments have suggested that a current like this would flow for at least 10,000 years, and that the conductivity of a superconductor is around 17 orders of magnitude larger than the conductivity of a typical metal.

Type I and Type II Superconductors

Superconductors can be divided into two classes—type I superconductors and type II superconductors—according to how they behave in a magnetic field. It only takes the application of a relatively low external magnetic field to a type I superconductor for it to have its superconductivity destroyed. In addition, the change between the superconducting state and the normal state for type I superconductors occurs abruptly when the external magnetic field, $\mathbf{H}$, (see Section 8.1.2) is at a value known as the *critical field*, $\mathbf{H}_c$, as shown in Figure 6.29(a).

So if a magnetic field with a value above that of $\mathbf{H}_c$ is applied to a type I superconductor when it is below its critical temperature, and therefore is in the superconducting state, it will revert to its normal state and no longer have zero resistance.

Just as every superconducting material has a unique value of T_c, every different type I superconducting material will have a different value of $\mathbf{H}_c$. $\mathbf{H}_c$ is zero at the critical temperature, and increases as the temperature is reduced. Figure 6.30 shows the dependence of $\mathbf{H}_c$ on temperature for several superconductors, which is approximately given by the following equation:

$$\mathbf{H}_c(T) = \mathbf{H}_c(0)\left[1 - \left(\frac{T}{T_c}\right)^2\right] \tag{6.27}$$

where $\mathbf{H}_c(0)$ is the critical field as the temperature approaches absolute zero.

It is not just an applied magnetic field that can destroy superconductivity. High currents passing through a superconductor can have the same effect. This is because, as the Danish physicist Hans Christian Oersted (1777–1851) discovered in 1820, any current flowing along a wire produces a magnetic field. So for a type I superconductor there is a "critical current" which produces a field equal to $\mathbf{H}_c$ when it flows through the superconductor. If the current remains below this critical value, the material can still superconduct, but if it exceeds it the superconductivity will be destroyed by the associated magnetic field, and the material will revert to its normal state.

With the exception of niobium, all of the elements that are superconducting are type I superconductors. By contrast, all alloys are type II superconductors. Instead of changing from a superconducting state to a normal conducting state at a single critical field (or critical current), type II superconductors change from one state to the other over the range of fields between $\mathbf{H}_{c1}$ and $\mathbf{H}_{c2}$ shown in Figure 6.29(b). $\mathbf{H}_{c1}$ is known as the "lower critical field" and $\mathbf{H}_{c2}$ as the "upper critical field". Between these two values—which can differ by as much as a factor of 100—the material is in

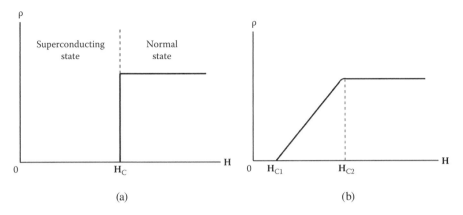

FIGURE 6.29 The resistivity of (a) type I superconductors and (b) type II superconductors in the presence of an external magnetic field. There is an abrupt change from the super-conducting state to the normal state at the critical field **H**$_c$ for type I superconductors, whereas the change in state for type II superconductors happens over a range of values of magnetic field between **H**$_{c1}$ and **H**$_{c2}$.

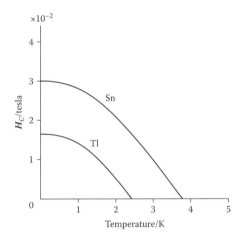

FIGURE 6.30 The experimentally measured dependence of the critical field, **H**$_c$, on temperature for some type I superconductors.

a so-called mixed state, in which some parts of it are in the superconducting phase and other sections are behaving like a normal conductor.

High-Temperature Superconductors

As we will see in the next section, there are already a number of important applications of superconductors. However, the ultimate goal for scientists is to discover a material that is superconducting at room temperature and that can be made into wires for carrying electric current.

In the meantime, any material that can superconduct at a temperature above 77K (the boiling point of liquid nitrogen) is considered to be a high-temperature

**EXAMPLE QUESTION 6.9   CRITICAL FIELD
OF A SUPERCONDUCTOR**

If a sample of tin with a value of $H_c(0) = 0.0305T$ is cooled below its super-conducting transition temperature of 3.72K, what is the approximate value of its critical field when it is at 1K?

ANSWER

The approximate relationship between the critical field and the temperature is given by

$$H_c(T) = H_c(0)\left[1 - \left(\frac{T}{T_c}\right)^2\right]$$

so inserting the values for T_c, T, and $H_c(0)$ gives us

$$H_c(1K) = 0.0305\left[1 - \left(\frac{1}{3.72}\right)^2\right] = 0.028T$$

(This value is about 400 times larger than the Earth's magnetic field strength.) If we now refer back to Figure 6.30, we can see that this equation does indeed provide a reasonable approximation to the real-life situation.

superconductor. Examples include yttrium barium copper oxide (YBCO) with its superconducting transition temperature of 92K, and bismuth strontium calcium copper oxide (BSCCO), which can have two different structures, one of which has a T_c of 93K, while the other has a T_c of 110K. These materials are easier to use than lower temperature superconductors because they can be kept in a superconducting state with liquid nitrogen rather than liquid helium cooling them. This produces a substantial financial saving, as liquid nitrogen is much more widely available than liquid helium. It is also easier to work with and so reduces the technical challenges involved in making functional devices for research or industry.

This does not mean that having a relatively high superconducting transition temperature automatically makes a material suitable for applications, however. Many of the high-temperature superconductors that have been discovered are brittle *ceramic* materials (compounds containing metallic elements together with non-metallic elements) that cannot be easily machined. If a superconducting material cannot be made into a tape or wire, it is of little commercial interest, and even if it can be, it needs to be able to be manufactured cheaply and easily enough to make its use viable. Of course different criteria apply for research, and some of the more expensive materials that could not make much of an impact in industry do still find their way into research laboratories around the world.

FIGURE 6.31   The Meissner effect allows a small magnet to levitate above a superconductor, providing an enjoyable laboratory demonstration. (Courtesy of Darren Peets, UBC Superconductivity Lab, Canada.)

The Meissner Effect

One of the best-known laboratory demonstrations of superconductivity is a small magnet being levitated above a superconductor (see Figure 6.31). This is possible because of the Meissner effect, which is the expulsion of magnetic flux from a metal in a magnetic field when it is cooled below T_c. This means that most magnetic fields cannot exist inside superconductors, although if the field is high enough the superconductor will revert back to its normal conducting state and allow the field to penetrate it.

The field is expelled because it induces electric currents on the surface of the superconductor, which in turn create a magnetic field that cancels out the magnetic field inside the superconductor. Small magnets can therefore be levitated above the surface of a superconductor, as the field produced by the surface currents opposes the field produced by the magnet and repels it from the surface.

Superconductors behave like perfect diamagnets, which also produce a magnetic field that opposes an external magnetic field, and are discussed in detail in Section 8.2.1. However, unlike the diamagnetic levitation described in Box 8.1, which is caused by the behaviour of the atoms in the levitating object, the levitation caused by the Meissner effect is solely produced by the electric currents flowing on the surface of the superconductor.

Thermal Properties

Unlike ordinary conductors, superconductors are not good conductors of heat. However, thermal conductivity is not the only thermal property that is altered by the onset of superconductivity. The specific heat of a metal alters completely below the superconducting transition temperature.

BOX 6.1   THE BIGGER THE BETTER

There are not many pieces of laboratory equipment that need to be delivered through a hole in the roof. But that's exactly how the William R. Wiley Environmental Molecular Sciences Laboratory—part of the Pacific Northwest National Laboratory (PNNL) in the United States—received a new magnet in March 2002 (see Plate 6.1). The 16-tonne superconducting magnet, which was designed by Oxford Instruments in the U.K., was at the time of delivery the largest of its kind in the world. It was built to provide the magnetic field in a nuclear magnetic resonance (NMR) spectrometer (see Plate 6.2).

NMR spectrometers enable the structure of molecules to be determined because the NMR signals emitted by hydrogen atoms in the molecule under study have different frequencies, depending on what type of atoms surround the hydrogen. In addition, the area under each frequency peak reveals the numbers of those atoms present. However, there is very little difference in frequency for the NMR signals from different parts of a molecule, and many spectrometers produce NMR signals so close together that they overlap—particularly when those signals come from biological molecules like DNA, which contains millions of atoms. To improve the resolution (separation of the signals), and so enable both smaller and more complex molecules to be identified, a stronger magnetic field is needed.

So in 1998—in direct response to "a market pull from scientists who use NMR on the human genome project," says project manager Mr. Martin Townsend—Oxford Instruments began designing an NMR magnet far more powerful than any in existence at the time. To provide enough resolution to study complex biological molecules effectively, the magnet had to produce a field of 21.14 tesla. This is over 400,000 times stronger than the Earth's magnetic field, and not surprisingly, building a magnet this powerful presented Mr. Townsend's team with a number of technical challenges.

For a start, the magnet's coil had to be made from superconducting wire because it is impossible to obtain DC magnetic fields this high with standard electromagnets. However, since there was no way of producing the 300 km of superconducting wire needed in one length, the team had to develop a special chemical bonding process to join the separate lengths of wire together into a single electrical circuit. These joints had to be superconducting or at least have very low resistance (of the order of 10–15 ohms), as any changes in resistance through the coil alter the value of the magnetic field, which must remain as constant as possible for such high-resolution experiments.

Another problem was caused by the large currents that run through this superconducting wire, which produce a lot of mechanical stress—over 200 tonnes of force—in the coils of the magnet. Because these forces were so much higher than those in any previous magnets, the team needed to develop a different method of holding the coil structure together. This involved a new way of manufacturing coils, used with the standard procedure of filling any air gaps between the turns

PLATE 6.1 This picture shows a 16-tonne, three-storey-high superconducting magnet designed by Oxford Instruments in the U.K. being delivered to the William R. Wiley Environmental Molecular Sciences Laboratory in the United States in March 2002. The magnet provides the magnetic field in a high-resolution nuclear magnetic resonance (NMR) spectrometer that is capable of revealing individual atoms in complex biological molecules. The magnet must not be jolted, so was transported from the U.K. to the United States by cargo ship and air-ride lorries that have their cargo beds supported by large air dampers instead of suspension springs, which limits the transmission of vibration from the road to the cargo. (Courtesy of Oxford Instruments, U.K.)

PLATE 6.2 Large magnets like the superconducting magnet shown in this picture form the basis of nuclear magnetic resonance (NMR) spectrometers. In order to be analysed, samples have to be accurately positioned in the centre of the magnetic field. So a platform and staircase is attached, which enables users to place samples in the top of the magnet and then lower them down to the centre. (Courtesy of Oxford Instruments, U.K.)

of the coil with an epoxy resin. They also had to design a new system for managing the huge amount of electrical energy (in the range of 17–27 megajoules) that becomes stored in the coil, and a new cryostat to keep the magnet at –271°C, making it cold enough to remain in a superconducting state.

Speaking in 2002, shortly after the first of these new magnets was delivered to PNNL, Mr. Townsend said there was already demand for even more powerful NMR magnets, and that in principle they could be built using similar techniques.

In a normal metal the specific heat decreases in a continuous way, but a super-conducting metal shows a very different behaviour below T_c, with the specific heat first jumping to a higher value, then decreasing slowly. It finally falls below the values a normal metal would have at that temperature.

Since the application of a magnetic field destroys superconductivity, it is possible to compare the low-temperature specific heat of a metal sample in both its normal and superconducting states by measuring the specific heat in the presence of a magnetic field, and also without a field present. (The magnetic field has a negligible effect on the normal specific heat, so a meaningful comparison can be made.)

BCS Theory

Like theories of conductivity for solids in their normal state, theories of conduction for superconducting solids are both complicated and still being worked on. However, the basics of the widely used BCS theory—named after its inventors, American physicists John Bardeen (1908–1991) (who also coinvented the transistor), Leon Cooper (1930–), and Robert Schrieffer (1931–)—are fairly straightforward.

In this theory, electrons with opposite spins become loosely bound together, forming so-called Cooper pairs that can move through a superconductor without encountering any resistance and so produce a current. It is also assumed that the lowest energy state that a superconductor could exist in at $T = 0K$ is one in which all the electrons are in Cooper pairs.

Electrons would normally repel each other because they have like electrostatic charges, but as we are about to see, in this case they are attracted because of their interaction with the positive lattice ions. If you imagine a situation in which there are free electrons moving about in a lattice of positive ions, as a single electron moves past one of these ions, they will be momentarily electrostatically attracted to one another. As Chapter 5 revealed, the atoms (or ions) that solids are composed of are constantly vibrating about their lattice positions, and the brief attraction between the oppositely charged electron and positive ion will alter the vibration of the ion slightly. This then allows the lattice ion to interact with another nearby electron, and since the positive ion and this second electron also attract one another, the overall effect is the same as the two electrons being attracted (see Figure 6.32). An alternative way of thinking about the interaction between the electrons in a Cooper pair is in terms of phonons. In this case, the first electron can be considered to have emitted a phonon, which is then absorbed by the second electron.

Although our simplified explanation suggests that the electrons in a Cooper pair are close together, they do in fact have a complicated quantum mechanical interaction and can have thousands of atoms in between them. The important point is, however, that the electrons are moving as pairs when they take part in conduction, and can only be scattered by defects or thermal vibrations if the energy they receive from "collisions" with these defects or lattice ions is enough to split the pair up. In most cases there is not enough energy to do this, so the Cooper pairs encounter no resistance when moving through the solid.

By contrast, a magnetic field can provide enough energy to break up a Cooper pair, and this is exactly what happens when superconductivity is destroyed by a magnetic

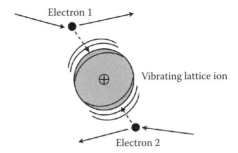

FIGURE 6.32 An electron moving through a superconductor modifies the lattice vibrations of a nearby ion when it is attracted to it. This change in the motion of the ion allows it to attract another electron, with the net result that the pair of electrons behave as if they are attracted to one another.

field. The energy needed to separate a Cooper pair is known as the superconducting energy gap. This energy gap decreases in size as the temperature increases towards T_c. In Section 6.1.1 we saw that the greater the amount of interaction between the conduction electrons and the lattice ions, the larger the resistivity is. The reverse of this situation occurs when a material is in the superconducting state, so the greater this interaction is, the stronger the attraction within the Cooper pairs and therefore the higher the energy needed to separate each pair.

6.5.2   SUPERCONDUCTOR TECHNOLOGY

While the search continues for superconductors with higher and higher critical temperatures, several superconducting materials are already in use, mainly in magnets. In this section we will look at some of the existing and planned applications of superconductors.

Superconducting Magnets

Superconductors make particularly good electromagnets (see Section 8.3.2) because, unlike the metals used in most electromagnets, they do not need a constant flow of electricity to maintain the magnetic field. Once a current is set up in a superconducting electromagnet, it will effectively flow indefinitely, and this property is made use of in both MRI machines and magnetically levitated (Maglev) trains (see next subsection). As well as requiring little power to operate, superconducting magnets can also generate particularly high fields.

One of the most common uses of superconducting magnets is in NMR (nuclear magnetic resonance) systems (see Box 6.1). These can either be used in medicine to view soft tissue inside the human body, in which case they are known as MRI (magnetic resonance imaging) systems, or in research to determine the compositions of complex biological molecules and other compounds. (Magnetic resonance will be discussed further in Chapter 8.)

Superconducting magnets are also used in particle accelerators to guide the charged particles along a particular path, and as storage devices for electric power.

FIGURE 6.33   The Shanghai Maglev Train (SMT) in China connects Longyang Road station with Pudong Airport station 30 km away in under 8 minutes, travelling at a maximum speed of 430 km/h. It was developed and delivered by Siemens AG. (Siemens press picture.)

Many of these superconducting electromagnets are made from niobium-titanium alloys because, unlike most of the high temperature superconductors discovered so far, they can be formed into wires.

Maglev Trains

Each carriage of a Maglev train has superconducting magnets fixed under it. Instead of running on rails, these trains run in a U-shaped guideway (just visible in Figure 6.33) that is lined with conducting wire coils. These coils are turned into temporary electromagnets as the train passes over them, because the moving magnetic field coming from the superconducting electromagnets on board the train sets up currents in them. The coils then produce a repulsive magnetic force field strong enough to levitate the train.

Propulsion is provided by a set of electromagnets in the guideway which are powered by an AC current. These electromagnets switch polarity as the current reverses direction, and this switching over is synchronised to the speed of the train, so that the electromagnets ahead of the train will be given an opposite pole to the onboard superconducting electromagnets and so will pull the train towards them. Conversely, the electromagnets behind the train will have the same polarity as the onboard superconducting magnets and so will repel them—pushing the train forward.

Future Applications

Being able to create transmission lines for electrical power which have no (or very small) losses is just one of the applications scientists have in mind for superconductors. This would both reduce power consumption worldwide—as several percent of the electricity generated is lost as heat because of the resistance in the wires it passes through—and allow much more powerful and efficient electric motors to be made.

These could, for example, be used to power the propellers of ships. Generators, transformers, and current limiters may also be made from superconducting materials in the future. Small loops of superconductor could even be used as the "bits" in a new generation of ultrafast computers.

FURTHER READING

Canfield, P. C., and Bud'ko, S. L. "Low-temperature superconductivity is warming up". *Scientific American* 62 (April 2005).
Dekker, C. "Carbon nanotubes as molecular quantum wires". *Physics Today* 52, no. 5 (1999): 22–28.
Matthews, R. "Super Train". *BBC Focus* p. 55–59, September 2008.
Schechter, B. "Ghost in the machine". *New Scientist* 29 May 2004: 38.
(More references for further reading, as well as Web links, are available on the following Web page: http://www.crcpress.com/product/isbn/9780750309721.)

SELECTED QUESTIONS FROM
QUESTIONS AND ANSWERS MANUAL

Q6.2

(a) Are ionic solids good or bad electrical conductors?
(b) Why is this?

Q6.6

(a) Does the resistivity of a metal increase or decrease with temperature?
(b) Why is this?
(c) What other nonthermal factors contribute to the resistivity?

Q6.8

(a) Which principle do electrons in a Fermi gas obey?
(b) Briefly, what does the principle state?
(c) In Pauli's quantum free electron model, what does the Fermi energy, E_F, represent?
(d) According to this theory, does an electron have to be (i) above or (ii) below the Fermi energy to take part in conduction?

Q6.14   What is the conductivity in a piece of silicon at 300K doped with (a) 6×10^{17} antimony atoms per cm^3 and (b) 5×10^{18} boron atoms per cm^3 if the hole mobility is 0.0458 m^2 V^{-1} s^{-1} and the electron mobility is 0.15 m^2 V^{-1} s^{-1}?

Q6.17   If a sample of indium with a value of $\mathbf{H}_c(0) = 0.0282$ T is cooled below its superconducting transition temperature of 3.41K, what is the approximate value of its critical field when it is at 2K?

7 Chips with Everything
Semiconductor Devices and Dielectrics

CONTENTS

7.1   INTRODUCTION TO SEMICONDUCTOR DEVICES

Imagine for a moment a world without any electronic equipment. In this hypothetical world there would be no mobile phones, no personal computers, no video recorders, no microwave ovens, and a complete absence of a host of other electronic gadgets that we tend to take for granted. This imaginary life would be our reality if it had not been for the invention of the transistor in 1947 by American physicists John Bardeen (1908–1991), Walter Brattain (1902–1987), and William Shockley (1910–1989), coupled with an increasing understanding of the properties of semiconductor materials.

The transistor, and almost every other electronic component, is made mainly from semiconductor materials, including silicon (Si), germanium (Ge)—from which many of the first electronic devices were made—and gallium arsenide (GaAs). The most widely used of these is silicon, as many electronic components contain silicon, and almost all integrated circuits are mounted on silicon chips. Silicon occurs naturally as silica (also known as silicon dioxide) in several different forms including quartz and sand. Unlike many of the Earth's resources, silicon is so abundant that it is unlikely to become used up. This is good news, as the semiconductor industry is one of the biggest industries in the world and more and more high-tech products come on the market each year.

In this chapter we will see how some of the most common electronic devices work, what they tend to be made from, and how they are manufactured. We will also discuss insulating materials, which, as we will discover, play an important role in semiconductor devices.

7.1.1   p-n JUNCTIONS

Forming a p-n Junction

If a layer of p-type semiconductor is sandwiched next to a layer of n-type semi-conductor, a *p-n junction* is formed. As Section 6.4.1 explained, both p-type and n-type semiconductors are electrically neutral overall, since intrinsic semiconductors

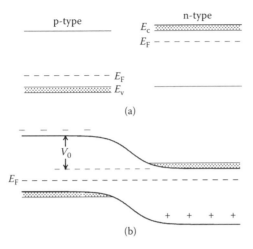

FIGURE 7.1   Energy band diagrams for (a) p-type and n-type semiconductors and (b) a p-n junction in thermal equilibrium made up from these semiconductors. (This diagram first appeared as Figure 8.10, p. 226, in *Quantum Mechanics, Foundations and Applications*, by D. G. Swanson, © 2007, Taylor & Francis Group, LLC. Kindly supplied by Gary Swanson.)

are electrically neutral, and so are any dopant atoms. So before they are joined together, the p-type and n-type sides of a p-n junction are electrically neutral, but there is a large concentration of electrons in the n-type material and a large concentration of holes in the p-type material. Figure 7.1(a) shows band diagrams for isolated n-type and p-type materials. The Fermi energy is near to the valence band in the p-type semiconductor, and near to the conduction band in the n-type semiconductor because in both cases there is a high dopant concentration. (Figure 6.25 shows how the Fermi level moves from the middle of the bandgap, depending on the amount and type of dopant present.)

As we saw in Section 6.4.2, whenever there is a concentration gradient of carriers within a semiconductor, diffusion occurs because it lowers the internal energy of the system (or material). So in this case, as soon as the two sides of the p-n junction touch, electrons diffuse into the p-side while holes diffuse into the n-side. This results in the region of the n-type material closest to the junction becoming positively charged as the electrons in that region leave behind ionised donor atoms. Meanwhile, the region next to the junction on the p-type side becomes negatively charged as holes move to the n-side, leaving negatively charged acceptor ions behind. The area either side of the junction—in other words, the region containing ionised donors plus the region of ionised acceptors—is known as the "depletion region". This is illustrated schematically in Figure 7.2.

As with many other topics in solid state physics, the ideas we have just seen are models of the true situation. It is helpful to assume that a p-n junction is formed by bringing two separate pieces of semiconductor together, as this provides a useful starting point for understanding the operation of the junction, but in reality p-n junctions are created as follows.

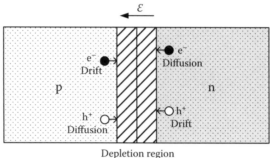

FIGURE 7.2   Schematic diagram of an unbiased p-n junction in thermal equilibrium.

When a p-n junction is made from one type of semiconductor, it can be doped with acceptor atoms on one side to make it p-type, and with donor atoms on the other side to make that side n-type. (In fact, more commonly n-type material has acceptors diffused into part of it to "cancel out" the donors and provide an excess of acceptors.) This type of junction is known as a homojunction. It is made from a single crystal because the grain boundaries (see Section 2.2.6) in a polycrystalline semiconductor would scatter the charge carriers, making the device inefficient.

It is also possible for a p-n junction to be made from two different types of semiconductor. In this case it is called a heterojunction, and is created by growing one type of semiconductor on top of the other. A single crystal serves as the *substrate* (base) for this growth, and the growing material—as Section 7.3.2 reveals in more detail—takes on the single-crystal nature of the substrate, and so will not contain grain boundaries.

Unbiased Junction in Thermal Equilibrium

The diffusion that occurs when a p-n junction is first created does not continue until the concentrations are uniform, as the charge from the uncompensated ions in the depletion region creates a potential difference (in other words, a voltage) between the two sides. This "junction voltage", which is a potential barrier for electrons to climb, is plotted as V_0 in Figure 7.1(b), and is typically within the range 0.6–0.8 V at room temperature for a silicon p-n junction. Once this voltage is established, any electrons wanting to move from the conduction band of the n-side into the p-side would then need to climb this potential barrier, and so their flow is inhibited. For the same reason, the flow of holes is also inhibited. (Remember that for a hole, the higher its energy, the lower it is from the edge of the valence band on a band diagram.)

The flow of electrons into the p-side and of holes into the n-side also produces an electric field that points from the n-side to the p-side. This field causes a drift current (see Section 6.4.2) of electrons and holes in the opposite direction to their respective diffusion currents, as shown in Figure 7.2. At thermal equilibrium, and with no voltage applied across the junction, the drift and diffusion currents for each type of carrier cancel one another out so that there is no net current flow across the junction. Having zero net electron and hole currents results in the Fermi energy being constant through the junction, as shown in Figure 7.1(b).

Unlike the flow of the diffusing majority carriers, the flow of minority carriers that make up the respective drift currents is not inhibited by the barrier voltage V_0. This can be illustrated most easily by looking at Figure 7.1(b) and imagining a minority electron trying to move from the conduction band of the p-type material into the conduction band of the n-type material. You can see that it can just "slide down" the slope that connects them.

Reverse Bias

If a battery is connected to the p-n junction so that the positive battery terminal is joined to the n-side and the negative terminal to the p-side, the p-n junction is said to be under "reverse bias" conditions. In order for the "reverse current" to flow, electrons will have to flow from the n-side into the p-side. Meanwhile, holes will need to flow from p to n. This leads to more ionised donors and acceptors being present than in the unbiased condition, which in turn leads to an increase in the size of the depletion region. Charge from these additional uncompensated ions creates a voltage that adds to the junction voltage, as shown in Figure 7.3(a). Almost no current flows through the junction because the increased size of the junction voltage acts as an even greater potential barrier (compare Figure 7.1[b] with Figure 7.3[a]) to the diffusion of majority carriers. The resistance (see Section 6.1.1) of a reverse-biased p-n junction is therefore very high.

In fact the only current flow—known as "reverse current" or sometimes "leakage current"—is due to drift of the minority carriers (see Section 6.4.1) across the junction. As Figure 7.4(a) reveals, the reverse current this produces is constant in value over a wide range of reverse-bias voltages. This is because the minority carriers that

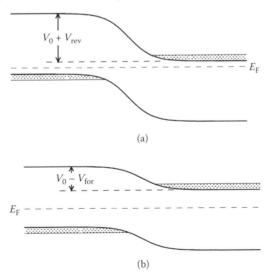

(a)

(b)

FIGURE 7.3 Band diagrams for (a) a p-n junction under reverse bias and (b) a p-n junction under forward bias. (This diagram is adapted from Figure 8.11, p. 227, in *Quantum Mechanics, Foundations and Applications*, by D. G. Swanson, © 2007, Taylor & Francis Group, LLC. Kindly supplied by Gary Swanson.)

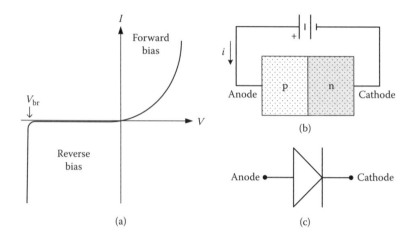

FIGURE 7.4 (a) Current-voltage characteristics for a p-n diode; (b) shows a forward-biased p-n junction, while (c) shows the circuit symbol for a diode and how this relates to the electrical connections shown in (b).

make up the current are created as a result of bonds breaking due to the temperature of the p-n junction. The reverse current is very small because there are so few minority carriers present.

Forward Bias

By contrast, if a battery is connected the other way around—in other words, with the positive terminal connected to the p-side, as shown in Figure 7.4(b)—the junction is said to be "forward biased". In this case the external circuit will supply majority carriers to both sides of the junction, so holes will be added into the p-side and electrons into the n-side. Some of these additional carriers will neutralise uncompensated ions in the depletion region, causing it to decrease from the size it is when the junction is unbiased. The reduction in size of the depletion region reduces the corresponding potential barrier across the junction (compare Figure 7.1[b] with Figure 7.3[b]), allowing a large diffusion current of majority carriers to flow. As with the reverse-bias case, the drift current of minority carriers remains similar to that of the unbiased junction.

Therefore many more electrons are diffusing from the n-side into the p-side than in the unbiased case. Once on the other side of the junction they then diffuse away from the depletion region and form a current through the device as they do so. From the conventions of electronics, this "forward current" flows in the opposite direction to the flow of electrons, and so flows from the p-side to the n-side, as illustrated in Figure 7.4(b).

The relationship between this current and the applied voltage is shown in Figure 7.4(a). As this reveals, the more the applied voltage is increased, the greater the amount of current passing through the junction, which means the depletion region becomes smaller as the applied voltage increases in size. Eventually the applied voltage is large enough to completely eliminate the depletion region, and beyond this point the current passing through the p-n junction is proportional to the voltage across it.

If we now look at the forward-bias situation in more detail, it becomes clear that the explanation above is oversimplified. Whilst this helps reveal the basic workings of the junction, it glosses over exactly how the large diffusion currents flow. In order to understand this we need to refer back to Section 6.4.1, where we saw that the movement of an electron into the conduction band of a semiconductor left behind a corresponding hole in the valence band. In certain circumstances electrons and holes can recombine and so effectively cancel each other out, and this is exactly what happens in a forward-biased p-n junction.

Looking first at an electron moving as part of a diffusion current from the n-side into the p-side, we know that once it is in the p-side it will become a minority carrier. It will also be surrounded by lots of holes—which are the majority carriers in p-type material—and is likely to recombine with one of these holes. In order for charge neutrality to be maintained, the external circuit must supply holes to the p-side to make up for the ones lost by recombining with electrons from the n-side. Meanwhile, the mirror of this process will be occurring for holes entering the n-side, which will recombine with electrons that in turn will be replenished by the external circuit. The result is that the current flowing through the external circuit is made up from the carriers flowing into the device to compensate for the *recombination* occurring within it.

So to summarise, applying a voltage to a p-n junction alters the balance achieved in unbiased thermal equilibrium between the drift and diffusion currents. In a p-n junction under forward bias, the drift current is much smaller than the diffusion current, whereas under reverse-bias conditions the diffusion currents are reduced to the point where the tiny drift currents dominate.

Rectification

We have just seen that connecting a p-n junction to an electric circuit turns it into a type of *diode* (an electronic device with two electrodes) which only allows current to flow in one direction through it. One of the most common uses of a p-n junction is as a rectifier. This is a device that converts AC current into DC.

Every half-cycle of the AC signal will reverse-bias the p-n junction and therefore will not pass through the diode, as Figure 7.5 illustrates. P-n junctions form the basis for many different types of semiconductor devices, as we will see in the next few sections.

Breakdown

Referring back to Figure 7.4(a), it can be seen that when a p-n junction is reverse-biased, the current is near zero until the applied voltage passes a critical value known as the "breakdown voltage," V_{br}. Beyond this point the current increases substantially. V_{br} is a function of the doping concentrations either side of the junction—the greater the doping concentration, the lower V_{br} is. Although in itself a large current passing through a reverse-biased p-n junction is not harmful to the junction, it can cause the junction to heat up beyond a safe level. This current therefore has to be limited by an external circuit.

There are two main mechanisms for breakdown in p-n junctions. For junctions with relatively low doping concentrations, "avalanche multiplication" occurs beyond V_{br}. This process is illustrated in Figure 7.6 and begins with a thermally

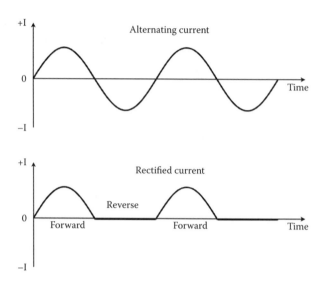

FIGURE 7.5 An alternating current and the same current once it has passed through a rectifier.

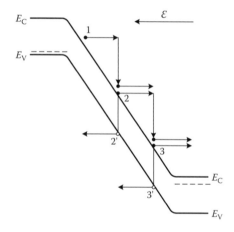

FIGURE 7.6 In this diagram of the avalanche multiplication process, a thermally generated electron labelled 1 gains so much kinetic energy from the electric field produced by the large applied voltage that it ionises an atom. This creates an electron-hole pair—labelled as 2 and 2′—the electron of which in turn creates a further electron-hole pair (pair 3), and so on. The large numbers of carriers produced by continual impact ionization allows a large current to flow.

generated electron gaining enough energy from the electric field produced by the applied voltage that it can cause impact ionization. In general, this process involves a high-energy collision between a particle and an atom which ejects one or more electrons from the atom. In a semiconductor, an electron-hole pair(s) is produced if the electron(s) released by impact ionization has enough energy to be promoted into the conduction band. The electrons from these electron-hole pairs will then create

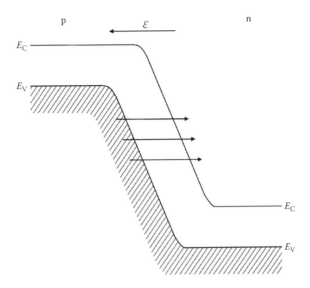

FIGURE 7.7 A schematic illustration of the tunnelling process in a highly doped p-n junction that occurs when a reverse voltage over a certain value characteristic of the junction is applied. This type of breakdown is sometimes referred to as "Zener breakdown".

more electron-hole pairs by impact ionization. Eventually there are so many charge carriers that a large current flows through the p-n junction.

The other process that can cause breakdown in p-n junctions is known as "tunnelling". Tunnelling is a quantum mechanical phenomenon and occurs when the wave function of a particle extends from one side of a potential barrier to the other. (The wave function of a particle is effectively a "probability wave" which shows the region of space over which the particle can reside. Section D1 of Appendix D explains more about wave functions.)

Tunnelling only occurs in p-n junctions with high levels of doping that have also had a reverse bias applied that is above a "critical value" that causes the top of the valence band on the p-side to lie above the bottom of the conduction band on the n-side. This situation is illustrated in Figure 7.7. An electron can then "tunnel" through the bandgap from the valence band on the p-side into the conduction band on the n-side that has states with the same energy available to the electron. Another way of looking at it is to say there is a probability the electron is in the valence band of the p-type material, but also a probability that it is in the conduction band of the n-type material because its wave function extends over both regions. Once in the conduction band, the electron can take part in conduction.

It is possible for the breakdown in a p-n junction to be a mixture of both the tunnelling and avalanche multiplication processes.

7.1.2 BIPOLAR JUNCTION TRANSISTORS

Modern electronics is based on the transistor because, in general, transistors can either be used to amplify electrical signals or as switches in electronic circuits which,

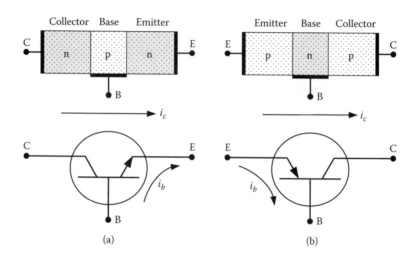

FIGURE 7.8   (a) An n-p-n bipolar junction transistor and its corresponding circuit symbol; (b) a p-n-p bipolar junction transistor with its circuit symbol.

for example, help computers process information. They can do this because computers handle information in the form of a binary code of 0's and 1's, and the "off" position of a switch can be used to represent binary 0 while the "on" position represents binary 1. Processing requires data to be stored while calculations and logic operations are carried out, and so a large number of transistors which can be switched on and off as required enable this to take place. All modern transistors belong to one of two main classes or categories: bipolar junction transistors and field effect transistors (which we will look at in the next section).

The bipolar junction transistor (BJT) has three regions, and can be thought of as two p-n junctions back to back, with a shared central section. That means the junction can either have an n-p-n structure or a p-n-p structure, as shown in Figure 7.8. As this diagram reveals, the three different regions of the transistor are called the emitter, base, and collector.

The n-p-n type of BJT is slightly more common than the p-n-p type, as it can be made more easily from silicon, can generally conduct larger currents, and can be made to switch between "on" and "off" more quickly. Figure 7.9 shows an n-p-n bipolar junction transistor connected to an external circuit that allows it to work in its normal mode of operation (sometimes known as "forward active" operation). This means that the base-emitter p-n junction is forward biased while the base-collector p-n junction is placed under reverse-bias conditions. Electrons flow across the forward-biased junction from the emitter into the base, where they are minority carriers. The emitter is so much more heavily doped than the base that, although some of these electrons recombine with the majority carrier holes in the base, many more diffuse across the base and are swept across the reverse-biased base-collector junction into the collector. From there they appear in an external circuit as a "collector current". (In fact in an ideal situation, the base would be narrow enough that almost no recombination would take place there, and so all the electrons injected into the

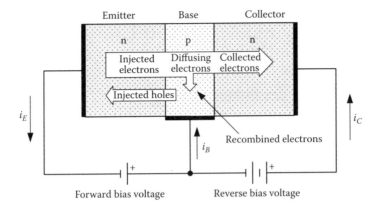

FIGURE 7.9 An n-p-n bipolar junction transistor in its "normal" mode of operation, with the base-emitter p-n junction forward biased, and the base-collector p-n junction reverse biased.

base could be assumed to be part of the collector current. In other words, the injected electron current would be equal to the collected electron current.)

Referring back to Figure 7.9, you will see that in addition to the collector current, there is also a base current. This is created by holes injected from the base into the emitter because the base-emitter p-n junction is forward biased. The base current is much smaller than the electron current flowing from emitter to base, because as we saw earlier the emitter is much more heavily doped than the base and so can provide more carriers. However, although the base current is small, it is the key to how the whole device works because if it is shut off, the flow of electrons from emitter to base (and hence the collector current) ceases. This is because the base-emitter p-n junction is no longer forward biased. The n-p-n BJT is therefore said to be a "normally off" device because it needs the application of a base current to switch it on. As well as acting as a switch, this transistor can amplify current because a small increase in base current produces a large increase in emitter-collector current.

Exactly the same types of process occur for the p-n-p BJT, but with holes diffusing into the base instead. The p-n-p BJT is a "normally on" device, with application of the base current needed to turn it off.

7.1.3 FIELD-EFFECT TRANSISTORS

Whereas the operation of a BJT is controlled by current, it is voltage that controls the action in a field-effect transistor (FET). Also in contrast to BJTs, the transport in FETs is via only one carrier type rather than both electrons and holes. This difference is indicated by saying that a FET is a "unipolar" device, while—as its name states—a BJT is "bipolar". Despite these differences, the current-voltage characteristics of the output from both types of transistor are similar.

Enhancement Mode MOSFETs

Figure 7.10 shows a type of MOSFET (metal-oxide semiconductor field-effect transistor), known as an "n-channel enhancement mode" device, and its corresponding

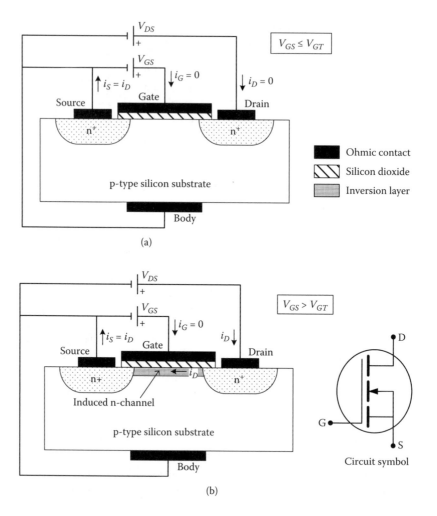

FIGURE 7.10 Schematic diagram of an "n-channel enhancement mode" MOSFET (a) with a gate voltage below the threshold (V_{GT}) and (b) with a voltage above V_{GT} applied to the gate. In (a) the device is switched off, while in (b) it is switched on. The circuit symbol for the device is shown on the right of part (b).

circuit symbol. Part (a) shows the device "off" (in other words with no current flowing through it), while part (b) of the diagram shows the device when it is "on". It is easiest to understand the device by looking at the structure with no current flowing through it, so as you will see from part (a), there are two islands of n-type semiconductor—one is called the "source" and the other the "drain"—sitting within a p-type silicon substrate. Metal connections known as terminals join the n-type semiconductor of the source and drain regions to an external circuit, while a layer of the insulator SiO_2 separates the gate terminal from the n-type semiconductor.

Every type of MOSFET has a threshold voltage for the gate (V_{GT}) that marks the divide between the device being on or off. For example, for the n-channel enhancement mode MOSFET, with a voltage applied to the gate that is less than the threshold voltage, there is no current flow through the device because the source is isolated from the drain. Figure 7.10(a) shows this situation. However, when a voltage above the threshold voltage is applied to the gate, as shown in Figure 7.10(b), the device is switched on. This is because the electric field the gate voltage produces will repel the holes—which are the majority carriers in p-type material—from the region in the substrate below the gate area, forming an area similar to the depletion region in a p-n junction.

This region is called the inversion layer because as the voltage is increased, a point is reached where electrons—which are minority carriers in a p-type semiconductor—accumulate there and so "invert" the charge carrier type. The accumulated charge forms an n-channel that links the n-regions of source and drain. If an external circuit like the one shown in the diagram is giving the drain a positive bias with respect to the source, electrons can then flow from source to drain via the n-channel, creating a current through the device. (In fact not all of the electrons that form this channel come from the substrate. Some also come out of the highly doped source and drain regions.)

As the voltage on the gate is increased, more electrons will move into the inversion layer, and so more current will flow through this MOSFET. In other words, increasing the gate voltage enhances the conducting channel, which is why the transistor is described as an "enhancement-mode" device.

Since a small variation in the value of the gate voltage produces a relatively large change in the current through the device, MOSFETs can be used to amplify very small currents. In addition, they can also be used as straightforward switches either in integrated circuits (as we will see later) or in much larger scale analogue electronic circuits (see Figure 7.11). This is because applying a gate voltage below the gate threshold or not applying a voltage at all will mean the device has no current flowing through it (as the drain is isolated from the source). Conversely, a gate voltage above the gate threshold allows current to flow (as the drain is connected to the source).

Example question 7.1 investigates a p-channel enhancement mode MOSFET, which like the n-channel device we have just looked at is a so-called "normally off" device because application of a gate voltage is required to make it conduct.

Depletion Mode MOSFETs

So-called "depletion-mode" devices make up another major family of MOSFETs. Like enhancement-mode devices, they can either be n-channel or p-channel, but we will not look at "p-type depletion-mode" MOSFETs as they are mostly theoretical and rarely made in reality. Instead we will concentrate on the "n-channel depletion mode" device shown in Figure 7.12. This device is "normally on". In the simplest case this means that as long as there is an external circuit making sure the drain is positive with respect to the source, current flows through the channel if the gate voltage (with respect to the source) is above the gate threshold (V_{GT}). For this device type the threshold is negative, so this will be the case if the gate is left unconnected from an external circuit, or is connected directly to the source or any voltage more positive than the source. Connecting the gate to a voltage below V_{GT} will turn the device off.

EXAMPLE QUESTION 7.1   ENHANCEMENT-MODE MOSFETS

The following diagrams labelled (i) and (ii) show a p-channel enhancement-mode device and its corresponding circuit symbol.

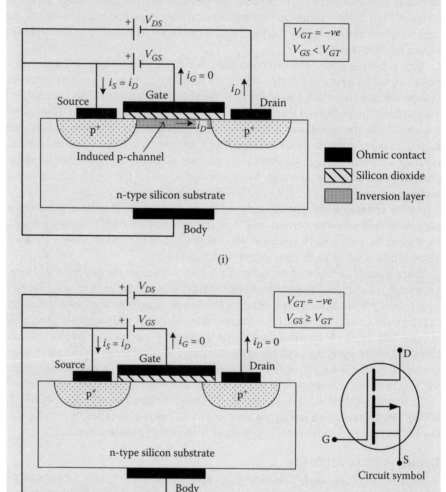

(i)

(ii)

(a) Which diagram shows the device "on"?

(b) What type of carriers are minority carriers in the n-type silicon substrate?

(c) What sequence of events needs to happen in order for the device to be "switched on"?

Answer on page 230.

FIGURE 7.11   The robots in this picture, which were built by Prof. Kevin Warwick and his team at the University of Reading in the U.K. in 2000, use ultrasonic sensors to detect objects. Like robot vacuum cleaners, they are programmed to avoid hitting things by moving out of the way whenever they sense an obstacle. This means their motors need to be turned on and off very quickly, and so MOSFETs—which can switch on and off particularly rapidly— are used as the switches for the motors. It is not just their high switching speeds that make MOSFETs suitable for applications like this, however. They are also chosen because the large currents needed to drive the motors can pass through them without causing any damage. Another version of these robots, which were designed by the Reading team in 2003, contain neural networks rather than standard electronic circuits. A neural network is an electronic network that mimics, in a very simplified way, the function of biological neurones. The hope is that by studying how robots and computers "learn" to carry out simple tasks—such as how to avoid obstacles—we might eventually be able to build machines that could "think" for themselves. (Courtesy of Kevin Warwick.)

As part (a) of Figure 7.12 shows, there are two islands of n-type semiconductor connected by a narrow n-channel, all sitting within a p-type semiconductor substrate. Metal connections join the n-type semiconductor of the source and drain regions to an external circuit, while a layer of the insulator SiO_2 separates the gate terminal from the n-type regions. Current flows because the n-type channel acts as a conducting channel between source and drain.

If a negative voltage is applied to the gate, the electric field it produces attracts holes from the substrate into the n-channel, where they deplete the numbers of electrons there by recombination. With fewer free electrons in the channel, the current through the device decreases. Another way of looking at this is to say the recombination creates a depleted region that reduces the area of the conducting channel, and so reduces the amount of current flow. If the gate voltage is made more negative than V_{GT}, as shown in Figure 7.12(b), the channel is totally "pinched off" by the depletion region and so no current flows.

MOSFETs are comparatively easy to manufacture and can be packed together very closely without affecting each other's operation. The power consumption of MOSFETs is also low, and they can be made smaller than BJTs. These features have

**ANSWER TO QUESTION 7.1 ON
ENHANCEMENT-MODE MOSFETS**

(a) Diagram (i).

(b) Holes.

(c) As this device has an n-type silicon substrate, electrons need to be repelled away from the region under the gate to create an inversion layer, so a negative gate bias must be applied. As this bias voltage is increased (i.e., made more negative), holes will accumulate in the inversion layer, creating a conducting p-channel. A hole current can then flow between source and drain, so the device is now "on" as a current is passing through it. (Note that since the gate must have a negative bias, the gate threshold voltage will also be negative, and to be bigger than this and so turn the device on, the gate voltage must be even more negative than the threshold voltage. In other words, $V_{GS} < V_{GT}$.)

helped make MOSFETs the most widely used semiconductor devices, and for this reason are the only type of FETs we will consider in this section.

CMOS

Integrated circuits based on MOSFETs are known as "MOS integrated circuits" and are used in both microprocessors and memory chips. In fact at the time of writing this book, the most common type of integrated circuit uses complementary metal oxide semiconductor (CMOS) technology. In CMOS, for every p-type transistor there is a corresponding (complementary) n-type transistor.

Thin-Film Transistors

When a MOSFET is made from a film just a few micrometres thick on an insulating glass or ceramic substrate, it is known as a thin-film transistor (TFT). The insulating substrate reduces the amount of charge the device stores, and so it can be operated at higher speeds than conventional MOSFETs. TFTs do not have to be made from standard semiconductors. In fact there are likely to be more and more so-called "plastic electronics" in the future. The rollable display screen shown in Figure 7.13, for example, shows just one use of organic TFTs.

7.2 OPTOELECTRONIC DEVICES

7.2.1 Interaction Between Light and Semiconductors

One of the reasons semiconductors are so useful is that the conductivity of some semiconductor materials is sensitive to light. When radiation falls on these semiconductors, if the photon energy hv is greater than the bandgap energy E_g, a photon can cause an electron in the valence band to be excited into the conduction band. The electron, and the hole it leaves behind, can then take part in conduction, so illuminating these semiconductors increases their conductivity.

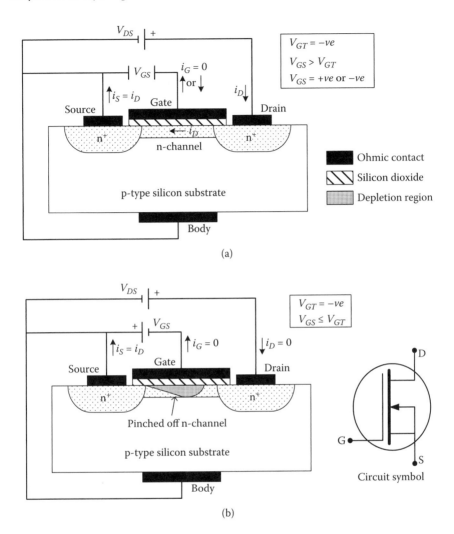

FIGURE 7.12 Schematic diagram of an "n-channel depletion-mode" MOSFET (a) with a gate voltage above the threshold (V_{GT}) and (b) with a voltage below V_{GT} applied to the gate. In (a) the device is switched on, while in (b) it is switched off. The circuit symbol for the device is shown on the right of part (b).

If the energy of the photons hitting the semiconductor is greater than the work function of the material, electrons will be liberated from it by the photoelectric effect rather than promoted into the conduction band. Photoconductivity therefore occurs when the energy of the incident radiation is in the range

$$\Phi > h\nu > E_g$$

where Φ is the work function, which is the minimum energy needed for an electron to be freed from a solid.

FIGURE 7.13   This flexible screen was made by Philips in 2005, and contains organic TFTs. (© Philips.)

Optical Absorption, Spontaneous Emission, and Stimulated Emission

There are three basic processes that can occur between a photon and an electron in a solid: absorption, spontaneous emission, and stimulated emission. Figure 7.14 shows these three mechanisms. E_1 is a state lower in energy than E_2, and so E_1 and E_2 can be considered to represent, respectively, the ground state and excited state of an atom. Equally, E_1 can represent the top of the valence band and E_2 the bottom of the conduction band in a semiconductor, and in this case the energy levels will therefore be electron energy levels. Given that this chapter is about semiconductor devices, we will assume them to be the latter.

Transitions can occur between these states with either the emission or absorption of a photon with an energy equal to the difference between these two energy levels. This means the photon will have an energy $h\nu_{12} = E_2 - E_1$, where the subscript on the frequency indicates that the photon is resulting from a transition from state 1 to state 2 (while ν_{21} would be the frequency for a transition from state E_2 to state E_1).

In thermal equilibrium, many more electrons will be in the lower energy level E_1 than in E_2. When a photon of energy $h\nu_{12}$ is input into the system, however, an electron in state E_1 will absorb it, and will be excited to the upper level E_2. This is called absorption and is shown in Figure 7.14(a).

As there is an unfilled lower energy state available, after a short period of time the electron will drop back down from its excited state into E_1, recombining with a hole and giving off a photon of energy $h\nu_{12}$ as it does so. This process is known as spontaneous emission and is illustrated in Figure 7.14(b).

Another type of emission, known as stimulated emission, can occur if a photon of energy $h\nu_{12}$ falls on a semiconductor whilst it has electrons in the excited state E_2.

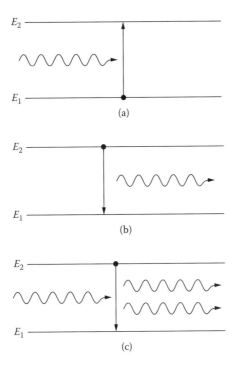

FIGURE 7.14   The three main types of transition process that can occur between two energy levels: (a) absorption, (b) spontaneous emission, (c) stimulated emission. (Adapted from Figure 5.2, p. 137 of *Quantum Mechanics, Foundations and Applications*, by D. G. Swanson, © 2007, Taylor & Francis Group, LLC. Original version kindly supplied by Gary Swanson.)

If we consider one electron sitting in E_2, instead of remaining there until spontaneous emission occurs, it may be forced to drop back down to E_1 by this photon falling on it, but it does not absorb the incident photon. As with spontaneous emission, the electron will recombine with a hole when it returns to state E_1, and so give out a photon of energy $h v_{12}$. However, in this case the emitted photon is in phase with, has the same state of polarisation, and travels in the same direction as the incident photon, which continues on its way as if it had not caused the transition. This process is shown in Figure 7.14(c), and this type of radiation is known as coherent radiation. Since one photon has gone into the semiconductor and two have come out, it is possible to see how an incident light wave can be amplified as it passes through a system like this.

Stimulated emission forms the basis for laser operation. (The word "laser" is actually an acronym for "Light Amplification by Stimulated Emission of Radiation".) In fact, in order to amplify light, there needs to be many more electrons in the higher energy level E_2 than there are in E_1, so that there are plenty available to be stimulated to fall back down to E_1 and so cause photons to be emitted. This is a nonequilibrium state known as a *population inversion*, and is essential for laser operation. (We saw earlier that in thermal equilibrium more electrons are in E_1 than E_2, and a system like this will act as a photon absorber. With same number of electrons in the two states, the amount of emission will exactly balance the absorption.)

One way of maintaining a population inversion is to shine electromagnetic radiation of a wavelength that can excite electrons from E_1 into an even higher level at E_3 so that when they relax down they are only falling as far as level E_2. This is how the ruby laser works, but the semiconductor laser (as we will see in Section 7.2.3) makes use of the properties of semiconductors to enable it to function with just two energy bands being involved (albeit plus the two energy pumping levels of the external voltage source).

Direct Gap and Indirect Gap Semiconductors

We saw in Section 6.4.1 what the band structures of hypothetical semiconductors look like. By contrast, Figure 7.15 shows the E-k relationship of two real semiconductors. (As we saw in Section 6.1.3, E-k diagrams are effectively graphs of energy vs. momentum for electrons.) Both styles of diagram show the conduction band and valence band with an energy gap (E_g) between them. It can be seen from Figure 7.15(a) that in some semiconductors the minimum energy level in the conduction band (usually referred to as the "conduction band minimum") and the maximum energy level in the valence band (the "valence band maximum") occur at different values of the wave vector, k. These semiconductors are therefore known as *indirect gap semiconductors*.

As the electrons will be in the minimum of the conduction band and the holes at the valence band maximum, any transitions must take place between the valence band maximum and the conduction band minimum. This means that transitions can only occur if there is a change in energy greater than or equal to the value of the energy gap, together with a change in the momentum of the electron. When electron-hole recombination occurs in such a material, phonons need to be generated to remove the extra electron momentum, and very little—if any—light is produced. Silicon, germanium, and diamond are examples of indirect gap semiconductors. In each of these materials, the top of the valence band and the bottom of the conduction band occur at different k values, so a change in electron momentum is required in any transitions occurring between the two bands.

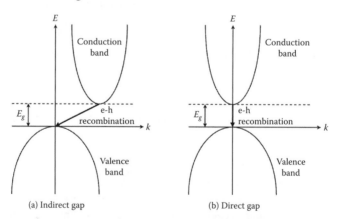

FIGURE 7.15 Energy band structure represented as a plot of energy against electron momentum for (a) an indirect gap semiconductor and (b) a direct gap semiconductor. (This diagram is based on Figure 2.8, p. 23 of *Introductory Semiconductor Device Physics*, by Greg Parker, © IOP Publishing Ltd., 2004. Reproduced with permission.)

By contrast, the top of the valence band and the bottom of the conduction band occur at the same value of electron wave vector for *direct gap semiconductors* like gallium arsenide, as Figure 7.15(b) shows. In this case, no phonons are generated during the emission process. As only two particles are involved (an electron and a photon), there is therefore a much higher probability of a radiative transition occurring in a direct gap semiconductor than in an indirect semiconductor where three particles (an electron, a phonon, and a photon) are involved in the transition.

EXAMPLE QUESTION 7.2   OPTOELECTRONIC DEVICES

You are a semiconductor manufacturer and have been asked to supply a research establishment. All you have been told is that they are trying to develop a new type of optoelectronic device. You make both direct gap and indirect gap semiconductors. Which type do you supply to the researchers?

Answer on page 236.

7.2.2   LEDs

Light-emitting diodes (or LEDs for short) are a familiar sight, as they are used as small indicator lights on many electrical and electronic devices. For example, automatic washing machines often have LEDs built into their control panels to enable users to see at a glance which program the machine is executing, and TV remote controls generally contain a red LED which lights up when a button on the remote is pressed. Meanwhile, the illuminated advertising signs in some sports stadiums are made up from thousands of LEDs, with each individual LED acting as one pixel of the picture, and architects and retailers have not been slow to make use of the technology, as Figure 7.16 shows.

FIGURE 7.16   New York department store Saks Fifth Avenue joined forces with Philips to produce this LED display of 50 snowflakes. (© Philips.)

ANSWER TO QUESTION 7.2 ON OPTOELECTRONIC DEVICES

In the absence of any further information, you would supply direct bandgap semiconductors to the researchers because, in general, the most efficient opto-electronic devices are made from direct gap semiconductors. This is because during emission all of the energy emitted will be released as light and not as phonons.

LEDs are a type of p-n junction diode (see Figure 7.17) made from a direct gap semiconductor (see Section 7.2.1). The p-n junction can be seen near the top of the device, with the crosshatched region in the diagram representing heavily doped p-type material. We saw in Section 7.1.1 that forward-biasing a p-n junction causes lots of electrons to diffuse into the p-side, where they recombine with holes (which are the majority carriers in p-type semiconductors). Conversely, holes diffusing from

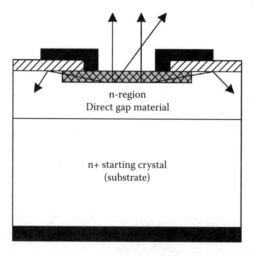

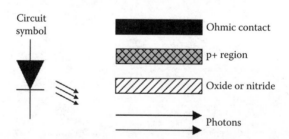

FIGURE 7.17 An LED consists of a p-n junction diode, which produces photons when it is forward biased. (This diagram is copied from Figure 7.1, p. 129, of *Introductory Semiconductor Device Physics*, by Greg Parker, © IOP Publishing Ltd., 2004. Reproduced with permission.)

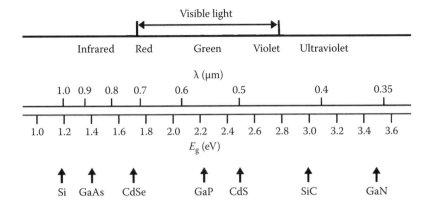

FIGURE 7.18 Some semiconductor materials that LEDs can be made from, and the wavelengths of light they emit.

the p-side into the n-side will recombine with majority electrons in the n-side. This recombination will produce photons, and if a standard p-n junction like this is made from a direct bandgap semiconductor, it will efficiently emit light whenever it is forward-biased and recombination is therefore occurring.

The greater the bias current, the more excess carriers are produced, and the greater the number of carriers that recombine, the brighter the light emitted. The colour of the light emitted is dependent on the type of material the LED is made from. This is because the energy of the emitted photons is equal to the bandgap energy. Figure 7.18 shows some of the semiconductors that are commonly used to make LEDs, and the wavelengths of light they emit. LEDs can be made from indirect gap materials, but they will have a lower efficiency than those made from direct gap materials.

LEDs are starting to replace conventional incandescent bulbs in a range of applications from domestic lighting to brake lights for road vehicles. One of the driving forces behind these efforts is the fact that the power consumption of LEDs is substantially less than that of any existing indoor room lighting. LEDs can also provide a more flexible solution than other light sources for certain applications, as Figure 7.19 reveals, and have a long life expectancy, which makes them attractive as a light source for use in environments where safety is paramount. For example, LEDs can be used for road traffic lights and to light landing strips in airports.

EXAMPLE QUESTION 7.3 LEDs

If an LED is made from GaP which has a bandgap of approximately 2.25 eV, what colour is the light it gives out?

ANSWER

Referring to Figure 7.18, we can see that GaP gives out green light (552 nm).

FIGURE 7.19 The locomotive shown in this picture is sold in 14 different European countries, each of which have different regulations for the use of the headlights. For example, a certain pattern of flashes in a particular colour may be required to indicate that the locomotive is reversing. In order to provide this flexibility, each headlight is made from 248 white, 66 green, and 102 red LEDs. (Courtesy of Siemens AG.)

7.2.3 SEMICONDUCTOR LASERS

Another device based on the p-n junction is the semiconductor or "diode" laser. As we saw in Chapter 1, diode lasers are used to read and write the data on CDs and DVDs, so are very widely used. A diode laser is very similar to an LED, but has the addition of two "mirrored" ends (as shown in Figure 7.20) that are formed by cleaving (in other words, splitting) the semiconductor crystal along its planes. For a GaAs laser, for example, the junction plane is (100), while the cleaved ends are (110) planes. (Section 2.2.7 explains the notation for crystal planes.) The difference in

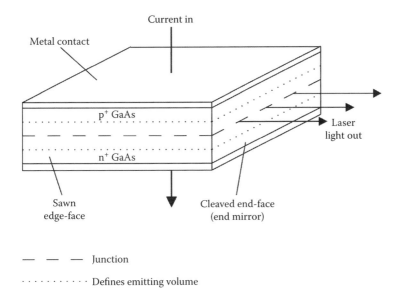

FIGURE 7.20 Schematic diagram of a semiconductor homojunction laser diode. (This diagram is copied from Figure 10.6, p. 232, of *Introductory Semiconductor Device Physics*, by Greg Parker, © IOP Publishing Ltd., 2004. Reproduced with permission.)

refractive index between the semiconductor and the air surrounding it causes these mirrored ends to reflect some of the light emitted by recombination within the p-n junction. The ends are therefore forming a laser cavity, which as we are about to see increases the light output.

Mirrored ends alone are not enough to turn an LED into a laser, though. As we saw in Section 7.2.1, a "population inversion", where there are more electrons in a higher energy state than a lower one, needs to be created in order for a material to lase, as the stimulated emission process has to dominate. For a diode laser, this means that the conduction band must be made to contain more electrons than the valence band. In addition, because it is recombination of electrons and holes that causes the emission of photons (with the same energy as the bandgap), a large number of holes must also be present in the valence band.

To achieve this, both the n and p materials are very heavily doped (indicated by the "+" superscripts in Figure 7.20), and a large forward bias is applied, which injects large concentrations of electrons and holes across the junction. This creates a population inversion in a narrow region on either side of the junction called the "active region", which then contains a large number of electrons in the conduction band and a large concentration of holes in the valence band. (The doping levels on both sides of the junction must be high enough that the Fermi level is below the top of the valence band on the p-side, and within the conduction band on the n-side.)

In a normal p-n junction, these carriers would be swept across the junction, where they would become minority carriers and recombine with the majority carriers to make up the current flow. In a direct bandgap semiconductor like those used for diode lasers and LEDs, however, the electrons and holes have a high probability of triggering photon emission when they recombine. These photons can either be absorbed by valence electrons or prompt stimulated emission from electrons in the conduction band. When the injected carrier concentration reaches a certain value, the stimulated emission exceeds the absorption, and so "optical gain" (in other words, light amplification) occurs in the active region.

Once stimulated emission is occurring, some of the emitted photons travelling along the axis of the system are reflected by the mirrored ends, and every time a photon is reflected it can pass back through the active region, producing yet more stimulated emission. This process is known as positive optical feedback. However, as the ends cannot reflect all the incident light, some of it escapes out from the device forming the laser beam. The emitted photons spread out a bit from the active region and so emerge from an "emitting volume", as shown in Figure 7.20. In this case the emitting volume is asymmetric about the junction because electrons diffuse further into the p-type GaAs—before recombining with the majority carriers there and so emitting photons—than the holes that diffuse into the n-side.

Diode lasers, just like other types of laser, emit a beam of light that is monochromatic (has only one wavelength). This is because the light they emit is characteristic of their bandgap energy. However, in contrast to other forms of laser, diode lasers are small—typically around 0.1 mm long—enabling them to be used as light sources for optical fibre communication, as well as in devices such as CD players/recorders and laser printers.

When a diode laser is made from just one semiconductor material, it is known as a homojunction laser, and when it is made from two or more different semiconductors, it is termed a heterojunction. Heterojunction lasers are made using epitaxial growth techniques (see Section 7.3.2) and consist of a thin layer of semiconductor (for example gallium arsenide), sandwiched between layers of a different semiconductor (such as aluminium gallium arsenide).

Since it is impossible for silicon to lase because it is an indirect gap semiconductor (see Section 7.2.1), diode lasers tend to be made from gallium arsenide and its alloys (called III-V compounds because of the groups in the periodic table from which they come). For example, the diode lasers used by the telecommunications industry to send data—in the form of pulses of light—through optical fibres are generally made from indium phosphide/indium gallium arsenide phosphide (InP/InGaAsP).

7.2.4 SOLAR CELLS

In Section 7.2.2 we saw that a forward-biased p-n junction can emit light. The opposite process can also occur, in other words, applying a reverse bias to p-n junction enables it to detect photons falling on it. Devices based on this idea are known as photodetectors (or photodiodes) and can only detect photons with energies greater than the bandgap of the material they are made from. This is because, as Section 7.2.1 revealed, photons can generate electron-hole pairs in a semiconductor if they have energies greater than E_g but less than the work function of the semiconductor. Since every semiconducting material has a particular value of E_g, it is immediately clear that a given semiconductor will only be able to detect light of certain wavelengths. It so happens that silicon is one of the best materials for detecting sunlight—as its

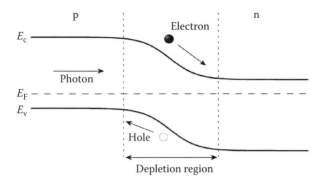

FIGURE 7.21 The production of a photocurrent in a solar cell. A solar photon generates an electron-hole pair, which is separated by the electric field generated by the wide depletion region. Once the charge carriers have been swept across the junction, they can form an electric current in an external circuit. All photodiodes work in this way. Solar cells are simply a type of photodiode optimized to work with radiation of the wavelengths the sun emits.

bandgap corresponds to infrared wavelengths present in sunlight—and so Si forms the basis for most solar cells. (Unlike LEDs, photodetectors can be made from either direct gap or indirect gap materials—like silicon.)

The basic operation of a solar cell is very simple. Photons from the incident sunlight fall onto the p-n junction and produce electron-hole pairs. Reverse-biasing the junction (see Section 7.1) creates a wide depletion region, and any electron-hole pairs that are formed in the depletion region, or near enough that they can diffuse into it, are quickly separated by the electric field generated by the depletion region. So electrons created on the p-side are swept towards an electrical contact to an external circuit at the end of the n-side, while holes drift from n to p and make their way to the contact on the p-side (see Figure 7.21). A current known as the "photocurrent" then flows in the external circuit, with a value proportional to the intensity of the incident radiation.

Solar cells are not particularly efficient. In most cases they convert only approximately 20% of the sun's energy falling on them into electricity. So to produce a useful level of power output, lots of individual solar cells are joined together to form a solar panel. Figure 7.22 shows a low-cost solar-powered garden light. The solar panel can be seen in the centre of the unit.

To provide larger levels of useful domestic electricity, solar panels can be fixed to the rooftops of houses, but the amounts produced will of course depend on the location of the house and hence how much sun it receives. Experimental solar-powered cars, which regularly feature in student engineering competitions, can also grind to a halt when the sun goes in, and so are not viable as a sole option for automotive power. By contrast, solar panels are a perfect choice for powering electronic equipment onboard satellites and spacecraft. In fact they are the primary source of power in most spacecraft, and any excess energy they produce is stored in rechargeable batteries. A special type of glass cover is used to protect these solar cells from micrometeorites, and high-energy particles that can create defects in the p-n junction materials that reduce the power output of the cell.

FIGURE 7.22 Solar-powered garden lights like the one shown here are a low-cost and straightforward option for domestic use.

Germany is currently leading the way in the widespread implementation of solar technology, with many homes and small businesses having rooftop solar panels that can feed the electricity they generate into the grid. By the end of 2007, all the solar panels in Germany combined together were generating approximately as much energy as three large conventional power stations.

7.2.5 MOS CAPACITOR

A variation on the MOSFET, known as the MOS capacitor, forms the sensing element of the charge-coupled device (CCD), which captures images in digital cameras. As Figure 7.23 reveals, a MOS capacitor consists of a metal or polysilicon electrode (gate) on top of a dielectric layer of silicon dioxide (SiO_2) that rests on a p-type silicon substrate.

When light falls onto the device, the energy from the photons creates electron-hole pairs within the p-type silicon. If a positive voltage is applied to the gate, the

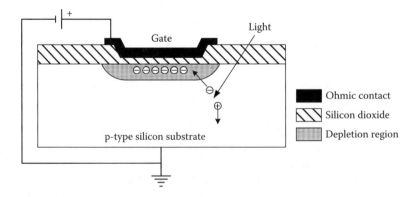

FIGURE 7.23 A MOS capacitor.

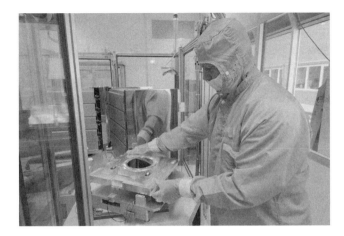

FIGURE 7.24 This employee at the Philips Microsystems Plaza at the High Tech Campus in Eindhoven in the Netherlands is wearing an antistatic suit while handling chips. (© Philips.)

electrons from these pairs will be attracted to the depletion region that forms immediately below the gate, where they will stay as a "surface charge" all the while the gate voltage remains positive. The amount of surface charge is proportional to the amount of light falling on the device.

7.3   DEVICE MANUFACTURE

In this section we will look at some of the main manufacturing processes used in the production of transistors, integrated circuits, and other semiconductor devices. One thing every device has in common is the need to be made from semiconductors, which—with the exception of impurities deliberately added to produce particular electrical characteristics—must be as free from defects and impurities as possible. This is partly achieved by manufacturing them in "clean rooms". These are rooms that have the air inside them constantly filtered to remove as much dust as possible. People enter clean rooms through an air lock, and must be dressed in an antistatic suit that stops any of their hair, or flakes of skin, from falling onto the circuits (see Figure 7.24).

There are several different standards of cleanliness that can be adopted, depending on the process being carried out. For example, a "class 100" clean room is specified as one that contains not more than 100 0.5-μm-diameter (or larger) particles of dust per cubic foot, which is approximately 3500 dust particles per cubic metre. This might seem a bit extreme, but even one speck of dust can spell disaster. A dust particle that becomes incorporated into the tiny oxide layer forming part of the gate of a MOSFET, for example, can increase the conductivity of the gate region and cause the device to fail. (There are various different notations used around the world to denote clean-room standards. For example, one other way for a class 100 clean room to be described is ISO 5.)

The semiconductors that form the substrates of devices not only need to be as pure as possible, they also need to be single crystals (see Section 2.2.6). Methods of growing high-purity single crystals of semiconductors are discussed in Section 7.3.1, while later subsections describe methods for creating the various layers that make up the operational part of semiconductor devices. (Some devices can be fabricated from amorphous semiconductors, especially amorphous silicon, but such devices are inevitably less efficient than corresponding single-crystal devices.)

Although this chapter is concerned with silicon-based devices, it is possible that plastic integrated circuits with diodes and transistors made out of polymers such as polyacetylene will replace conventional ICs in the future for some applications (involving only low power that causes little temperature rise) and enable entirely new uses for electronics. At the time of writing this book, potential commercial applications of plastic electronics range from electronic paper and wearable computers to chemical sensors. Although plastic chips may never be able to process information as fast as their silicon counterparts, one of their main attractions is that they should not require the sorts of complex fabrication processes described in this section. Instead they could be produced by simple, quick, low-cost methods such as a variation on standard ink-jet printing, which squirts layer upon layer of the polymers onto a substrate in the patterns needed to build up circuit components.

7.3.1 CRYSTAL GROWTH

Although they sometimes occur in nature, most single crystals (see Section 2.2.6) are artificially created for use in the semiconductor industry as the substrate for semiconductor devices to be built up on. For electronic devices and integrated circuits to work properly, these crystals need to be as pure as possible. In fact, there is often not more than one "electrically active" impurity atom for every 10^6 atoms in these semiconductors. This is a concentration about 100 times less than there can be of impurities like carbon and oxygen, which do not affect the electronic properties of silicon or germanium.

Once a high-purity single crystal has been obtained, it is sliced up into wafers. Each wafer is polished and then cleaned to give it a surface perfect enough to have a device created on top of it using methods such as epitaxy (discussed in Section 7.3.2) and deposition (covered in Section 7.3.3). The wafers are, however, much too big to form the substrate of any device, and so are cut up into much smaller pieces known as "chips" before device fabrication occurs. (Any of the three terms "wafer", "substrate", or "chip" is regularly used to describe the piece of semiconductor in which or on which the device itself is fabricated.)

The following three techniques are the most widely used for growing "perfect" semiconductor crystals.

Czochralski Technique

This involves growing a large single crystal from its molten form with the help of a "seed" that prompts crystallisation. The seed is a small single crystal of the same semiconductor, and one end of it is dipped into the polycrystalline molten material (usually referred to as "the melt"), which is held just above its melting point in a crucible often made of silica (SiO_2). Although the dipped-in end of the seed melts, the other end of it—which is attached to a rod—remains solid. The rod is then slowly

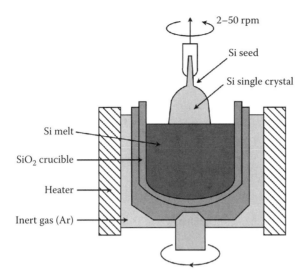

FIGURE 7.25 The apparatus used for the Czochralski method of crystal growth. Around 90% of the silicon crystals used by the semiconductor industry are grown in this way. Gallium arsenide can also be grown using the Czochralski method, but is generally produced by the Bridgman technique, which is discussed in the next subsection. (This diagram originally appeared as Figure 1.9, p. 16, of *The Physics and Chemistry of Solids*, by S. R. Elliott, © 1998 by John Wiley & Sons Limited. Reproduced with permission.)

pulled away from the melt, usually at a rate of around a few millimetres per minute. This allows the part of the melt in immediate contact with the seed crystal to cool enough for it to crystallise on the seed. This new crystal grows into a longer and longer single crystal as the rod is pulled further and further from the melt, with crystallisation continuously occurring at the solid–liquid interface.

Figure 7.25 shows a typical setup for the Czochralski (CZ) technique. In this case the rod and the crucible containing the melt are rotated in opposite directions to one another. This helps maintain a constant temperature at the interface between the solid crystal and liquid melt, and keeps compounds properly mixed. Rotating the seed also gives the crystal a circular cross-section. The resulting single crystal has the same crystallographic orientation as the seed crystal, and also has a cylindrical shape. As crystal diameters of 30 cm are not uncommon, this is a perfectly satisfactory shape for cutting up and then using for the manufacture of electronic devices. (In fact the single-crystal "boules" produced by the CZ technique can be almost 100 cm long, and Si boules weighing up to 100 kg can be made.)

The Czochralski technique can be carried out in an inert-gas atmosphere to reduce the likelihood of the growing crystal becoming contaminated with impurities, and the chance of any volatile constituents escaping from the melt. Alternatively, a layer of liquid can be used to contain—and so prevent decomposition of—molten compounds such as gallium arsenide and indium phosphide. The latter technique is known as liquid encapsulated Czochralski (LEC).

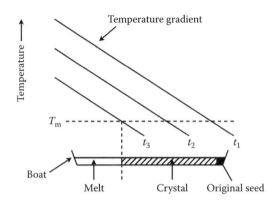

FIGURE 7.26   The Bridgman technique, where T_m is the temperature at which crystallisation occurs, and $t_1 > t_2 > t_3$. The melt crystallises as the boat temperature is slowly lowered starting from the end containing the seed. (This diagram originally appeared as Figure 1.10(a), p. 17, of *The Physics and Chemistry of Solids*, by S. R. Elliott, © 1998 by John Wiley & Sons Limited. Reproduced with permission.)

Bridgman Technique

Like the Czochralski technique, the Bridgman technique uses a seed crystal to grow a single crystal from a molten form of the material. In this case, however, a drop in temperature is used to prompt crystal growth, as illustrated in Figure 7.26. Other variants on this growth method include moving the furnace that heats the melt slowly away from the seed crystal along the length of the "boat"-shaped crucible that the melt sits in, or alternatively pushing the boat into a furnace and then slowly pulling it out.

In all cases the temperature at the interface between the growing crystal (initially the seed) and the melt—indicated by T_m in Figure 7.26—drops to the solidification point of the melt. The melt then crystallises—firstly onto the seed crystal, then onto the growing crystal. Eventually a large single crystal is formed that fills—and so has the same shape as—the boat containing it.

Floating Zone Method

Rather than growing a crystal from a melt, the float zone process produces a single crystal from a polycrystalline rod or "ingot". This rod is held vertically and has a seed crystal at its lower end. A heater is gradually moved up the rod (or alternatively the rod is moved relative to the heater) starting from just above the seed crystal. This heats the rod to above its melting point, and as the heater moves away from that area or "zone" of the rod up to the next zone, the zone that has just been heated cools down and crystallises into part of a growing single crystal. Figure 7.27 shows how the floating zone method works.

This method produces purer crystals than the Czochralski process because the molten material does not have to be contained in a crucible, which can contaminate it. In order to reduce contamination from any other sources, the rod and heater are enclosed in a small chamber that contains the inert gas argon. In fact, the technique

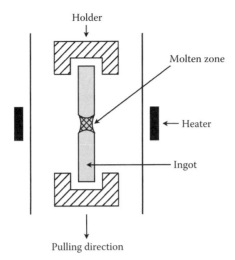

Holder

Molten zone

Heater

Ingot

Pulling direction

FIGURE 7.27 This schematic diagram shows the float zone process in action. (This diagram originally appeared as Figure 1.10(d), p. 17, of *The Physics and Chemistry of Solids*, by S. R. Elliott, © 1998 by John Wiley & Sons Limited. Reproduced with permission.)

can also be used to increase the purity of existing crystals. This is because impurities within crystals form a "solid solution" but dissolve out of this solution—in other words, move away from their positions amongst the host atoms—when the host material is molten. As the heater moves up through the crystal, the molten zone it produces moves upwards too, taking the impurities along with it. The end of the crystal containing most of the impurities is then cut off to leave behind a crystal with a much higher purity.

7.3.2 EPITAXIAL GROWTH METHODS

As we saw in Sections 7.1 and 7.2, most semiconductor devices are composed of layers of different materials (heterostructures) or layers of the same material that have different levels of doping. One way of making these layered structures is to use *epitaxy* or "epitaxial growth", which is a method of growing thin layers of crystal. (The term "epitaxy" is formed from the Greek words "epi", meaning "upon", and "taxis", meaning "arrangement".)

There are several different epitaxial growth techniques, and the three methods used most frequently are described below. In every case, there is a substrate wafer that acts as a seed crystal for the subsequent layers, which attempt to take on the same crystal structure as the substrate. (The substrate wafer will in turn have been cut from a single crystal created by one of the processes detailed in Section 7.3.1.)

Ideally heterostructures should consist of semiconductors with the same crystal structure that have lattice constants less than 0.1% different. However, in reality, some technologically important structures do not have such a good match between their lattice constants, which leads to strain and sometimes to the formation of dislocations. For example, Figure 3.9 shows a GaSb film grown by molecular beam epitaxy on a GaAs substrate. In this case the lattice mismatch between the film and the substrate has caused dislocations to form, which then act as a growth site for new atomic planes and so create spiral structures.

Vapour Phase Epitaxy

This is the most commonly used method of epitaxy for silicon devices, and involves depositing atoms from a gas onto a substrate. The process takes place inside a "deposition furnace", which contains a heater that keeps the substrate just below its solidification temperature. As the atoms from the gas reach the surface they become adsorbed (attached to the surface). Then as we saw in Section 3.1.2, because they will condense much more readily at sites where there are nearest neighbours present, the atoms move across the surface to take up positions that allow the underlying crystal structure to be preserved. Dopants can also be introduced in vapour form.

Liquid Phase Epitaxy

Epitaxial layers can also be grown from molten materials by placing them in contact with a substrate and so allowing atoms from the melt to solidify on the surface of the substrate. This technique is known as liquid phase epitaxy (LPE), and is used for the growth of devices such as long-wavelength photodetectors (see Section 7.2.4) and AlGaAs for heterojunction diode lasers (see Section 7.2.3). The most common setup for the technique involves the substrate being contained in a small "boat", while the melts are housed in wells in a graphite block. To enable epitaxial growth, the substrate is slid under the surface of either one melt, or a succession of melts if additional layers are to be grown. As with VPE, dopants can be introduced via LPE.

Molecular Beam Epitaxy

Molecular beam epitaxy (MBE) is a very precise technique, and allows the growth of almost every type of semiconductor. It is carried out under ultrahigh vacuum (UHV) conditions and can be used to control chemical compositions, doping profiles, and layer thickness to such an extent that it can be used to make single-crystal multilayer structures that have layers just a few atoms thick. It can also facilitate the manufacture of quantum dots (see Box 4.1), and is often used to deposit the semiconductor layers of diode lasers (see Section 7.2.3).

Figure 7.28 is a schematic diagram of a typical MBE system that enables hot beams of atoms or molecules to be deposited onto a heated crystalline substrate. Alternatively, a single beam can be directed at the substrate, allowing a single type of atom or molecule to be deposited. Rotating the substrate helps improve the uniformity of the epitaxial layers produced.

Each element to be deposited is heated in a separate oven. The resulting vapour is turned into a molecular beam by allowing it to escape from the oven through a small hole. A special type of entrance known as a "load lock" allows any of the constituents to be changed to another type whilst UHV is maintained, helping to make this a versatile technique. For example, a dopant can be added to a particular layer of a device, then a swap made to another type of dopant when a subsequent layer with different electrical properties is required. The doping profile is controlled by adjusting the dopant beam flux relative to the silicon or gallium flux of atoms.

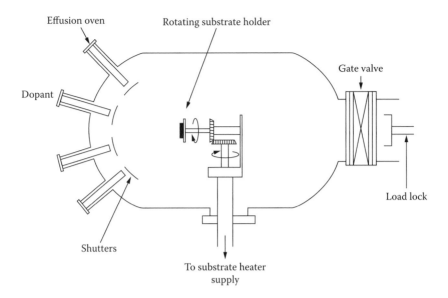

FIGURE 7.28   The arrangement of effusion ovens and substrate in a typical MBE system.

A quantum dot—a tiny group of semiconductor atoms buried inside another type of semiconductor material—can be made by directing a beam of the atoms to be incorporated into the dot at a substrate of the host material. Once "dots" have formed on the substrate surface, the beams are changed to atoms of the host semiconductor, which then fills all the spaces between the dots, and covers over the top of them.

7.3.3 Deposition

Semiconductor devices and integrated circuits can contain all or some of four main types of thin films: metal films, polycrystalline silicon, thermal oxides, and dielectric layers. In general, all types of thin films are created using various "deposition" techniques, which means—quite literally—that the atoms making up the thin film are deposited onto either the substrate or a previously created layer. Although the methods vary slightly, they are all based on the following main technique.

Chemical Vapour Deposition

This technique is basically the same as the vapour phase epitaxy (VPE) technique described in Section 7.3.2. The source materials are in the form of a gas, and they react on the surface of the substrate to form a thin layer. The quality of the thin films produced by chemical vapour deposition (CVD) varies depending on the type of source material, the temperature of the substrate, and the pressure of the gases inside the tube in which the process takes place. CVD-grown semiconductor layers are generally polycrystalline at high temperatures and amorphous at lower temperatures, while the dielectric layers produced are amorphous and the metal layers tend to be polycrystalline. Careful adjustment of the conditions allows the VPE process to take place, resulting in layers with the same crystal structure as the substrate.

BOX 7.1   GROWING WHISKERS THAT WON'T NEED SHAVING

As manufacturers try to incorporate more and more functions into electronic gadgets like mobile phones and laptop computers, and at the same time decrease their size, the need for smaller electronic circuit components increases. One contender in the race to develop the next generation of circuit components is Prof. Lars Samuelson from Lund University in Sweden. His team is building minute electronic devices within wires just 20 billionths of a metre wide, then studying how well they operate.

Wires this small are known as nanowires, or nanowhiskers. Scientists have been able to grow semiconductor nanowires for several years, but until recently had no way of mixing different materials together within one wire. However, Prof. Samuelson's team was one of three groups who, early in 2002, simultaneously reported the ability to grow single nanowires made from layers of different semiconductors (see Plate 7.1). Since conventional electronic devices are made from layers of different types of semiconductor, this breakthrough in growing methods meant that tiny electronic devices could potentially be made as an integral part of a nanowire.

A few months later, the Swedish scientists took this work a step further. They combined the semiconductor materials indium arsenide and indium phosphide in layers of particular thicknesses to produce a commonly used electronic device known as a double-barrier resonant tunnelling device, and then investigated its electronic properties.

The team's nanowire devices are made using a variation of a standard technique for forming semiconductor devices known as chemical beam epitaxy. A substrate for the nanowire to grow on is put inside an ultra-high vacuum chamber. This substrate is heated up, and small particles of carefully controlled size that contain gold atoms are placed on its surface. The semiconductor materials that the wire will be made from are then injected into the chamber in the form of gases. These gases react with the gold atoms and condense into a solid crystalline structure. Layers are formed by introducing each gas separately and in turn; the more gas let in at any one time, the thicker that particular layer will be.

The first solid layer forms between the gold atoms and the substrate, while each subsequent layer builds up between the substrate and the last layer grown. This means the gold atoms end up sitting at the end of the nanowire that is furthest from the substrate surface. Growing the nanowires in this way allows the atoms of each layer to line up perfectly with those of the layer below, creating an ideal heterostructure (composite material made from layers of different types of semiconductor) as Plate 7.2 reveals.

As well as enabling tiny versions of existing electronic devices to be manufactured, entirely new types of devices could be developed using this technology. This is because the nanowires are so small that they only allow electrons—which produce electric currents in metals and semiconductors by moving steadily through them—to move in just one direction. "In general, all heterostructure devices have their specific functions because we can design the 'landscape' in which electrons can move. The one-dimensionality of nanowire devices offers many more exciting opportunities, such as creating ultra-small light sources for the safe transfer of information via quantum cryptography," said Prof. Samuelson shortly after his team produced their first nanowire device. Nanowire devices could also be used in displays and solar cells.

30 nm

PLATE 7.1 This transmission electron microscope image shows a nanowire made from layers of indium arsenide (light grey) and indium phosphide (dark grey). The gold atoms that induce the nanowire's growth can be seen capping the top of it. New types of electronic devices, and miniaturised versions of existing electronic components, could be made using the same technique that produced this nanowire. (Image taken by Torsten Sass and Reine Wallenberg. Courtesy of Lars Samuelson, Lund University, Sweden. Plates 7.1 and 7.2 reprinted (in part) with permission from M. T. Bjork, et al. "One-dimensional Steeplechase for Electrons Realized", *Nano Letters*, Feb 1, 2002, Vol 2 issue 2. Copyright 2002 American Chemical Society.)

(This box is based on a press release the author wrote for the Institute of Physics to accompany the 26th International Conference on the Physics of Semiconductors in 2002.)

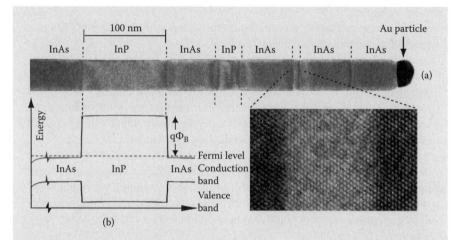

PLATE 7.2 The transmission electron microscope image (a) shows a nanowire made from layers of indium arsenide (InAs) and indium phosphide (InP), with the gold (Au) atoms that induced its growth capping the end. The magnified image (below right) shows one of the InP layers in more detail. The regular crystalline arrangement of the atoms can be seen, as can the abrupt changeover from one region to the next. Unlike conventional growth techniques, this method allows materials with very different lattice constants to be sandwiched together without the formation of defects or changes in the crystalline structure.

Diagram (b) shows the difference in bandgap width between InAs and InP. The increase in height of the conduction band means each region of InP acts as a potential barrier to electrons trying to move along the nanowire. The electrons do not have enough energy to cross this barrier in a classical way, but they can "tunnel" through it. Tunnelling is a quantum mechanical effect that occurs when the wave associated with a particle—which represents the probability of finding the particle in a particular place—spreads across a potential barrier to the other side. So if the wave associated with an electron moving along the nanowire extends across an InP barrier as the electron approaches it, there is a small but finite chance that this electron will appear on the far side of the barrier. (TEM images taken by Torsten Sass and Reine Wallenberg. Courtesy of Lars Samuelson, Lund University, Sweden.)

Different types of layers require different deposition rates and reaction times for best results, so the basic CVD method is altered accordingly. Some common variations include plasma enhanced or "plasma assisted" chemical vapour deposition (PECVD or "PCVD") and low-pressure chemical vapour deposition (LPCVD). Figure 2.34 shows polycrystalline diamond films that have been grown using the plasma enhanced CVD process.

7.3.4 DOPING SEMICONDUCTORS

Earlier in this chapter (and in Chapter 6) we saw that the semiconductor industry is effectively based on the fact that semiconductors can be deliberately doped with impurity atoms to give them the required electronic properties. For example, adding boron to silicon makes it p-type, while a dopant of selenium will make gallium arsenide n-type. In the previous section we also saw that dopants can be added during MBE. Alternatively they can be introduced into a crystal melt, or added at the same time as the silicon tetrachloride ($SiCl_4$) gas that produces layers of silicon in the VPE process.

Interestingly, the same type of dopant atom can behave differently depending on which of the different epitaxial processes is used. For example, in VPE growth of GaAs, the deposited layer of GaAs dissociates (breaks down into its component ions) unless there is an overpressure of arsenic. That means that during the GaAs growth there is formation of Ga vacancies because there is more arsenic present. If silicon is the dopant element, the silicon atoms go onto the Ga vacancy sites, and therefore the silicon acts as a donor, as it has one more electron than gallium.

The same situation applies during MBE growth of GaAs. Again the growth process at the GaAs surface requires an excess of arsenic, and so a greater flux of arsenic atoms compared with gallium atoms is provided. Therefore, Ga vacancies are formed and the silicon atoms go into these sites, where they act as donors.

By contrast, in LPE growth of GaAs the starting GaAs material is dissolved in molten gallium. The excess of gallium leads to the formation of arsenic vacancies. Because in this case dopant silicon atoms will go into arsenic sites, they act as acceptors. (The excess Ga and As vacancies produced by these methods are not stable at any of the GaAs growth temperatures and so are not part of the final epitaxial layers.) Referring back to Figure 6.26 will reveal the different energy levels caused by silicon acting as a donor (uppermost Si level) and as an acceptor in GaAs.

Despite the ability of these methods to add dopants, the precisely controlled amounts required for devices are usually incorporated into solid semiconductors using one of the following methods instead.

Diffusion

Dopant atoms can diffuse into intrinsic semiconductors if the semiconductors are heated in a gas of the dopant atoms. The diffusion begins after dopant atoms condense on the surface of the semiconductor. The number of atoms entering the material and the depth they diffuse to depends on both the temperature and the concentration gradient inside the specimen. A high temperature (different for each type of impurity, but approximately in the range of 800°C–1200°C for silicon and 600°C–1000°C for gallium arsenide) is required for diffusion into the semiconductor to take place.

In addition, once a concentration gradient exists—in this case, as the diffusion process begins there will be a greater dopant concentration just below the surface than deeper inside the intrinsic semiconductor—diffusion will occur from regions of high dopant concentration into areas with fewer dopant atoms. This type of diffusion stops when a uniform concentration of the dopant atoms is reached, while lowering the temperature halts diffusion into the semiconductor.

Ion Implantation

This technique consists of firing high-energy ionized dopant atoms at the surface of a semiconductor so that they are forced inside it. The depth these impurities reach

FIGURE 7.29 This MEMS micro chain was made in 2002 at the Sandia National Laboratories in the United States. It was designed to rotate several drive shafts simultaneously, and the centre of each link of the chain is only 50 μm (which is around 70% of the width of a human hair) from the centre of either of its neighbouring links. (© Sandia National Laboratories.)

is controlled very accurately, as it is proportional to the electric potential used to accelerate the ions ready for firing. As the energies involved in ion implantation are high enough to cause lattice atoms to be displaced into interstitial sites (see Section 3.1.1), the process is usually followed by thermal annealing treatment (see Section 4.2.2). In this case annealing removes the damage to the lattice by providing enough energy for the lattice atoms to move back to their correct positions.

7.3.5 MEMS

Many of the processing steps used to make semiconductor devices (including deposition and ion implantation) are also used to manufacture microelectromechanical systems (MEMS). These are miniature mechanical components (see Figure 7.29) integrated in with the electronics on a silicon substrate. MEMS components include temperature sensors, motors, and gears, and have a wide range of potential uses.

At the time of writing this book, one of the most common uses of MEMS is in vehicle air bags, where they act as a detection system for rapid deceleration. Each individual sensor is a flat piece of circuitry that can sense deceleration in two mutually perpendicular directions, but not the third direction. So that deceleration can be detected along the three axes directions of up and down, left and right, and forwards and backwards, there are normally two sensors mounted at right angles to each other within each air bag.

7.4 DIELECTRICS

7.4.1 INTRODUCTION TO DIELECTRICS

As anyone affected by the blackouts in New York, Detroit, Toronto, Ottawa, and London in August 2003 will no doubt testify, power cuts cause major disruption.

Whilst there are many different causes, one problem that can lead to power failures is the breakdown of the plastic insulation surrounding the conducting wires of a power cable.

This insulation is, of course, essential. Without insulators we would not be shielded from the harmful currents carried by any electrical cables—from the massive power lines of electrical grids to the much smaller wires that supply electricity to household devices like kettles and televisions. Meanwhile, semiconductor devices and integrated circuits would be useless without the insulating materials in between the conducting channels that stop them from touching one another and so prevent short-circuiting. Many devices also contain dielectric layers that provide electrical insulation between conducting layers. In addition, dielectrics are used as masks during doping (see Section 7.3.4) and for passivation. Passivation is the protection of electronic components and circuits from moisture, scratching, and impurities. The surfaces of most silicon chips, for example, will be covered with a layer of silicon dioxide (SiO_2) or silicon nitride (SiN), which helps protect them from harmful environmental conditions such as a humid atmosphere.

Physicists often refer to insulators as "dielectrics", although strictly speaking a *dielectric* is in fact an insulator that becomes "polarised" in the presence of an electric field. This means its atoms either acquire an electric dipole (see Section 2.1.5), or if they already have one, tend to align themselves with the field. Before looking at the various ways in which solids can become polarised, we will first discuss electric dipoles in more detail.

Electric Dipoles

Whilst the electrical behaviour of conductors and semiconductors is produced by moving electric charges, the electric properties of dielectrics depend on static electric charges. Whenever a dielectric is in an electric field, the positively charged nucleus and negatively charged electrons move relative to one another, giving each atom an electric dipole moment as illustrated in Figure 7.30. As we saw in Chapter 2 when we were looking at van der Waals bonding, any dipole consists of two equal and opposite charges separated by a small distance.

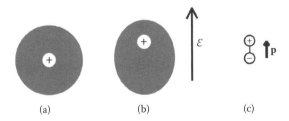

(a)   (b)   (c)

FIGURE 7.30   When an atom is placed in an external electric field, as shown in (b), the negatively charged cloud of electrons moves relative to the positively charged nucleus that it surrounds. By contrast, (a) shows the relative positions of the electrons and nucleus when there is no applied electric field. In reality, the amount of displacement is very small. It has been exaggerated in this diagram to illustrate the process more clearly. In (c) the vector **p** is shown alongside a schematic representation of an electric dipole.

The electric dipole moment, **p**, is a vector pointing from the negative charge towards the positive charge as shown in Figure 7.30(c), and is defined as the product of one of the electric charges and the distance between them, that is,

$$\mathbf{p} = q\,\mathbf{s} \tag{7.1}$$

where q is the electric charge and **s** the distance between the two charges. (The SI unit of **p** is the coulomb metre, C m.)

EXAMPLE QUESTION 7.4   DIPOLE MOMENTS

What is the magnitude of the dipole moment of a pair of point charges 5 nm apart from one another if one charge has the value +3 pC and the other −3 pC?

ANSWER

The dipole moment $\mathbf{p} = q\,\mathbf{s}$, where q is the charge and **s** the distance between them, so

$$\mathbf{p} = 3 \times 10^{-15} \times 5 \times 10^{-9} = 1.5 \times 10^{-23} \text{ C m}$$

Dipole moments can be much smaller than this in real life. The dipole moment of the water molecule, for example, is 6.2×10^{-30} C m, and this is considered to be relatively large. Many of us make regular use of the electric dipole moments of water molecules, as it is their ability to vibrate in an oscillating electric field that allows microwave ovens to cook food (see Figure 7.31).

Polarisation

For any dielectric material, the polarisation **P** is the dipole moment per unit volume and, if all the dipoles are assumed to be the same, is given by

$$\mathbf{P} = n\,\mathbf{p} \tag{7.2}$$

where n is the number of atoms in a unit volume of the material.

There are three different ways in which materials can become polarised: electronic polarisation, ionic polarisation, and orientation polarisation.

Electronic polarisation is the process already illustrated in Figure 7.30. In the absence of an electric field, the electrons and nucleus of an individual atom are arranged in such a way that there is no electric dipole moment. However, if the atom is placed in an electric field, the cloud of electrons is distorted by the field and so moves relative to the nucleus, producing a dipole moment as shown in Figure 7.30(b). Electronic polarisation occurs in all types of dielectric materials when they are in an electric field.

By contrast, ionic polarisation can only occur in ionic materials, that is, in materials composed of ions instead of neutral atoms. When ionic materials (including ionic

FIGURE 7.31 The microwaves produced by microwave ovens subject anything they pass through to a rapidly oscillating electric field. All foods contain water, and the electric dipoles of water molecules will always try to line up with an external electric field, and so will vibrate as they attempt to align themselves with the electric fields of microwaves, which reverse direction about 10^9 times per second. The heat created by the vibrating water molecules spreads throughout the food by thermal conduction and so cooks it.

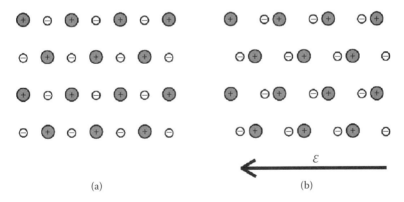

FIGURE 7.32 In the absence of an electric field, the ions of an ionic solid are arranged in such a way that their charges cancel each other out. This results in an electrically neutral solid overall as shown in (a). When an ionic solid is placed in an electric field, however, the negatively charged anions move towards the positive region of the field and the positively charged cations move towards the negative part, creating dipole pairs as shown in (b).

solids) are subjected to an electric field, the negatively charged anions will move in the opposite direction to the field direction. Meanwhile, the positively charged cations move in the field direction, so in the opposite direction to the anions. This shifting of the position of the ions turns each pair of ions into a dipole pair, as shown in Figure 7.32, and so polarises the ionic solid.

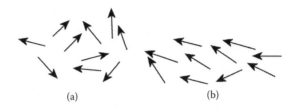

(a)   (b)

FIGURE 7.33 Without an electric field present, a solid containing permanent dipole moments will be electrically neutral overall, as each dipole is randomly oriented with respect to its neighbours, as shown in (a). When an electric field is applied, however, the dipoles tend to line themselves up with the field, and the solid acquires a net dipole moment, as shown in (b).

The third type of polarisation, orientation polarisation, occurs in materials that are made up from molecules known as "polar molecules" that have a permanent dipole moment because there is charge asymmetry within the molecule. In other words, the charges in the molecule are distributed so that the centres of positive and negative charge are separated.

Dielectric solids with permanent dipole moments on the atomic scale include PVC (polyvinyl chloride), amorphous polymers, and some ceramics. When these solids are placed in an electric field, there is a tendency for their dipole moments to line up in the direction of the field, with the positive end of each dipole pointing towards the negative region of the field. The solid therefore acquires a dipole moment as a whole and so becomes polarised. (In fact this tendency for the dipole to rotate and align itself with the field is counteracted by the randomising effect of temperature, and so the degree of alignment is determined by the classical Boltzmann distribution [see Section E2 of Appendix E for a reminder of classical statistical mechanics].)

Without a field present, there is no overall dipole moment, as each dipole is randomly oriented with respect to its neighbours, so over the whole solid the dipoles cancel each other out. Figure 7.33 shows the dipoles in a solid with a permanent dipole moment with and without an applied electric field.

As we saw earlier, electronic polarisation occurs in all types of dielectric materials when they are in an electric field, and while some dielectric materials only exhibit electronic polarisation, others show ionic and/or orientation polarisation as well. In every case, the contributions from the different mechanisms add together as follows:

$$P = P_e + P_i + P_o \qquad (7.3)$$

where P is the total polarisation, P_e the electronic polarisation, P_i the ionic polarisation, and P_o the orientation polarisation.

EXAMPLE QUESTION 7.5   POLARISATION

What types of polarisation contribute to the total polarisation of a diamond crystal?

Answer on page 260.

Sometimes polarisation can be localised. For example, a polarisable surface is a surface whose atoms or molecules are likely to have a temporary dipole induced in them by the nearby dipole of another atom or molecule. Such surfaces tend to then attract atoms and molecules via the van der Waals force. It is this method of electrostatic attraction that allows geckos to hold on to ceilings (see Box 2.1).

7.4.2 FERROELECTRICITY

Some types of dielectric materials have an electric dipole moment when they are below a certain critical temperature even when they are not in an electric field. These materials are known as *ferroelectrics*. If the temperature is raised above this critical temperature, which is known as the Curie temperature, T_C, the material loses its electric dipole moment unless it happens to be in an external electric field. In other words, below the Curie temperature the material is ferroelectric, whereas above the Curie temperature it is an ordinary dielectric. The Curie temperature for several different ferroelectric materials is shown in Table 7.1.

When crystals are in the ferroelectric state, the lattice ions actually move slightly from their normal positions so that the electric charges in the crystal no longer cancel each other out. This allows the solid to acquire a dipole moment. Figure 7.34 shows

TABLE 7.1
The Curie Temperature for a Range of Ferroelectric Crystals

Ferroelectric Crystal	T_c (K)
Potassium dihydrogen phosphate (KH_2PO_4)	123
Barium titanate ($BaTiO_3$)	408
Potassium niobate ($KNbO_3$)	708
Lithium niobate ($LiNbO_3$)	1480

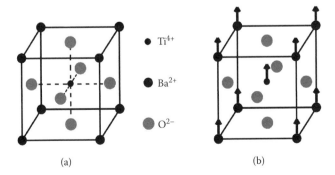

(a)   (b)

FIGURE 7.34   The crystal structure of the ferroelectric crystal barium titanate (a) above and (b) below the Curie temperature.

ANSWER TO QUESTION 7.5 ON POLARIZATION

The electronic polarization is the only contribution to the total polarization, because diamond (Figure 2.25) is neither an ionic solid nor a solid with a permanent dipole (as all the atoms are the same and so there are not positive and negative ions present).

how the crystal structure of the ferroelectric material barium titanate changes when it is below the Curie temperature.

As can be seen from the figure, below the Curie temperature both the Ti^{4+} and the Ba^{2+} ions (at the centre and corners of the unit cell, respectively) move upwards relative to the c-axis of the unit cell. At the same time, the O^{2-} ions can move slightly downwards as well. This movement causes the unit cell to elongate slightly in the c direction, so whilst barium titanate has a cubic structure above the Curie temperature, it has a tetragonal structure below it (see Section 2.2 for more information on crystal structures). The opposite movement to that shown in the diagram can also occur; that is, the Ti^{4+} and the Ba^{2+} ions can move downwards and the O^{2-} ions slightly upwards when the material is below the Curie temperature. Alternatively, the movement can be along any of the remaining four cube-edge directions. No matter which of the six possible directions the ions move in, though, an ionic dipole moment is produced.

Ferroelectric Domains

Of course a crystal of barium titanate will contain far more than one unit cell's worth of atoms, and when there is no additional influence from an external electric field, unit cells that are polarised in the same direction are found together in regions known as *domains*. Neighbouring unit cells are likely to have the same polarisation. This is because some of the atoms from a given unit cell, namely those at the corners and face-centres, will be shared with neighbouring unit cells, and so any movement of atoms in one cell has a knock-on effect in the surrounding cells.

Neighbouring domains can have polarisation in any cube-edge direction. So if, for example, a crystal contains domains which are each polarised in an opposite direction to their neighbour, and the total volume of all the domains polarised in one direction equals the total volume of the domains polarised in the opposite direction, then the overall (net) polarisation of the crystal is zero. However, if an external electric field is applied to the crystal, this situation changes.

An applied field causes any of the domains that are already lined up parallel (or almost parallel) to the field to grow larger at the expense of domains not pointing in this direction. As the intensity of the electric field is increased, these parallel domains become larger and larger until eventually every dipole moment in the crystal is aligned parallel to the field. It is possible to observe these domains growing by etching a crystal (removing part of its surface with chemicals).

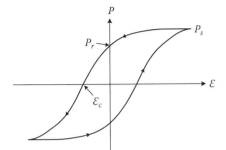

FIGURE 7.35   A hysteresis curve for a ferroelectric material.

Hysteresis

Although the overall polarisation of the crystal clearly increases as the applied electric field increases (until the crystal cannot be polarised any further because it is completely polarised), this increase in polarisation is not proportional to the increase in electric field intensity. Instead, if the relationship between the electric field, ε, and the polarisation, P, in a ferroelectric material is plotted as a graph, a so-called hysteresis loop is formed, as shown in Figure 7.35.

At the point P_s, known as the saturation polarisation, all the dipoles in the material are aligned with the ε field. If this field is removed, the material loses some, but not all, of its polarisation, which drops back slightly to the value P_r, known as the remanent polarisation. This remanent polarisation remains unless the crystal is subjected to an electric field in the opposite direction to the one that polarised it. In other words P_r can only be removed by reversing the electric field. At a value of electric field called the "coercive field", ε_c, the polarisation in the specimen is completely removed. However, if the reverse field continues to be applied, the dipoles will turn to face the new field direction and so be in the opposite direction to the one they were pointing in at point P_s. The remainder of the curve is negotiated by reversing the field again so the dipoles begin to turn back the other way. The area within a hysteresis loop is proportional to the energy dissipated during a full hysteresis cycle, and this process is analogous to the hysteresis seen in ferromagnetic materials and discussed in Section 8.2.3.

Capacitors

A capacitor consists of two conducting plates separated by a dielectric. With the plates connected up to a battery and an air gap left between them, charge will remain stored on the plates even when the battery is switched off. Although dry air acts as a dielectric, inserting another dielectric material between the plates will increase the capacitance. This is because the electric field caused by the charge on the plates will create electronic polarisation of the inserted dielectric, and this in turn increases the amount of stored charge, known as the capacitance.

The higher the value of the dielectric constant for a dielectric material, the greater the increase in capacitance that material can produce if it is inserted between the plates of a capacitor. So capacitors often contain ferroelectrics, as they can have dielectric constants 1000 times higher than those of other dielectric materials when subjected to an alternating electric field. For example, while silicon has a dielectric constant of 12, the dielectric constant of barium titanate at room temperature can have a value as great as 5000. Because of their vastly increased ability to store charge, capacitors containing ferroelectrics can be much smaller than capacitors with other dielectric materials between their plates.

7.4.3 Piezoelectricity

All ferroelectric materials are also *piezoelectric*, which means that when they are subjected to stress, their surfaces acquire a dipole moment, and so opposite surfaces will become oppositely charged (see Figure 7.36). Changing the electric field that the material is subject to also affects the polarisation as well as causing a strain. There is a linear relationship between applied stress and polarisation for small values of stress.

Although all ferroelectric materials are piezoelectric, materials can be piezoelectric and not ferroelectric. Quartz is a well-known example of the latter, and one of the best-known applications of quartz is in clocks and watches, where it acts as an oscillator. All clocks and watches must have some form of oscillator to help them keep time. The oscillator produces regular vibrations that are used to keep the hands turning or digits changing at a constant rate. In the case of quartz, an electric field is applied to the crystal, which then starts to vibrate. With the quartz mounted into a suitable circuit, the electricity produced by the vibration can be fed back into the crystal so that the vibration continues indefinitely.

Other applications of piezoelectric materials range from strain gauges—which use the amount of electricity produced by a piezoelectric crystal attached to the gauge as a measure of the strain (see Figure 7.37)—to damping out vibrations in sports equipment. Tennis players and skiers are among the first people to have benefited from the technology. In the case of tennis, a large amount of stress is put on a player's arm as the shock and vibration from the racket coming into contact with the ball travels into their body. To counteract this effect, piezoelectric fibres can be embedded into the handle of the racket. Then, when a player serves or returns a ball, the fibres bend along with the rest of the racket's frame. The stress on the fibres causes them to generate electric charge, which then flows through a tiny circuit that sends the current

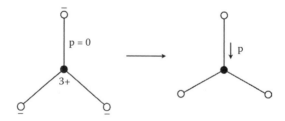

FIGURE 7.36 How a uniaxial strain causes a molecule to acquire a dipole moment. (This diagram originally appeared as Figure 9.13, p. 276, of *Solid State Physics*, 2nd edition, by J. R. Hook and H. E. Hall, © 1991 by John Wiley & Sons Limited. Reproduced with permission.)

FIGURE 7.37   This carbon fibre test wing—which has been deliberately flown to the point where it has suffered structural failure—was used in a NASA experiment in 2001. Surface-mounted piezoelectric strain actuators measured the strain it suffered during a series of flights at increasing speeds and altitudes, including the final flight that led to the failure. (NASA image, taken by Tony Landis.)

back through the fibres. This current is sent back out of phase so that, when the fibres start to vibrate, their motion interferes destructively with the vibrations of the racket and so reduces the overall level of vibration felt by the player.

Skis with piezoelectric fibres integrated into their design can work in the same way, with the expansion and contraction of the fibres reducing the vibration caused by the rigid skis travelling at high speeds over the tiny bumps in snow. Alternatively, the current produced by the vibrating fibres can just flow away via a resistor.

Piezoelectricity is also important in medicine, as it is piezoelectric crystals that allow the ultrasound waves used to scan unborn children to be produced. The technique is similar to radar (see Figure 7.38), but instead of measuring the time between sending out a radio wave and the time it takes to return to see how far away the object that has reflected it is, ultrasound uses pulses of sound of ultrasonic frequencies (generally in the range 1–15 MHz) to create a picture.

The emitter/receiver in a sonar system also contains a piece of piezoelectric ceramic. To detect the depth of underwater objects, the emitter/receiver is placed in the water and an electric signal is applied to it. The oscillations set up in the crystal by the signal are transmitted through the water as high-frequency vibrations. Any objects underwater and in the path of these vibrations reflect them back towards their source. The piezoelectric material then acts as a detector, converting the reflected vibration into an electrical signal. The depth of the reflecting object can be worked out from the time delay between the emitted and reflected vibrations and the frequency of those vibrations.

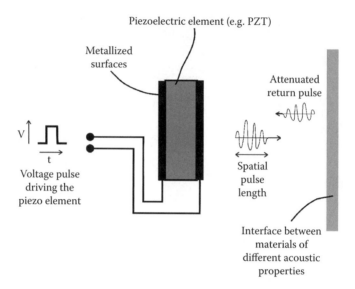

FIGURE 7.38 The basic principles of ultrasound detection.

FURTHER READING

Coghlan, A. "It's lights out for a household classic". *New Scientist* 26 (31 March 2007).

Daviss, B. "Our solar future". *New Scientist* 32 (8 December 2007).

Goosey, M., "Eco-electronics". *Materials World* 17, 26–28, March 2009.

"The Silicon Pioneers". *Pictures of the Future* (Siemens AG) Fall 2005.

Sze, S. M. *Semiconductor Devices: Physics and Technology.* 2nd ed. New York: John Wiley & Sons, 2001.

(More references for further reading, as well as Web links, are available on the following Web page: http://www.crcpress.com/product/isbn/9780750309721.)

SELECTED QUESTIONS FROM
QUESTIONS AND ANSWERS MANUAL

Q7.4 Is the depletion region in a p-n junction largest when the junction is under (a) no bias, (b) reverse bias, or (c) forward bias?

Q7.5 When a p-n junction is under the condition given in your answer to question 7.4, do the drift currents or diffusion currents dominate?

Q7.11

(a) Name the three basic processes that can occur between a photon and an electron in a solid.

(b) Which of these processes forms the basis for laser operation?

Q7.13

(a) What type of diode forms the basis for an LED?

(b) In order for the LED to emit light, does the diode in your answer to part (a) have to be forward biased or reverse biased?

(c) What determines the colour of the light emitted by an LED?

Q7.18   What is the magnitude of the dipole moment of a pair of point charges 6 nm apart from one another if one charge has the value +4 pC and the other −4 pC?

8 Living in a Magnetic World
Magnetism and Its Applications

CONTENTS

8.1   INTRODUCTION TO MAGNETISM

Magnetism has been put to use for centuries. In fact it is possible that the ancient Chinese were making compasses from magnetite—a naturally occurring magnetic mineral which is also known as lodestone—as early as the 11th century. Whilst compasses remain important instruments to this day, there are countless other modern applications of magnetism, including medical imaging, electric motors, and audio, video, and data recording that we have become heavily reliant on. New magnetic materials are being developed all the time that could improve the performance of devices like these and may even enable others to be created.

As with many physical phenomena, magnetism was observed and used far earlier than a satisfactory explanation for it had been developed. In fact it took roughly six centuries before anyone knew *why* compass needles pointed north. This breakthrough was made by the English physician and physicist William Gilbert (1544–1603), who—when he wasn't taking care of the medical needs of Queen Elizabeth I—carried out experiments on electricity and magnetism. He published his results in 1600 in a book called *On the Magnet, Magnetic Bodies, and the Great Magnet Earth*. As the title suggests, one of the ideas Gilbert introduced was the fact that the Earth behaves like a giant bar magnet.

In this chapter we will look at the various magnetic properties that solids can have, and also consider some of the applications of magnets and magnetism. It may seem surprising, but absolutely everything, whether it would normally be classed as magnetic or not, will have some response—however small—to an applied magnetic field. So every object and living thing in the world is magnetic, and in the next section we will see how this magnetism comes about.

8.1.1   THE ORIGINS OF MAGNETISM

All magnetic moments are produced by the angular momentum of electrons in the atoms of solids, and there are two types of angular momentum for electrons in atoms: spin and orbital.

In the simplest rendition of Bohr's model of the atom, a single electron orbits the nucleus, as in Figure 8.1(a). The electron going round its orbit produces a magnetic moment at right angles to the plane of the orbit in the same way that an electric current flowing in a closed loop of wire produces a magnetic moment at right angles to the plane of the loop.

Electrons also have an intrinsic spin, which for simplicity can be thought of as spinning about their own axes, as shown in Figure 8.1(b), and this adds to the total magnetic moment of each electron. (This is in fact a classical model being used to describe a quantum mechanical effect, as an electron is actually a point object and so does not have an axis.) As discussed in more detail in Appendix C, the spin axis can only point in two directions. These are called "up" and "down". This means the spin magnetic moment—which is directed along the spin axis—can only be in an "up" direction (represented by a plus [+] sign) or a "down" direction (represented by a minus [−] sign). The spin magnetic moment for an electron in an atom is given by a constant called the Bohr magneton (μ_B), which has the value 8.27×10^{-24} A m^{-2}.

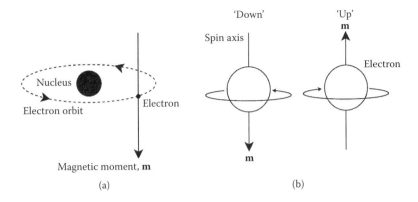

FIGURE 8.1 (a) In Bohr's model of the atom, electrons orbit the nucleus. The negatively charged electron moving round its orbit is similar to an electric current flowing round a closed loop of wire in that, in both cases, a magnetic moment is produced perpendicular to the plane of motion. (b) The two directions that electron spin can have.

As every atom except hydrogen contains more than one electron, there is the opportunity for some of the magnetic moments from the electrons to cancel one another out. For example, if one electron has a spin magnetic moment of $+\mu_B$ and another in the same atom has a spin magnetic moment of $-\mu_B$, then these will cancel each other out. Orbital magnetic moments can also cancel each other out, so the total magnetic moment of an atom is made up from all the orbital and spin magnetic moments of the individual electrons that have not cancelled each other out.

In fact, any atomic shells and subshells (see Appendix C) that are completely full have no overall magnetic moment because all the contributions from the various electrons have cancelled each other out. So solids composed of atoms that have completely filled shells (see Section 2.1.1)—like those of the inert gases—are unable to be permanently magnetised. This is because (as we will see in Section 8.2) the different types of magnetism that a solid can possess are governed by how its atomic magnetic moments respond when a magnetic field is applied to the solid.

8.1.2 Magnetic Properties and Quantities

Before discussing the different types of magnetism that solids have, it is useful to be familiar with different magnetic properties and quantities because the different magnetic phenomena are usually explained in terms of these properties and quantities. A very large number of magnetic properties—which in turn can be described by an enormous variety of units—can be defined, but in this section we will look at a selection of the most commonly used terms in magnetism, and will keep to using SI units.

Magnetic Flux Density

If we imagine a coil of wire in a vacuum that has a fixed current of 1 A running through it, the electric current will cause a magnetic field to be produced. One way of indicating the strength of a magnetic field is via the magnetic flux density (sometimes known as the "magnetic induction"). The magnetic flux density, **B**, is a vector proportional to the strength of the magnetic field, and pointing in the same direction as the magnetic field. In this particular case, the magnetic flux density is given by $\mathbf{B}_{vac}$. (This will not have the same value everywhere because, just as the magnetic field lines cluster around the poles of a bar magnet, the field produced by the wire will be stronger in some areas than others.)

If we now imagine that the entire universe surrounding this wire is made from just one single type of material, for example oxygen, the magnetic flux density would be different to $\mathbf{B}_{vac}$. This is because the surrounding material would give a contribution to the overall magnetic flux density from the magnetic moments of the atoms it is made up from. So

$$\mathbf{B} = \mathbf{B}_{vac} + \mathbf{B}_m \qquad (8.1)$$

where $\mathbf{B}_m$ is the contribution to the field that originates from the material the field is in. The SI unit of **B** is the tesla (T). (Vectors like **B** are often referred to as "field vectors" simply because they are a vector quantity representing a type of field such as a magnetic, gravitational, or electric field.)

Permeability

We have just seen that **B** in a material is different from **B** in a vacuum, and there is no reason why all of magnetism cannot be explained in terms of **B**, which is as fundamental to magnetism as the electric field $\boldsymbol{\varepsilon}$ is to electrostatics. However, for mathematical convenience, another vector **H**, known as the magnetic intensity, is used as well which is related to **B** as follows:

$$\mathbf{B}_{vac} = \mu_0 \mathbf{H} \qquad (8.2)$$

where μ_0 is a constant known as the permeability of free space, which is the magnetic permeability of a vacuum and has the value $4\pi \times 10^{-7}$ Hm^{-1}. (We first encountered **H**, which has units of Am^{-1}, in Section 6.5.1 when discussing the difference between Type I and Type II superconductors.) The permeability has units of henries per metre (Hm^{-1}) and is a "defined constant" as opposed to a measured constant, in this case defined by the definition of the ampere. An **H** field can be produced by a solenoid, which is a coil of wire much longer than its width, and will be discussed in more detail in Section 8.3.1. The field strength will then depend on the length of the solenoid, the number of "turns" or loops in the coil, and the current passing through it. This means that **H** only describes current and geometry, so its units are amperes per metre (Am^{-1}).

The relative permeability of a material is defined as

$$\mu_r = \frac{\mathbf{B}}{\mathbf{B}_{vac}} = \frac{\mathbf{B}}{\mu_0 \mathbf{H}} \qquad (8.3)$$

where μ_r is unitless because the units of $\mathbf{B}$ and $\mathbf{B}_{vac}$ are identical. The relative permeability, μ_r, is the simplest measure of the magnetic properties of a material because it indicates the degree to which it can be magnetised by an externally applied field. The relative permeability can have either positive or negative values and is not necessarily single-valued. For example, ferromagnetic materials (see Section 8.2.3) such as iron have relative permeabilities that vary with the applied field strength.

Magnetisation

By using equation 8.2, we can rewrite equation 8.1 as

$$\mathbf{B} = \mu_0\mathbf{H} + \mathbf{B}_m \qquad (8.4)$$

As we have already seen, $\mathbf{B}_m$ is due to the atomic effects that we discussed in Section 8.1.1. It is also related to $\mathbf{M}$, which is often referred to as the magnetisation of a material, but is perhaps more accurately described as the magnetic dipole moment per unit volume, and given by the following expression:

$$\mathbf{B}_m = \mu_0\mathbf{M} \qquad (8.5)$$

Combining equations 8.4 and 8.5 gives

$$\mathbf{B} = \mu_0\mathbf{H} + \mu_0\mathbf{M} \qquad (8.6)$$

One way of thinking of the magnetisation, $\mathbf{M}$, is as the ratio of the total magnetic moment of the whole material to the total volume of the material. Alternatively, it can be considered to be the ratio of the number of atomic moments multiplied by the value of an individual moment to the total volume of the material. Whichever way it is thought of, the magnetisation links the microscopic atomic effects with the macroscopic measurable magnetic effects by

$$\mathbf{M} = n\,m \qquad (8.7)$$

where n is the number of atomic moments per unit volume and m is the moment of each.

Magnetic Susceptibility

Like the relative permeability, the magnetic susceptibility, χ, provides another fairly simple way of describing the macroscopic magnetic properties of a material, as it is also a measure of how a magnetic material responds to an applied magnetic field. The magnetic susceptibility is related to the magnetisation as follows:

$$\chi = \frac{\mathbf{M}}{\mathbf{H}} \qquad (8.8)$$

and to the relative permeability by

$$\chi = \mu_r - 1 \tag{8.9}$$

As equation 8.9—containing the unitless μ_r—suggests, χ is also unitless. The values of χ (like those of μ_r) can be either positive or negative, and can vary in ferromagnetic materials. In some anisotropic crystalline solids, χ can vary depending on the direction of the applied field with respect to the crystal orientation.

8.2   TYPES OF MAGNETISM

8.2.1   DIAMAGNETISM

Diamagnetism is the weakest form of magnetism there is, and only exists in the presence of an applied magnetic field. Every type of atom has a diamagnetic component in its response to a magnetic field, but in materials that exhibit ferromagnetism (see Section 8.2.3), the diamagnetism is masked by the ferromagnetism, which is a much stronger—and permanent—effect. As we will see in the next subsection, paramagnetic effects can be either stronger or weaker than diamagnetism, and so it will only be masked by paramagnetism in certain materials.

There are some solid objects, however, which only show a diamagnetic response. In the presence of a magnetic field, the atomic electrons of these objects change their velocities and orbits around the nucleus in such a way that the atoms produce a magnetic field that opposes the external field. In metals, the conduction electrons also contribute to the overall diamagnetism by changing their motions to produce a magnetic field that opposes the external field. This means that the magnetic susceptibility, χ, (see equation 8.8) will be negative, as the magnetisation, **M**, is opposite to the applied field, **H**. The relative permeability, μ_r, (see equation 8.3) is fractionally less than 1 (by about 1 part in 10^5) for diamagnetic materials because the **B** field within them is less than the **B** field in a vacuum. Electrons in a metal also contribute a positive (paramagnetic) component to the susceptibility that may make the total susceptibility positive or negative.

Under certain circumstances, it is possible to levitate diamagnetic objects, as Box 8.1 describes. This is a different form of levitation to the floating of a magnet above the surface of a superconductor, which we discussed in Section 6.4.1. In that case, the levitation was caused by electric currents rather than an atomic response, but superconductors—just like plastics, bismuth, copper, hydrogen, and living creatures, including the frog in Box 8.1 as well as ourselves—are diamagnetic. (This is why the rather obscure term "solid objects" has been used in this section, so that it encompasses everything from elemental solids to living, organic objects that are made from more than one type of material.) In fact, as we saw in Chapter 6, superconductors are perfect diamagnets because they totally expel any magnetic flux lines within them.

Since the individual atomic fields oppose the external field, diamagnetic substances are very slightly repelled by large magnetic fields. This is in contrast to all other types of magnetism, in which substances are attracted to magnetic fields to a greater or lesser extent. Although other forces would come into play if large diamagnets like humans were placed in a magnetic field that was stronger in some areas than others, small diamagnets would tend to move to the weaker regions of the field. This is because the strongest regions of the field would repel them.

EXAMPLE QUESTION 8.1   THE DIAMAGNETIC PROPERTIES OF SUPERCONDUCTORS

What is the value of the magnetic susceptibility for a superconducting material when it is in the superconducting state?

ANSWER

The magnetic susceptibility, χ, is given by

$$\chi = \frac{\mathbf{M}}{\mathbf{H}}$$

where $\mathbf{M}$ is the magnetisation, and $\mathbf{H}$ is the applied magnetic field. $\mathbf{M}$ and $\mathbf{H}$ are in turn linked to the magnetic flux density $\mathbf{B}$ by

$$\mathbf{B} = \mu_0\mathbf{H} + \mu_0\mathbf{M}$$

where μ_0 is the permeability of free space.

When a superconductor is below T_C, it expels all magnetic flux from its interior, so $\mathbf{B}$ within it must be zero. This means we have

$$-\mu_0\mathbf{M} = \mu_0\mathbf{H} \equiv \mathbf{H} = -\mathbf{M}$$

Combining this result with the equation for χ reveals that the magnetic susceptibility for a superconductor in its superconducting state is -1.

By contrast, other diamagnetic materials have values of χ in the order of $-x \times 10^{-5}$ (see Table 8.1).

Unlike all other forms of magnetism, which we will discuss in the following sections, diamagnetic behaviour is totally independent of temperature.

8.2.2 PARAMAGNETISM

Like diamagnetism, *paramagnetism* is only apparent when a magnetic field is applied, and is caused by the motion of electrons. The similarity ends there, however, as rather than being repelled, paramagnetic materials have a weak attraction to magnetic fields.

BOX 8.1   TOAD IN THE HOLE

When scientists make discoveries, it is not often that their colleagues think their work is a hoax. But that's exactly what happened to Prof. Andrey Geim from the University of Manchester in the U.K., whilst he was working at the University of Nijmegen in Holland in 1997. "I had come across a scientific paper which reported a rather strange phenomenon. The authors had placed a cuvette filled with water in a horizontal magnet and observed that the water level split into two walls, leaving a completely dry bottom," recalls Prof. Geim.

Intrigued by the fact there seemed to be no explanation for this effect, he poured water inside the bore (central hole) of a powerful electromagnet. "To my surprise, the water did not end up on the floor but stuck in the middle. And some droplets were floating freely in mid-air," says Prof. Geim, who soon realised both the reported splitting of water, and the flotation of water droplets that he was seeing was due to diamagnetism.

Diamagnetism is the weakest form of magnetism, and exists at all temperatures in every substance. Because it is such a weak effect, it is often masked by other types of magnetism, so is only observed in substances that do not show any stronger magnetic effects—like living creatures and plants, water, and plastics. When a diamagnetic substance is placed in a magnetic field, every atom of that substance produces its own tiny magnetic field that expels the external field. This makes diamagnetic substances and strong magnetic fields repel each other as if they were two magnetic north poles.

If a small diamagnetic object is placed in a large magnetic field and the repulsive force pushing the object upwards matches the force of gravity that it normally feels pulling it towards the ground, then the object will levitate. This same repulsion will also cause water to be split in half if it is placed in a horizontal magnetic field, as the centre of the field will push the water away from it off to the left and right.

Encouraged by his floating water, Prof. Geim began levitating "anything else that I could find. The next day, I proudly explained to several colleagues that I could levitate things because of diamagnetism, and no-one believed me!" he laughs. This was despite the fact that other scientists had previously reported diamagnetic levitation, albeit of materials like bismuth and graphite, which exhibit very large diamagnetism.

To illustrate his results for colleagues and students, Prof. Geim began setting up demonstrations of levitation with various diamagnetic objects, including a live frog (see photographs). Strong magnetic fields are not believed to produce any adverse effects on living organisms, but although the frog was fine, the experiment did produce an adverse effect on Geim. "It truly took all my experimental skills to levitate the frog, which was rather sticky and jumped! I am not tempted to try this again," he quips.

However frogs jumping around proved to be the least of Geim's problems. Having now convinced his immediate colleagues, the picture of the levitated frog was first published in the April 1997 edition of *Physics World*. "Colleagues

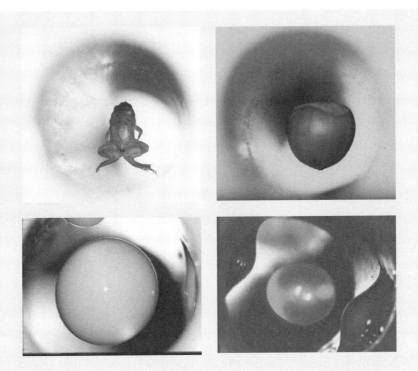

The frog, hazelnut, milk drop, and water droplet shown here have all been photo-graphed whilst being levitated inside the bore (central hole) of a strong magnet. It is only possible to levitate diamagnetic objects, and although every substance on earth shows some diamagnetic response to a magnetic field, only living organisms, wood, plastics, and other substances that would not normally be considered as magnetic are diamagnetic overall. This is because diamagnetism is so weak that it is masked by stronger forms of magnetism. If a diamagnetic substance is placed in a magnetic field, the electrons of every atom of the substance change their orbits slightly and produce a weak magnetic field that opposes the external field. So diamagnetic substances and strong magnetic fields repel each other. If a diamagnetic object like a water droplet or a frog is placed in a large magnetic field—such as the 10 tesla field within the bore of the magnet shown here—it becomes a small magnet that generates a magnetic field of about 1 gauss. (This is roughly the same strength as the Earth's magnetic field.) The force of repulsion between the 10 tesla and 1×10^{-4} tesla magnets pushes the diamag-netic object upwards and compensates the force of gravity that would normally cause it to fall to the ground, so the object levitates. In principle, anything diamagnetic, including a human being, could be levitated in this way if the magnetic field was strong enough. (Images courtesy of Andrey Geim.)

from abroad sent me messages congratulating me on a nice April Fool's Day hoax!" says Prof. Geim, who finally saw his work taken seriously after further coverage in both the popular and scientific press. Since then, diamagnetic levi-tation has become commonly used in research for replicating the microgravity conditions experienced in space.

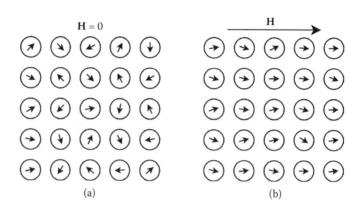

FIGURE 8.2 In (a) the general situation is shown in which atomic magnetic moments in a paramagnetic solid are randomly orientated, while in (b) the material has been placed in an external magnetic field so the moments tend to line up with the field.

Atomic Paramagnetism

We saw in Section 8.1.1 that the total magnetic moment of an atom is made up from all the orbital and spin magnetic moments of the individual electrons that have not cancelled each other out. When there is no external magnetic field, the magnetic moments of the atoms in paramagnetic materials are randomly orientated so that there is no overall magnetic moment from them. However, with a field applied, this situation changes and a net magnetisation is produced, as the individual magnetic moments partially line up with the external field, as shown in Figure 8.2. Referring back to equation 8.6, it is the term $\mu_0\mathbf{M}$ that represents the contribution this aligning gives to the overall **B** field inside the material.

Completely full shells of electrons have zero net angular momentum and therefore zero magnetic moment. In most atoms it is only the outermost shell that is not completely filled, and in solids, atoms bond—either by releasing, sharing, or exchanging electrons—so that their outer electron shells are full (as we saw in Section 2.1). However, "transition metal" atoms, for example, those of the iron group and rare earth group, have an incomplete inner shell (the $3d$ and $4f$ shells, respectively) and so have a nonzero magnetic moment even when their outer shell is complete. It is therefore transition metals and their compounds that show atomic paramagnetism.

Paramagnetism is temperature dependent, as heat tends to randomise the direction of the magnetic moments. In fact, in moderately sized magnetic fields, and at everyday temperatures, paramagnetic gases roughly obey "Curie's Law" in which

$$\chi = C/T \tag{8.10}$$

where C is known as the Curie constant, and T is the temperature. This law is named after the French physicist Pierre Curie (1859–1906), who although most famous for

TABLE 8.1

The Susceptibility for a Range of Materials

Material	Magnetic Type	Susceptibility χ
Copper	Diamagnetic	-0.96×10^{-5}
Sodium chloride	Diamagnetic	-1.41×10^{-5}
Zinc	Diamagnetic	-1.56×10^{-5}
Manganese sulphate	Paramagnetic	3.70×10^{-3}
Aluminium	Paramagnetic	2.07×10^{-5}
Sodium	Paramagnetic	8.48×10^{-6}

his work on radioactivity also carried out research into solid state physics, including studying the magnetic properties of materials for his doctorate.

Paramagnetic solids and liquids, and ferromagnetic solids in their paramagnetic region (see Section 8.2.3), tend to obey a variation on this law known as the Curie–Weiss law. This states that

$$\chi = \frac{C}{(T - \theta)} \tag{8.11}$$

where the Weiss constant θ can have either a positive or negative value.

So paramagnetic substances have a positive susceptibility, albeit with a small value, that is temperature dependent. They also have a relative permeability greater than 1. (If they did not have $\mu_r > 1$, as equation 8.9 reveals, then χ would have to be negative—as it is for diamagnetic materials.) Table 8.1 compares values of χ for a selection of paramagnetic and diamagnetic materials.

Pauli Paramagnetism

There is also a type of paramagnetism known as "Pauli paramagnetism" or free-electron paramagnetism. This comes from the magnetic moments of free electrons—or nearly free electrons—in the conduction band, and so is seen in all metals, albeit as a weak effect.

Electrons can be considered to be spinning about their own axes, and the magnetic moments of the free electrons in the conduction band of a metal are caused by their intrinsic spins. Quantum mechanics dictates that these spins can only point in one of two directions (known as "up" and "down" for convenience) which are antiparallel to each other (see Appendix C).

In the absence of an applied magnetic field, the spins of the free electrons in a metal will be orientated in such a way that there is no overall magnetic moment. When a magnetic field is applied, however, electrons with spins parallel to the field have their energy reduced, while those antiparallel to the field increase in energy. (This is because the energy of a dipole, μ, in a field, B, is equal to $-\mu B \cos \theta$, and cos $180° = -1$, while cos $0° = 1$, so there is a positive value for the energy of antiparallel spins, and a negative value for parallel spins.) To keep the overall energy of the system as low as possible, some electrons change their spin orientation, giving a net result of more electrons with spins aligned with the field than against it. This creates an overall magnetic moment—the Pauli paramagnetism—parallel to the applied field.

This effect is named after the Austrian-Swiss physicist Wolfgang Pauli (1900–1958). He showed that applying Fermi–Dirac statistics (see Appendix E) to conduction electrons was able to correctly predict that the probability of electron spins being parallel to an applied field was greater than the probability of the spins being antiparallel.

Overall Magnetism

It is worth bearing in mind that a paramagnetic material can have a contribution to its magnetic moment from several sources. For example, aluminium behaves as a paramagnet. However, its overall magnetic moment is made up from a diamagnetic contribution from the ion cores as well as both a paramagnetic and a diamagnetic contribution from the conduction electrons. (The latter is a result of the conduction electrons changing their motions to produce a magnetic field that opposes the external field.) So in this case the paramagnetic effects are stronger than the diamagnetic ones, and so the material behaves as a paramagnet overall.

Many other metals are diamagnetic, and in their cases the diamagnetic contribution of the conduction electrons coupled with the diamagnetism of the ion cores is stronger than the Pauli paramagnetism.

8.2.3 FERROMAGNETISM

Ferromagnetism is a much stronger effect than paramagnetism, and *ferromagnetic materials* are the materials that are thought of as magnetic in everyday life. Even weak magnetic fields will magnetise these materials, and attract other ferromagnetic materials to them because they have a permanent magnetic moment known as "spontaneous magnetisation" that exists even in the absence of an applied field.

Iron, cobalt, and nickel, as well as alloys containing these elements, are ferromagnetic as a consequence of a quantum mechanical force known as the exchange interaction. The exact nature of the exchange interaction is not known, but we do know that it exists between the spins of electrons, and that it can make those spins line up parallel or antiparallel to one another. In ferromagnetic materials the exchange interaction causes the spins on different atoms to align parallel with one another, and so create—in conjunction with a component from the orbital angular momentum—a large overall magnetic moment which gives the material its ferromagnetic character.

Formation of Domains

Inside ferromagnetic materials, it is energetically favourable for the magnetic moments of the individual atoms to be arranged in areas called *domains* within which all the moments point in the same direction. Domains reduce the overall magnetic energy of a ferromagnet mainly because they minimise the "magnetostatic" energy of the solid. Maximum magnetostatic energy would be created in a ferromagnet if all the atomic magnetic moments were to line up parallel to one another as shown in

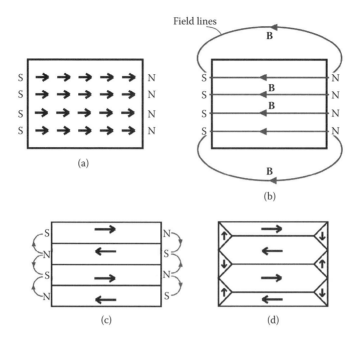

FIGURE 8.3 How the formation of domains reduces the magnetostatic energy. In (a) all the atomic magnetic moments are aligned, producing an internal and external magnetic field as shown in (b). The energy stored in this arrangement is known as the magnetostatic energy, and it can be minimised by the formation of domains as shown in (c) and removed completely by the domain arrangement shown in (d).

Figure 8.3(a). As the diagram reveals, this arrangement gives rise to magnetic poles on the surface of the solid, and these produce a magnetic field both inside and outside the material in the same way a bar magnet would (see Figure 8.3[b]). This magnetic field then tends to turn the atomic moments round as it opposes their alignment, and the magnetostatic energy is the energy stored within the arrangement of magnetic moments pointing in the opposite direction to the field they are creating.

Magnetostatic energy is greatly reduced if domains like those shown in Figure 8.3(c) form. It can be seen that the domains point in different directions to reduce the energy. In fact the magnetostatic energy can be completely removed if small domains called "closure domains", which are represented by the triangular shapes shown at the sides of the solid in Figure 8.3(d), form.

There are also other magnetic energies within a ferromagnetic solid that the detailed structure of the domains works to minimise. For example, the spontaneous magnetisation of a ferromagnet—and also the direction of magnetism within its domains—will tend to lie in a particular crystal direction known as the "easy" direction in which the "magnetocrystalline anisotropy energy" is lowest. This energy is minimised in iron (Fe) if its overall magnetisation is along the <100> axes, whereas the "easy" direction in nickel (Ni) is <111> axes, and for cobalt (Co), which has the close-packed hexagonal structure shown in Figure 2.26, it is the c axis.

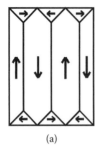 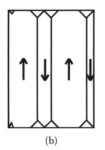 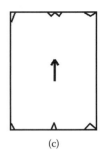

(a)   (b)   (c)

FIGURE 8.4 The changes in domain structure in a ferromagnetic material as the applied magnetic field is increased. (a) In zero applied field, there is no net magnetisation of the sample, as the volume of the domains magnetised in the up direction is equal to that of domains magnetised in the down direction, and similarly the volume of left- and right-magnetised domains is the same. (b) When an external magnetic field is applied in the upwards vertical direction, domains magnetised close to that direction grow at the expense of those magnetised in the opposite direction. There is then an overall magnetisation in the direction of the applied field. (c) Eventually, when the applied field is increased to a large value, the domain walls (indicated by lines) have almost all been removed, and the sample is magnetised to saturation.

Changes to Domain Structure Due to the Application of a Magnetic Field

In the absence of an external magnetic field, ferromagnetic domains point in different directions, as shown in Figure 8.4(a), so the overall magnetisation of the material is zero. However, when an external magnetic field is applied to a ferromagnet, domains which are in line or almost in line with the direction of the field grow, while those domains magnetised in other directions become smaller, as shown in Figure 8.4(b). Each of the lines shown in Figure 8.4 represents a domain wall, which is the region between adjacent domains, and this process of domains changing in size takes place via the movement of domain walls. The net result is an overall magnetisation in the direction of the applied field.

As the external field increases, the direction of magnetisation within any remaining unaligned domains rotates to align with the field. Eventually, once it is completely magnetised, a ferromagnetic material will consist of just one huge domain, with its axis parallel to the external applied field as shown in Figure 8.4(c). There are now almost no domain walls left, and the sample is said to be "magnetised to saturation".

It is possible to use a microscope to see where the domains are in ferromagnetic crystals by polishing the surface of the crystal then coating it with ferromagnetic particles in a suspension. Since the magnetic field is stronger at the domain boundaries than within the domains, the particles become attracted to the domain walls and so reveal them. The patterns produced by the particles are known as Bitter patterns, and this technique can also be used to detect imperfections in ferromagnetic materials.

Figure 8.5 shows the gradual change in atomic dipole moments across a domain wall. In general, domain walls are a few hundred atoms thick and tend to be straight because energy is stored in the walls, and keeping them straight keeps their total volume and therefore energy minimised. Domains range in size from about a tenth of a millimetre to a few millimetres across, and can be several centimetres long in a large crystal.

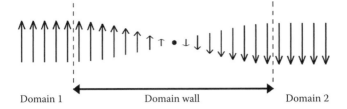

FIGURE 8.5   The direction of the atomic dipole moments changes gradually across a domain wall. In this example, the direction of magnetisation changes by 180° between domain 1 and domain 2, so this type of wall is called a "180° wall". In reality, the change of angle between adjacent dipoles is much smaller, so that the domain wall is several hundred atoms wide.

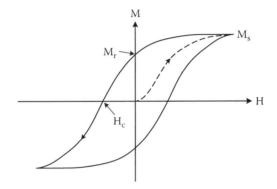

FIGURE 8.6   An M vs. H hysteresis loop.

Hysteresis

In the last subsection we saw that ferromagnetic solids have domains just like ferro-electric solids (see Section 7.4.2). They also show hysteresis like ferroelectric solids. Figure 8.6 shows a typical hysteresis loop for a ferromagnetic material. Both M vs. H, and B vs. H hysteresis loops are common, but B is the quantity measured in exper-iments. (In fact although B, M, and H are vector quantities, it is just their magnitudes that we are concerned with here, hence they do not appear in boldface.) B also deter-mines the effects of the magnetic field (e.g., forces), so a B vs. H hysteresis loop is most commonly used when discussing applications of magnetic materials (as we will see later). However, the following explanation for the physics behind the loop shape holds for both types.

A ferromagnetic material will follow the values of M and H on a hysteresis loop when it is subjected to a magnetic field that is cycled from zero up to some maximum value, decreased back to zero again, then reversed down to another value before being brought back up to zero again. When a magnetic field is first applied to a ferromagnet, it follows the path of the dotted line in Figure 8.6. If the applied field, H, is strong enough, a ferromagnetic material can reach a value known as the *saturation magnetisation*, M_s. This is the point at which it cannot be magnetised any further by an increase in H.

On a microscopic level, applying a magnetic field to a ferromagnetic material first moves the domain walls so that domains magnetised in almost the same direction as the field orientation grow at the expense of domains magnetised in other directions (as we saw earlier). Once a point is reached where there are no domains left pointing in a direction different to that of the field, any further increase in field will just line up the magnetisation within the domains so that it is as parallel as possible to the applied field.

If the applied field is fairly small, the change in domain sizes will be reversible, but a further increase in applied field will produce an irreversible change in the domain wall positions. This is because in order to move further, the walls will have to have been pulled over various defects including impurities and dislocations. Once these defects become incorporated in the domain wall, they reduce the volume of the wall, and so reduce the amount of energy stored in the wall. The wall will then favour its "low energy" position over moving further.

So when H is reduced to zero, the material follows the values on the hysteresis curve, and as Figure 8.6 shows, it retains some of its magnetisation when the external field is zero. This retained magnetisation is known as the "remanence" or remanent magnetisation and is indicated in the diagram by M_r. It is this feature that allows ferromagnets to act as permanent magnets. The remanence can be removed by reversing the magnetic field to a value known as the "coercive force", H_c (sometimes known as the "coercivity" and also indicated in the Figure).

The area inside a hysteresis curve is proportional to the energy dissipated during a full hysteresis cycle, so the magnets used for the cores of transformers and electric motors have high and thin hysteresis curves, as they must not suffer many losses as they go round their cycle. By contrast, the hysteresis curves of materials used to store information magnetically tend to be wide and square. This is because, in order to store information in binary form (in other words as 0 or 1), there needs to be a very clear difference between the spins pointing one way, and then pointing the other when the applied field is reversed.

Paramagnetic Region

The spontaneous magnetisation in a ferromagnet decreases as the temperature increases, as Figure 8.7 shows, becoming zero at the *ferromagnetic Curie temperature*, T_C. This is because thermal energy opposes the exchange interaction, and once the temperature is as high as T_C or above it, the ferromagnetism disappears completely.

Above the Curie temperature, ferromagnetic solids become paramagnetic and obey the Curie–Weiss law introduced in Section 8.2.2. Table 8.2 gives the Curie temperature, T_C, for a range of ferromagnetic materials.

8.2.4 ANTIFERROMAGNETISM AND FERRIMAGNETISM

Both these types of magnetism can be considered to be a type of ferromagnetic behaviour.

Antiferromagnetism

In the same way that the magnetic moments in a ferromagnetic material are only aligned below the Curie temperature (T_C), in an antiferromagnetic material the

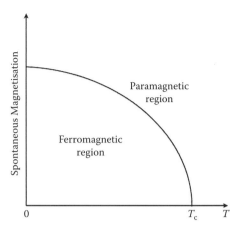

FIGURE 8.7   The spontaneous magnetisation in a ferromagnet decreases as the temperature increases, as this graph shows. The temperature at which ferromagnetism disappears and the material becomes paramagnetic is known as the Curie temperature, T_C.

TABLE 8.2
The Curie Temperature for a Range of Ferromagnetic Materials

Metal	Curie Temperature (K)
Gd	289
Ni	631
Tungsten steel	1033
Fe	1043
Co	1404

moments are only in an ordered arrangement below a critical temperature known as the Néel temperature (T_N). Also a ferromagnet becomes paramagnetic above T_C because the magnetic moments no longer align, and in an analogous way antiferromagnetic materials become paramagnetic above T_N.

When they are below T_N, antiferromagnetic materials (like ferromagnetic materials) have the magnetic moments of all their atoms pointing in a particular direction. In this case, the moment on any given atom will be pointing in an opposite but parallel direction—in other words, antiparallel—to the moment on its neighbour. This is illustrated in Figure 8.8 for manganese oxide. Since the moments are almost equal, this has the effect of making the magnetic susceptibility—the ratio of the amount of magnetisation of a magnetic material to the strength of the magnetic field magnetising it—almost zero.

French physicist Louis Felix Néel (1904–2000) suggested the existence of antiferromagnetism around 1930, and that above a certain temperature—which has now taken his name—antiferromagnetic materials cease to show the ordered arrangement of

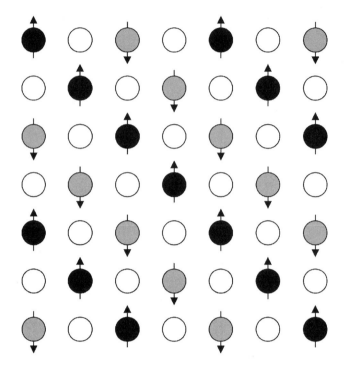

FIGURE 8.8 The spin alignments in the (100) plane of MnO. The O^{2-} ions represented by the open circles in this diagram do not have a magnetic moment associated with them because their spin and orbital magnetic moments cancel each other out. By contrast, the Mn ions have a spin magnetic moment—either "spin up" shown by the black circles or "spin down" illustrated by the grey circles. The overall solid has no net magnetic moment because the spin moment of each ion is opposite to that of its neighbours within the same plane and in adjacent planes, so all the spin moments cancel each other out.

their magnetic moments, and hence the antiferromagnetism stops. Most antiferromagnetic materials are ionic compounds like MnO, and include FeO and NiO.

Ferrimagnetism

Néel also suggested a similar phenomenon in which the magnetic moments of neighbouring atoms were antiparallel, but were also unequal in strength. This is known as ferrimagnetism. (Néel won the 1970 Nobel Prize in Physics for his extensive work on magnetism.) Figure 8.9 compares the directions of the electron spins for ferromagnetic, antiferromagnetic, and ferrimagnetic materials.

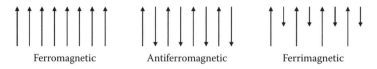

    Ferromagnetic   Antiferromagnetic   Ferrimagnetic

FIGURE 8.9 The directions of the electron spins for ferromagnetic, antiferromagnetic, and ferrimagnetic materials.

There is permanent magnetisation in ferrimagnets like there is in ferromagnets, and the macroscopic properties of ferro- and ferrimagnets are similar—the most noTable difference being the saturation magnetisation, which is not as high in ferrimagnets. Ferrimagnetic materials are ceramics and include ferrites and garnets. In fact the first magnetic material discovered—magnetite (in other words, lodestone)—is a ferrite with the chemical composition Fe_3O_4. Garnets have complicated crystal structures and a general composition of $M_3Fe_5O_{12}$, where M represents a rare earth ion. For example, the most common material of this type is yttrium iron garnet ($Y_3Fe_5O_{12}$). Because ceramic materials, being oxides, are good insulators, they are ideal for certain magnetic applications, such as providing the magnetic core between the two coils in high-frequency transformers.

8.3 TECHNOLOGICAL APPLICATIONS OF MAGNETS AND MAGNETISM

We have come a long way since ancient Chinese mariners floated a piece of lodestone in a container of water, and in so doing made the first compasses. In a world without magnetism, we would not only have problems with navigation, but we would be without one of the most useful methods of investigating molecules and imaging the human body, as well as a host of electronic devices relying on electromagnetism, such as modems and telephones. We would also have no electrical motors or generators. In this section we will look at some of the main technological applications of magnetism.

8.3.1 SOLENOIDS

Coils of current-carrying wire that are much longer than they are wide are known as *solenoids*. They were used by Faraday when he first demonstrated his discoveries in the early 19th century, and are often used in schools' laboratory demonstrations of the principles of electromagnetic induction.

These days, however, the most common use for solenoids is in relays and switches. Relays are electrical circuits that can control another electrical circuit without being electrically connected to it. Electromechanical relays use the magnetic force created by the flow of an electrical current in one circuit to close a switch in another circuit. (Alternatively, this task can be performed by a semiconductor device known as a solid-state relay.)

Solenoids also come in handy if you are intending to build a robot, as they can be used to produce linear motion. This motion can, for example, push keys or buttons, operate a valve, or even allow a robot to jump. This is achieved by placing a cylindrical piston made from iron or steel inside the coil, which shoots out of the coil when current flows round the coil. The movement is caused by the piston either being repelled by or attracted to the magnetic field produced by the current-carrying coil. So that it can return to its position inside the coil as soon as the current is switched off, one end of the piston is attached to a spring.

Much larger solenoids can also be produced, such as those within the 4000-ton PHENIX particle detector shown in Figure 8.10.

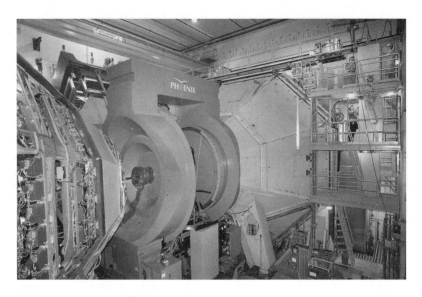

FIGURE 8.10 The PHENIX detector shown here allows particles produced in collisions in the Relativistic Heavy Ion Collider at the Brookhaven National Laboratory in the United States to be analysed. Three magnets in the detector bend the trajectories of charged particles, enabling their charge and momentum to be measured. The north and south muon magnets, which, as their name suggests, are responsible for guiding muons along a curved path, can be seen either side of the two-part central magnet that bends the paths of a variety of particles, including electrons, protons, and deuterons. Each magnet is a solenoid made up from copper coils wound around a steel core. (Courtesy of Brookhaven National Laboratory.)

As we saw in Section 8.1.2, an **H** field can be produced by a solenoid, and the field strength will depend on the length of the solenoid, the number of "turns" or loops in the coil, and the current passing through it. In terms of an equation, this is expressed as follows:

$$\mathbf{H} = \frac{NI}{l} \tag{8.12}$$

where N is the number of turns, l is the length of the solenoid, and I is the current passing through it.

8.3.2 Electromagnets

When a solenoid has the space between its coils—often referred to as "turns"—filled with a core of ferromagnetic material such as iron, it is known as an *electromagnet*. The addition of the core increases the strength of the magnetic field that the current flowing through the coils can produce.

Electromagnets have many uses including heavy lifting, reading and writing data in magnetic recording (see Section 8.3.5), allowing bells to work, telephone calls to

---

EXAMPLE QUESTION 8.2   SOLENOIDS

Calculate (a) the magnetic field strength and (b) the magnetic flux density for a solenoid 25 cm long, with 70 turns and carrying a current of 0.15 A in a vacuum.

ANSWER

(a) The magnetic field strength, $\mathbf{H} = \dfrac{NI}{l} = \dfrac{70 \times 0.15}{0.25} = 42 \text{ Am}^{-1}$

(b) $\mathbf{B}_{vac} = \mu_0 \mathbf{H}$, where $\mathbf{B}_{vac}$ is the magnetic flux density in a vacuum, and μ_0 is a constant known as the permeability of free space, which is the magnetic permeability of a vacuum and has the value $4\pi \times 10^{-7} \text{ Hm}^{-1}$.

So in this case, $\mathbf{B}_{vac} = \mu_0 \mathbf{H} = 4\pi \times 10^{-7} \times 42 = 5.28 \times 10^{-5} \text{ T}$

---

be heard, and Maglev trains to move. When the coil is made from superconducting wire, very high magnetic fields can be produced (see Box 6.1).

8.3.3   PERMANENT MAGNETS

Most of the technological applications of magnetism involve electromagnets, but permanent magnets also have their uses. As their name suggests, permanent magnets remain permanently magnetic unless they are deliberately subjected to a large demagnetizing field. All permanent magnets, which are also known as hard magnetic materials, have a large area within their hysteresis curve. By contrast, a narrow hysteresis curve indicates a soft magnetic material, which is a material that is readily magnetized and demagnetized.

Permanent magnets tend to be made from alloys, and along with steel, the most widely used alloy is known as Alnico. In fact Alnico is not just one alloy but a series of iron (Fe) based alloys with a slightly varying composition of nickel (Ni), aluminium (Al), cobalt (Co), and copper (Cu). Heat-treating these alloys using a process which causes their grains to line up will increase their magnetic strength. Mechanically, however, Alnicos are not very strong, and because they are so hard and brittle, they can only be formed into a particular shape by casting or by pressing them in powder form.

In 1983, hard magnetic materials containing rare earth elements were discovered. These materials, called "neomagnets", as they generally contain neodymium (Nd), are so magnetic that when they are used in motors, these motors can be at least five times lighter—and much smaller—than a motor with the same power that contains another type of permanently magnetic material. This is because much less magnetic material needs to be used to provide the same magnetic strength if it is neomagnetic. Referring back to Figure 1.2 reveals the decrease in size of permanent magnets over the last 50 years.

As we also saw in Chapter 1, devices such as hard disks, floppy disks, CDs, and DVDs are spun by tiny motors during reading and writing, and it was the discovery of these powerfully magnetic rare earth materials that enabled such small motors to be produced.

Another widely used group of permanent magnetic materials are ceramic ferrite magnets. They are relatively inexpensive compared with other magnetic materials, and can be mixed with plastics in their powdered form. This enables flexible magnetic strips to be formed, such as those found round the edges of refrigerator doors.

8.3.4 MAGNETIC RESONANCE

There are some experimental techniques that are used throughout the sciences, and nuclear magnetic resonance (NMR) is one of them. Biologists, chemists, and physicists use the technique to analyse many different types of molecule (as we saw in Box 6.1). Meanwhile, the medical profession use NMR—under the name of magnetic resonance imaging (MRI) to avoid unnecessary panic by misinterpretation of the word "nuclear" by patients—to reveal information about the soft tissues of the body at a resolution unobtainable by any other technique.

During an MRI scan, the patient must lie very still inside the bore of an MRI scanner containing a powerful permanent magnet or electromagnet that produces a magnetic field typically at least 1.5 T in strength. (This is 30,000 times greater than the Earth's magnetic field, and enables clear, high-quality images to be produced.) Water makes up around 65% of the human body, and the MRI image is obtained from the spins of protons within the hydrogen atoms present in this water. When they are not subjected to a high magnetic field, these proton spins are randomly aligned. However, once subjected to the field created by the scanner magnet, they align their spins within the applied field, precessing (wobbling) on their axes as they go round. (In fact some protons align with the field, while others line up antiparallel to it.)

Bombarding the spinning protons with radio waves that have the same frequency as this precession then causes the spins to precess in phase. The next step is for the radio waves to be switched off, which allows the protons to "relax back" to how they were spinning before it was on. This relaxation produces small changes in the magnetic field that are detected by electronics within the scanner. The amount of time the relaxation takes is also detected. As the relaxation time of a proton depends on the type of tissue surrounding it, computer software can then be used to analyse and process the detected signals and create a detailed image of the body that enables different types of tissue to be identified.

As Figure 8.11 reveals, MRI scans cannot show up bones, but they are extremely useful for looking at the central nervous system and musculoskeletal system.

At the time of writing this book, the main clinical applications of MRI are in planning for certain types of surgery including brain surgery, imaging sports and spinal injuries, and helping to diagnose cancer. Functional MRI (fMRI), which reveals the parts of the brain activated when a patient in the scanner carries out a task such as looking at a picture or moving a joystick, is also a growing area.

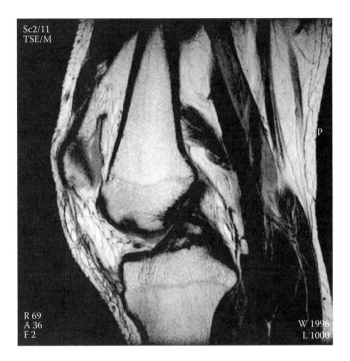

FIGURE 8.11 An MRI image of a knee. The black X-shaped lines behind the knee joint are ligaments that are commonly damaged by sporting injuries. (Courtesy of Guy's and St. Thomas' Hospital NHS Trust.)

8.3.5 MAGNETIC RECORDING

All magnetic storage systems work in roughly the same way. The media can be "particulate", which means they consist of very small particles of a magnetic material such as chromium oxide (CrO_2) bonded onto a polymeric film (in the case of magnetic video or audio tapes) or onto a polymer or metal disk. Media can also be "thin film", which is a polycrystalline film on a chromium substrate. Each particle of a particulate medium can be magnetised in either of two directions (to represent 0 or 1), while thin films store the information within each of their domains. (In a particulate medium, the individual particles are effectively single domains because each one is about the same size as the thickness of a domain wall, and the energy involved in forming a wall would be greater than the magnetostatic energy of the single domain.) More data can be stored within the same space on a thin film compared with particulate media because, unlike separate particles, the domains in a thin film have no unusable spaces between them.

Tape Recording

Magnetism was first used to record sound in the late 19th century by the Danish engineer Valdemar Poulsen. Poulsen's magnetic recording device worked on the same principle as a modern cassette recorder, but instead of using tape, it recorded sound as variations in magnetism on a steel wire.

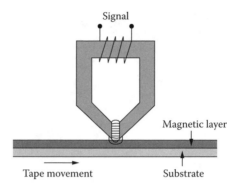

FIGURE 8.12 Writing sound data on a magnetic tape.

In modern systems, sound data is stored as variations in the amount of magnetisation on a tape surface. As we have just seen, the tape is usually a polymer film with tiny particles of a magnetic material such as chromium oxide (CrO_2) bonded to it. Writing and reading from the tape is achieved via a read/write head. This consists of a core of soft magnetic material with a gap cut into it. There is a coil wound round part of this core so that it acts as an electromagnet, producing a magnetic field across its gap when electricity flows through the coil. The tape is fed past the read/write head, and if the head is being used to write data, the field across the gap magnetises the section of tape directly below it. The magnetisation produced is proportional to the current flowing through the coil, and remains in place even when the read/write head is removed. Figure 8.12 shows a schematic diagram of the writing process.

To read data rather than write it, the tape is again moved past the read/write head, but this time a voltage is induced in the coil whenever it encounters a change in the magnetic field of the tape. The varying electrical signal from the coil therefore contains the audio information stored on the tape, and can be fed to a loudspeaker to enable the sound to be heard.

MiniDisc

Another device that makes use of magnetism to record sound is the MiniDisc (MD). There are two types of MDs. Premastered MDs containing music by professional artists are similar to CDs, with pits on their surface representing digital 1 signals, and areas with no pits representing 0's. When a "playback beam" from a diode laser (see Section 7.2.3) is shone at the surface of the MD, it is scattered by the pits and strongly reflected by the areas with no pits, enabling the data to be decoded.

The second type of MD, which we can record on, stores audio data on a magnetic layer. To record, the playback laser operates at a higher power than when it is reading data, and heats whichever spot of the magnetic layer it is shining at above a critical temperature (approximately its Curie temperature of 180°C). This demagnetises that region of the disc and so removes any data stored there. At the same time, a magnetic recording head induces either a North or South magnetic polarity at the spot being heated, which the spot takes on as it cools. A digital 0 signal is represented by areas left with a South pole, while all the regions with a magnetic North pole represent 1.

During playback the magnetic North poles rotate the polarisation of the playback laser's beam in one direction, while the South poles rotate it in another direction. These two polarisations are separated by a polarised beam splitter, then sent to different light detectors, which register a corresponding 0 or 1.

Computer Disk Drives

Disk drives in a computer read and write data in a similar way to tape recorders. However, the magnetic storage medium is a disk rather than a tape, and the read/write head is moved to the required place over the disk to read or write information, vastly speeding up data retrieval. Also the data is stored digitally on a disk in the form of bits which occupy very small regions less than 1 μm long. Each bit can either be magnetised to represent a 1, or magnetised in the opposite direction to represent a 0. The data is written in "sectors" on concentric tracks that run from the innermost diameter of the disk to its outer edge.

A read/write head can be seen on the end of a suspension arm that looks like a pair of paper scissors in the photograph of an uncovered hard disk drive shown in Figure 8.13. A suspension arm is able to move its corresponding head into the correct position for retrieving saved data, or writing new data on an unused sector of the disk. As the picture reveals, disk drives usually contain more than one disk. Since data is stored on both sides of every disk, there will be six read/write heads for the disk drive shown.

FIGURE 8.13 A hard drive containing three disks. The suspension arms for the read/write heads are visible on the left of the picture, and the uppermost read/write head can be seen above the top disk of the stack. (Photo taken by Marcin Barłowski.)

Although they perform the same function as those in a tape recorder, the read/write heads in a hard disk are different in that the "reading" parts of them are made from a metal or semiconductor that changes its resistance to electricity when it is subjected to a magnetic field. This effect is known as magnetoresistance (MR).

So in a disk drive the reading section of the read/write head can decode the stored information because the stronger the magnetisation is in a particular area, the stronger the corresponding magnetic field is around that point and the greater the increase in read/write-head resistance it produces. This increase in resistance reduces the amount of current flowing through the head, which is detected by an electronic circuit that decodes any changes in current flow to reveal the data.

By contrast, data is written using exactly the same principles as tape recording. The data is sent in the form of electrical pulses to a coil in the writing part of the head, which produces a representative pattern of magnetisation on the disk.

Giant Magnetoresistance

We all want to squeeze more and more data into the same space, so researchers are constantly working on developing higher density magnetic recording. If more information is to be stored on a standard-sized disk, the heads need to be more sensitive in order to read the more closely packed data. In fact the latest computer hard drives, as well as iPods and Xboxes, can store so much information mainly thanks to a new generation of sensitive heads. These contain a variation on a type of structure discovered in 1988 to exhibit larger resistance changes (when subjected to a magnetic field) than the conductors used in magnetoresistance heads. This magnetic effect was termed giant magnetoresistance (GMR), and was independently discovered by German physicist Peter Grünberg (1939–) and French physicist Albert Fert (1938–), who went on to share the 2007 Nobel Prize for Physics for their work.

A basic GMR structure consists of ultrathin metal layers (only around a few atoms thick) of a ferromagnetic material separated by nonmagnetic layers. These layers form a continuous metal sandwich of one magnetic layer followed by one nonmagnetic layer, repeating again and again.

Applying a magnetic field to a GMR structure changes the relative orientations of the magnetisation in neighbouring magnetic layers, which each act like a single magnetic domain. An antiparallel configuration gives the material a relatively high resistivity, whereas successive magnetic layers with parallel magnetisation allow the material to conduct well.

FURTHER READING

Heinrich, B. "Nickel reveals a new magnetic face". *Physics World* (June 2005): 26–27.
Keevil, S. F. "Magnetic resonance imaging in medicine". *Physics Education* 36 (2001): 476–485.
(More references for further reading, as well as Web links, are available on the following Web page: http://www.crcpress.com/product/isbn/9780750309721.)

SELECTED QUESTIONS FROM
QUESTIONS AND ANSWERS MANUAL

Q8.1   Which statement describes the Bohr magneton?

(a) The total magnetic moment of a free electron
(b) The spin magnetic moment of a free electron
(c) The spin magnetic moment for an electron in an atom
(d) The total magnetic moment for an electron in an atom

Q8.3

(a) What unit is the magnetic susceptibility, χ, measured in?
(b) How does χ relate to the magnetisation?

Q8.5   Which type of magnetism is being described by the following statement?

The weakest form of magnetism, that only exists in the presence of an applied magnetic field.

Q8.7

(a) Which way do the magnetic moments of the individual atoms of a ferro-magnetic material point within a single domain?
(b) Does the existence of domains reduce or increase the overall magnetic energy of a ferromagnet?
(c) What is the region between adjacent domains known as?
(d) Describe the main changes in domain structure in a ferromagnetic material as an applied magnetic field is increased.

Q8.10   Calculate (a) the magnetic field strength and (b) the magnetic flux density for a solenoid 20 cm long, with 60 turns and carrying a current of 0.16 A in a vacuum.

Appendix A: Some Useful Maths

A1   VECTORS

VECTOR VS. SCALAR

The distance and the direction between two points in space can be represented by a quantity known as a vector. Vectors are drawn as lines that have a "magnitude" (in other words a size—which in this case is the length) which is proportional to the real-life distance between the points, as well as a direction the same as that of the real-life direction. By contrast, quantities known as "scalar" quantities only have magnitude, so they can be described using just a real number. The length of a vector, which is completely represented by a single number, is therefore an example of a scalar quantity.

The magnitude of a vector is given mathematically by the modulus. So the magnitude of vector **a** shown in Figure A1 is $|\mathbf{a}|$, which can also be denoted in normal type—that is, as "a"—instead. (When writing by hand, it is difficult for most of us to make a clear distinction between boldface and normal, so it is conventional to underline vectors. In this case vector **a** would be written as a̲.)

Another vector will only be equal to **a** if both its magnitude and direction are identical. If the magnitude of another vector is identical to **a** but it points in a direction exactly opposite to **a** (as shown in Figure A2), then this vector is known as a "negative vector".

Vectors can be multiplied by scalars; for example, **a** can be multiplied by 2 to become 2**a**. The magnitude of this new vector is $2|\mathbf{a}|$ and it is in the same direction as the original vector **a**. Figure A3 shows an example of a vector **a** multiplied by a scalar.

UNIT VECTORS

Any vector with a magnitude of 1 is known as a "unit vector". Unit vectors are often denoted by a small pointed symbol (commonly called a hat) above the letter labelling the vector, for example $\hat{\mathbf{w}}$.

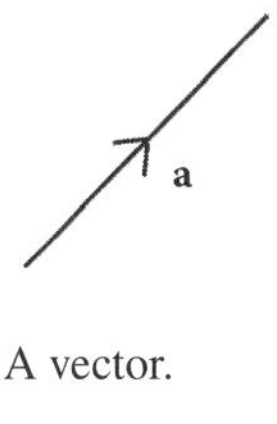

FIGURE A1   A vector.

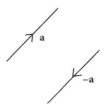

FIGURE A2   Two vectors with the same magnitude, but opposite directions.

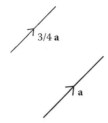

FIGURE A3   A vector **a** multiplied by a scalar.

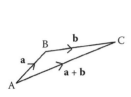

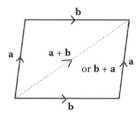

FIGURE A4   Vector addition.

FIGURE A5   The commutative nature of vector addition.

ADDITION OF VECTORS

Vectors can be added together by placing them so that the end of one vector coincides with the start of the vector it is to be added to. For the example shown in Figure A4, the vector **a** connecting points A and B is being added to the vector **b** that connects points B and C. The vector joining points A and C is known as the "resultant" or vector sum and is written as **a** + **b**.

Of course real situations tend to get a bit more complicated than this, but the addition procedure is the same. An important feature of vector addition is that it is "commutative". This means that the result will be the same no matter what order you add the vectors in. So **a** + **b** ≡ **b** + **a** as Figure A5 illustrates.

A2   CARTESIAN COORDINATES

It was the French mathematician and philosopher René Descartes (1596–1650) who introduced the Cartesian coordinate system into mathematics; in fact, the word *Cartesian* is taken from his name. The system allows the positions of points to be determined by how they relate to reference axes. In three dimensions, the Cartesian frame of reference consists of three mutually perpendicular axes (axes that are each at right angles to the other) as shown in Figure A6.

The position of any point is given by three coordinates (x,y,z) that represent how far along each of the respective axes the point lies. The point where the axes intersect is known as the origin, and is the point $(0,0,0)$. In a right-handed system, distances along the x-axis that are to the right of the origin are taken to be positive, while those to the left of the origin are taken to be negative. In a similar way, distances to the right of the origin on the y-axis are positive, while those to the left

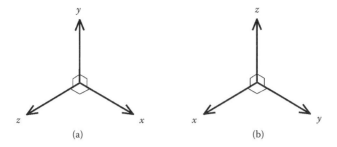

(a)   (b)

FIGURE A6 The Cartesian frame of reference. The axes shown in (a) and (b) are said to be right-handed because holding the thumb, first and second fingers of your right hand so that all three fingers are as close to being mutually perpendicular as you can get without straining will create the axes. If you first look at part (a) if you line up your thumb so it points along the x-axis, your first and second fingers will then point along the y- and z- axes respectively in the directions indicated. The axes shown in (b) are also a right-handed set because they are the set shown in (a) rotated clockwise by 90°.

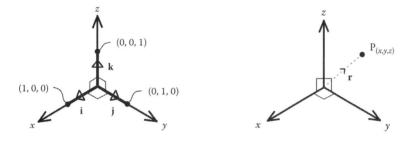

FIGURE A7 The Cartesian unit vectors $\mathbf{i}$, $\mathbf{j}$, and $\mathbf{k}$ with the coordinates of each of their end points marked.

FIGURE A8 The position vector, $\mathbf{r}$, of the point P.

are negative, and distances above the origin on the z-axis are positive, while those below are negative. (If the system is left-handed the opposite is true.)

For example, three different coordinates are indicated in Figure A7, which shows how Cartesian unit vectors (sometimes known as Cartesian base vectors) are defined. As the diagram reveals, the unit vector—that is, the vector with a magnitude of 1—along the x-axis begins at the origin and ends at the point (1,0,0). Meanwhile, the unit vectors along the y- and z-axes both start at the origin and end at points (0,1,0) and (0,0,1), respectively.

Points do not have to be represented by coordinates. They can also be located via a position vector like the one shown for the point P in Figure A8. In general, any position vector joins a single point to a fixed reference point. In this case the position vector $\mathbf{r}$ joins point P to the origin of the Cartesian frame of reference.

Position vectors can be described in terms of their "Cartesian components", that is, in terms of the Cartesian unit vectors. This means that any distance along a particular axis will be given as multiples or fractions of the unit vector along that axis. So in general, for any point with coordinates (x,y,z), the position vector $\mathbf{r}$ is given by $\mathbf{r} = x\mathbf{i} + y\mathbf{j} + z\mathbf{k}$. For example, the position vector of the point $(5, 1, -2)$ is $5\mathbf{i} + \mathbf{j} - 2\mathbf{k}$.

A3   DERIVATION OF SELECTED EQUATIONS

DERIVATION OF EQUATION 6.17

In an intrinsic semiconductor, $n = p$, as equation 6.14(a) shows; so if we want an expression for the Fermi energy in an intrinsic semiconductor, we can equate equations 6.15 and 6.16 as follows

$$n = N_C \exp\left[-\frac{\left(E_C - E_F\right)}{k_B T} \right] \tag{6.15}$$

$$p = N_V \exp\left[-\frac{\left(E_F - E_V\right)}{k_B T} \right] \tag{6.16}$$

So

$$N_C \exp\left[-\frac{\left(E_C - E_F\right)}{k_B T} \right] = N_V \exp\left[-\frac{\left(E_F - E_V\right)}{k_B T} \right] \tag{A.1}$$

$$\therefore \frac{N_V}{N_C} = \frac{\exp\left[\dfrac{-E_C + E_F}{k_B T} \right]}{\exp\left[\dfrac{-E_F + E_V}{k_B T} \right]} \tag{A.2}$$

$$\therefore \frac{N_V}{N_C} = \exp\left[\frac{-E_C + E_F + E_F - E_V}{k_B T} \right] \tag{A.3}$$

Taking logs gives us

$$\ln\left(\frac{N_V}{N_C} \right) = \frac{-\left(E_C + E_V\right) + 2E_F}{k_B T} \tag{A.4}$$

Re-arranging gives

$$k_B T \ln\left(\frac{N_V}{N_C} \right) = -\left(E_C + E_V\right) + 2E_F \tag{A.5}$$

Dividing through by 2 gives

$$\frac{k_B T}{2} \ln\left(\frac{N_V}{N_C}\right) = \frac{-(E_C + E_V)}{2} + E_F \tag{A.6}$$

A final rearrangement then gives us equation 6.17 as follows:

$$E_F = \frac{E_C + E_V}{2} + \frac{k_B T}{2} \ln\left[\frac{N_V}{N_C}\right] \tag{6.17}$$

DERIVATION OF EQUATION **6.18**

The intrinsic carrier concentration can be obtained by multiplying equations 6.15 and 6.16 together as shown:

$$n = N_C \exp\left[-\frac{(E_C - E_F)}{k_B T}\right] \tag{6.15}$$

$$p = N_V \exp\left[-\frac{(E_F - E_V)}{k_B T}\right] \tag{6.16}$$

So

$$np = N_C N_V \exp-\left[\frac{E_C - E_F + E_F - E_V}{k_B T}\right] \tag{A.7}$$

Tidying up gives

$$np = N_C N_V \exp-\left[\frac{E_C - E_V}{k_B T}\right] \tag{A.8}$$

and as $(E_C - E_V) = E_g$,

$$\Rightarrow np = N_C N_V \exp\left(-\frac{E_g}{k_B T}\right) \tag{A.9}$$

Since equation A.9 does not involve the position of E_F, which moves depending on the amount of doping in a semiconductor, it is independent of whether there is doping or not, and so it is valid for both intrinsic and extrinsic semiconductors.

However, as $n_i^2 = np$ (equation 6.14[b]) in an intrinsic semiconductor, equation A.9 can be rewritten as

$$n_i^2 = N_C N_V \exp\left(-\frac{E_g}{k_B T}\right) \tag{A.10}$$

Taking the square root of each side of this equation leads to equation 6.18 as follows:

$$n_i = \sqrt{N_C N_V} \exp\left[-\frac{E_g}{2k_B T}\right] \tag{6.18}$$

PROOF THAT $n_i^2 = np$ **(EQUATION 6.14[B]) IS VALID FOR EXTRINSIC AS WELL AS INTRINSIC SEMICONDUCTORS.**

We have already seen that equation A.9 is valid for both extrinsic and intrinsic semiconductors. In fact, the form in which it is written would be particularly suitable for an extrinsic semiconductor, but if we wanted to express it for use with an intrinsic semiconductor we could use equation 6.14(b) to replace the np term, giving us equation A.10.

As the right-hand side of both equations A.9 and A.10 are equal, their left-hand sides must be equal too, showing that $n_i^2 = np$ for both intrinsic and extrinsic semiconductors.

Appendix B: Vibrations and Waves

B1 PROPERTIES OF WAVES

ELECTROMAGNETIC AND ELASTIC WAVES

When a wave travelling through a medium (solid, liquid, or gas) is made up from the vibrations of the particles of the medium, it is said to be elastic. Elastic waves travel along by these vibrations being passed on from one particle to the particle next to it and so on. If, however, the wave is not disturbing the particles of the medium, but is instead made up from oscillations of the electric and magnetic fields in that region of space, the wave is an electromagnetic wave.

Sound waves are elastic waves, and so whilst they can be carried through solids, liquids, and gases, they cannot pass through a vacuum because there are no particles present to do the vibrating. (Since the density of particles in space is low enough that it can be considered to be a vacuum, it is true that in space nobody can hear you scream.) By contrast, electromagnetic waves—including light and X-rays (see Figure B1 for a reminder of the contents of the electromagnetic spectrum)—can travel through a vacuum. (So at least you would see the enemy spaceship approaching.)

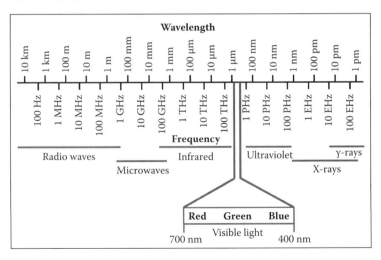

FIGURE B1   The electromagnetic spectrum.

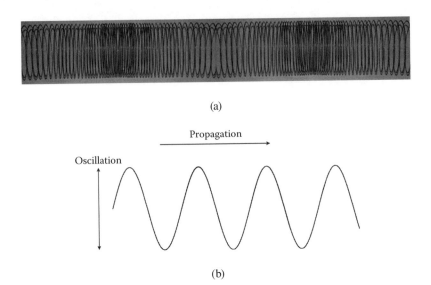

(a)

(b)

FIGURE B2   (a) A longitudinal wave can be set up in a slinky spring by first laying it on the floor and gently stretching it out. Then whilst keeping hold of both ends, move just one hand to and fro so you continually push some of the loops together at one end of it. You should be able to see regions where the coils are closer together travelling down the spring. By contrast, trapping one end of a piece of string in a desk drawer and then holding the other end and flicking your hand rapidly up and down will—with some practice—set up a transverse wave like the one shown in (b).

Longitudinal and Transverse Waves

There are two basic ways in which waves can travel along. Longitudinal waves, such as sound waves, move in the same direction as the vibration making up the wave—as Figure B2(a) illustrates. By contrast, ripples on a pond and electro-magnetic waves are examples of transverse waves (see Figure B2[b]), which are waves that travel in a direction perpendicular to the direction of motion of the oscillations that form them.

Describing Waves

Many of the effects that physicists study are dependent in some way on the behaviour of various types of wave, so in order to explain these effects and to carry out calculations on them, an accurate method of describing waves is needed. Both transverse and longitudinal waves can be represented by a sine wave like that shown in Figure B3. As the figure indicates, the wavelength, λ, is the distance from one peak to the next (or one trough to the next), and the amplitude, A, is the maximum displacement from the undisturbed position that the oscillations or vibrations can have. (The real-life shapes of waves [waveforms] are not, of course, restricted to sine functions. Other waveforms include square waves, pulses, ripples, and sawtooth waves. However, sinusoidal waves are important because other waveforms such as square waves that cannot simply be written as a single mathematical function can be

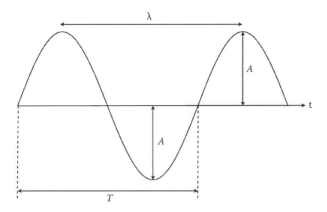

FIGURE B3   Some of the basic properties of waves.

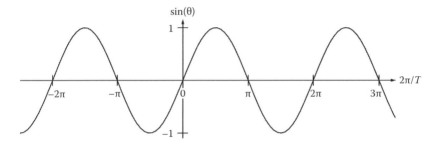

FIGURE B4   The sine function.

represented by a Fourier series, which is a mathematical series of sinusoidal waves that is often infinitely long.)

Other wave properties include the frequency (which can be represented either by f or by v), which is the number of oscillations per unit time, and the period, T, which is the time it takes for one complete oscillation to take place. In both cases an oscillation is taken to mean a complete cycle of movement in which the transmitting medium of a wave has moved to and fro once and then returned back to the point it was at before the disturbance took place. The period is related to the frequency as follows

$$T = 1/v \qquad \text{(B1.1)}$$

Figure B4 shows another sine wave, which in this case is the graph of the mathematical sine function. The sine function is known as a circular function because the shape of its graph repeats every 2π radians (= 360°). Since each oscillation of a wave is a complete cycle of movement, and every real-life wave can be represented in terms of sine waves, the period of any elastic or electromagnetic wave can also be thought of as an angle and is therefore 2π.

This means that any two waves can be compared with one another by representing each as a vector tracing out a complete circle every cycle. (See Appendix A for a

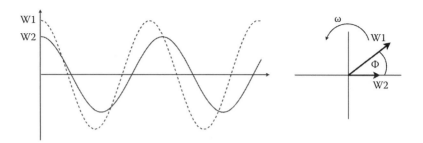

FIGURE B5   Two ways of representing the waves W1 and W2 that have a phase difference Φ.

reminder of vectors.) For example, Figure B5 shows two waves W1 and W2 that have the same frequency, but are at a different point in their respective cycles. The length of each vector is equal to the amplitude of the wave it represents, and the vectors rotate at an angular frequency, ω, through 2π every period. The angle between W1 and W2 is known as the phase angle, Φ.

In general, the phase can be thought of as the fraction of the period that has elapsed between a chosen origin and another point in the cycle. As with the period, the phase difference between two waves can either be expressed as a time difference or as a difference in angle. When one wave is further through its cycle than another, this first wave is said to "lead" the second wave. For example, wave W1 in Figure B5 leads wave W2. By contrast, a wave which is behind another in its cycle is said to "lag" behind, so another way of describing the phase difference between these two waves is to say that wave W2 lags behind W1. Whenever the phase of two waves is exactly the same, they are said to be "in phase". In all other circumstances the waves are "out of phase", and when one wave is exactly π radians ahead (or behind) another, the waves are described as being "exactly out of phase" or "in antiphase".

It is often easier to describe the frequency of a wave in terms of its angular frequency, ω, which is related to the ordinary frequency as follows

$$\omega = 2\pi\nu \qquad\qquad (B1.2)$$

The wave number is another regularly used term. Strictly speaking, this is defined as the number of waves per unit length, or the reciprocal of the wavelength, and has the symbol σ, but in solid state physics it is taken to mean the magnitude of the wave vector of a wave and is denoted by k. The wave vector, **k**, (sometimes known as the "propagation vector") represents the magnitude and direction of a wave (see appendix A for a revision of vectors). (Although the vectors in Figure B5 are being used to represent waves W1 and W2, they are not wave vectors, as they do not indicate the direction the waves are travelling. Instead, these vectors are simply helping provide a convenient way of thinking about phase geometrically.) The wave number k is related to the wavelength as follows

$$k = \frac{2\pi}{\lambda} \qquad\qquad (B1.3)$$

Velocity of Propagation

The velocity of an elastic wave depends on the elastic properties and the density of the material it is travelling through. In general, the less dense the medium is, and the greater the value of the elastic modulus that best describes the part of the medium that a given elastic wave moves through, the faster this elastic wave can travel. (See Section 4.1.4 for more about the various moduli of elasticity.) It is important to note that the velocity of propagation of an elastic wave is not related to its "particle velocity", which is the velocity at which the particles of the elastic medium vibrate as they transmit the wave.

Elastic waves travel at much lower speeds than electromagnetic waves, which move through a vacuum at the speed of light (approximately 3×10^8 m/s), and at speeds slightly slower than this when travelling through a medium.

B2   WAVE BEHAVIOUR

Interference

If any two waves of the same frequency and similar amplitudes and intensities line up as shown in Figure B6(a) when they meet, so that the peaks of one wave coincide with the peaks of the other, these waves are said to be "in phase" and they add together to produce a larger wave. This is known as constructive interference.

If the peaks line up with the troughs instead, the waves are exactly "out of phase"—as shown in Figure B6(b)—and they cancel each other out. (See Section B1 for a reminder of phase.) This is known as destructive interference.

Of course, any two or more waves—whatever their respective frequencies, amplitudes, or phases—will interact with one another. In general, the resulting wave in any region where waves meet is the vector sum of all the waves that are interacting. We will not bother to revise this here, however, as in this book we are mainly concerned with the destructive and constructive interference that occurs between light, sound, and electron waves that have come from the same source.

Diffraction

Light, and in fact all other forms of electromagnetic radiation, travels in a straight line unless it encounters some sort of obstacle in its way. If it passes through a hole or slit, or around the outside of an object, a light beam breaks up into a number of smaller light beams—which has the effect of spreading the beam out. This process is known as diffraction, and "diffraction patterns" are produced when diffraction occurs and the broken-up parts of the light interfere with one another. (Sound waves and other waves, such as electron waves, can also be diffracted in a similar way, but for simplicity we will only discuss the diffraction of light—as an example of an electromagnetic wave—in this appendix.)

When diffraction of light occurs, some of the light beams created interfere constructively while others interfere destructively. This produces a diffraction pattern consisting of light and dark areas. The light areas are the result of constructive

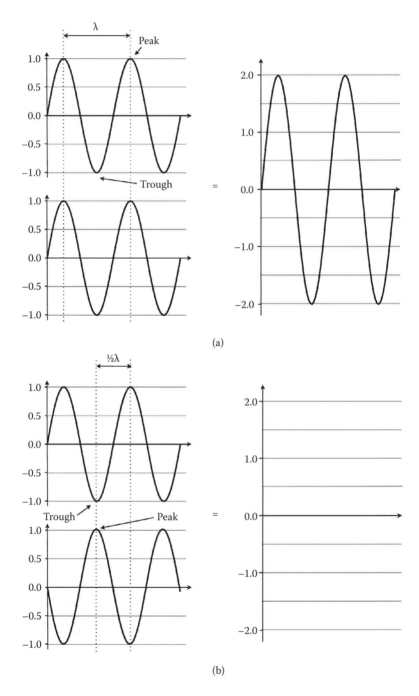

FIGURE B6 Constructive interference occurs when any two waves are in phase when they meet as shown in (a). If, however, two waves that are out of phase meet as shown in (b), destructive interference takes place, and instead of adding together like the waves in (a), they cancel each other out.

interference taking place—which causes a bright beam to be produced, whereas the dark appears wherever light has destructively interfered and therefore been cancelled out.

If you pick up a CD in daylight (or other "white" light sources) and look at the side the music or data is saved on, you are likely to notice some coloured bands of light. The tiny concentric rings of data are the cause of these colours—which change as you tilt the CD towards and then away from the light—because they are diffracting any light beams that fall on them and separating them into a large number of distinct beams with different wavelengths. Wherever they meet, these beams then interfere with one another, and any of the colours of this diffracted light that we can see are produced when two or more beams of that particular wavelength have interfered constructively. Any colours missing from the spectrum have been cancelled out by destructive interference.

DIFFRACTION GRATINGS

One way to deliberately produce a diffraction pattern is to shine a beam of light through a "diffraction grating". A diffraction grating is a large array of very narrow slits (usually in the order of several thousand slits per centimetre) which can separate out different wavelengths of light by diffracting each different wavelength by a different angle.

Figure B7 shows what happens in one small section of a diffraction grating when a monochromatic beam of light (in other words, a beam of light consisting of a single

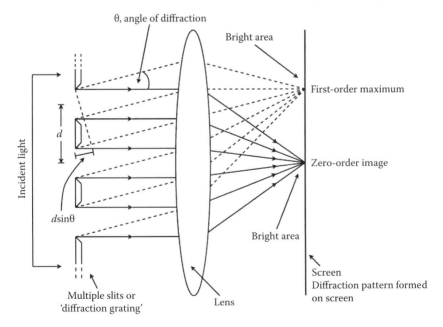

FIGURE B7   The bright fringes of a diffraction pattern are produced when the path difference of light beams from neighbouring slits is equal to a whole number of wavelengths.

wavelength) falls on it. As the light passes through each slit, it is diffracted, and while some of the light continues to travel straight on through the grating (shown by the solid lines in the diagram), many other diffracted beams are produced. (The dotted lines in Figure B7 show one such set of these diffracted beams.) The diffracted light beams from all these slits then interfere to produce a diffraction pattern on a screen, as indicated.

Each of the bright areas of a diffraction pattern appears whenever diffracted light beams from two neighbouring slits have interfered constructively. This happens when the path difference (difference in distance travelled by each beam from the diffraction grating to the screen) between two adjacent beams is a whole number of wavelengths—in other words when the path difference is equal to $n\lambda$, where n is an integer. (Referring back to Figure B6[a], you can see that this situation will cause the two little light beams to be in phase.) As Figure B7 reveals, the path difference can be obtained by simple trigonometry and is equal to $d \sin \theta$, where d is the distance between the centres of two neighbouring slits (although, as the slits are equally spaced, it can sometimes be more convenient to think of d as the distance from the bottom of one slit to the bottom of the slit either above or below, or even the distance between the tops of two neighbouring slits), and θ is the angle between the incident beam and a diffracted beam which has constructively interfered with another and will produce a bright region when it falls on the screen. This means a bright region will only occur in a diffraction pattern produced by a diffraction grating when the condition

$$d \sin \theta = n\lambda \qquad\qquad\qquad (B2.1)$$

is met.

The central bright line in a diffraction pattern like this is known as the zero-order image as in this case $\theta = 0$ and $n = 0$. The lines on either side of this central maximum are produced when $n = 1$ and θ is some angle not equal to zero, and are known as first-order maxima. Meanwhile, the lines to the outside of these are known as second-order maxima and correspond to a situation where $n = 2$ and θ is some angle greater than the angle of the first-order maxima, and so on through the pattern. As Figure B7 indicates, each of the maxima is in fact composed of contributions from more than one pair of slits—as all of the beams emerging from successive slits are in phase—although a single pair of slits would still produce a maximum.

If light composed of more than one wavelength falls on a diffraction grating, first-order, second-order, third-order (and so on) maxima are produced for each of the constituent wavelengths. It was mentioned at the start of this subsection that different wavelengths of light are diffracted through different angles. So this means colours are separated out within a diffraction pattern, as, for example, blue light will produce a first-order maximum at a different angle than the first-order maximum for red light.

Unknown wavelengths can be calculated by measuring the spacing between the slits of a diffraction grating and the angles of the bright lines in the diffraction pattern it produces. (Conversely, if the angles of the bright lines or spots in a diffraction pattern of a known wavelength of light are measured, the spacing between the slits of the diffraction grating is revealed.) The central maximum (where $n = 0$) is easy to spot because it is much brighter than the other lines.

Appendix C: Revision of Atomic Physics

There are various topics in solid state physics—including bonding and energy band formation—that can only be properly understood if you are armed with a basic knowledge of atomic physics and the notations used to describe atoms. The following sections should provide a quick reminder of the atomic physics that is necessary for understanding the material covered in this book.

C1 ATOMIC STRUCTURE AND PROPERTIES

ATOMIC STRUCTURE

Atoms contain negatively charged electrons that can be considered to be orbiting around a positively charged nucleus. For every atom other than hydrogen, the nucleus contains protons and neutrons. (Hydrogen has just a single proton and no neutrons in its nucleus.) While neutrons are electrically neutral, protons have a positive charge exactly the same size (1.602×10^{-19} C) as the negative charge on electrons. Since isolated atoms have no overall electric charge, the number of protons is equal to the number of electrons.

ATOMIC NUMBER

The number of protons in the nucleus is known as the atomic number, and is indicated by Z. (For an electrically neutral atom, Z will also represent the number of electrons circling the nucleus.)

The atomic number Z determines where each different element appears in the periodic table—as this lists the elements in order of increasing atomic number. If you look at the periodic table, you can see that, for the natural elements, Z ranges in value from 1 for hydrogen to 92 for uranium. (Section C3 discusses the periodic table in more detail.)

MASS NUMBER AND ISOTOPES

The total number of protons and neutrons in the nucleus of an atom is known as the mass number, and is denoted by A. (The mass number can also be called the "nucleon number".) Most of the elements have different variants known as isotopes, which all have the same atomic number but have different values for the mass number because they contain different numbers of neutrons in their nuclei. For example, hydrogen has three known isotopes. Ordinary hydrogen has no neutrons at all, while deuterium has one neutron. It is also possible to artificially make a third isotope of hydrogen called tritium, which contains two neutrons and is radioactive.

Different isotopes of the same element can be indicated by a superscript representing the mass number A and a subscript representing the atomic number Z preceding the chemical symbol for that element. For example, the ordinary isotope of hydrogen would be written as ^1_1H. Alternatively, just the mass number can be stated, as in ^{12}C and ^{14}C, which can also be written as carbon-12 (or C-12) and carbon-14 (or C-14).

ATOMIC MASS UNITS

Atomic weight or "relative atomic mass" (one of the few areas of physics where there is no need to make a distinction between mass and weight) is denoted by A_r and is the average weight of an element, taking into account the different weights of all its naturally occurring isotopes. For example, chlorine has an atomic weight of approximately 35.5, which allows for the fact that there are two isotopes, ^{35}Cl and ^{37}Cl, and that for every atom of ^{37}Cl that exists there are three atoms of ^{35}Cl.

Atomic weight is listed on the periodic table in atomic mass units (amu), which are defined so that one atom of carbon-12 has an atomic weight of 12u, where u is the symbol for the atomic mass unit and $1u = 1.6606 \times 10^{-27}$ kg (to five significant figures). In other words, one amu is 1/12th of the atomic mass of the C-12 isotope. Using this unit makes the atomic weight of any atom approximately equal to its mass number. (As an electron is about 1836 times lighter than a proton and around 1839 times lighter than a neutron, it is not surprising that electrons do not contribute much to the weight of an atom.)

Most calculators will have u as a preset physical constant button, which is useful, as we need to multiply any atomic weight value given in amu by 1u before inserting it into calculations. For example, if you look up the atomic weight of Li on the periodic table, you will find it is 6.94 u, that is, $6.94 \times (1.6606 \times 10^{-27}) = 1.15 \times 10^{-26}$ kg (to three significant figures).

C2   ELECTRON SHELL NOTATION

ATOMIC SHELLS

Any electrons that are approximately the same distance away from the nucleus are said to be occupying the same "atomic shell". These atomic shells are separated out in space and have larger radii the further they are from the nucleus, as Figure C1 shows. (In reality, not all the shells are spherical; for example, some are dumbbell shaped. However, for ease they are usually depicted with the same basic shape.) They can be thought of as "energy levels" or "energy states" that electrons can reside in. The further an atomic shell is from the nucleus, the greater its energy—and consequently the greater the energy of any electrons in it—will be. (Having just said that, we are about to see that the energies of the electrons in a given atomic shell are not in fact identical. However, they are very similar and are far, far closer in value than the energies of any two electrons when one is in a different atomic shell to the other.)

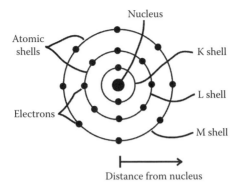

FIGURE C1   This schematic diagram shows how atomic shells increase in radius the further they are from the nucleus. As well as having a larger radius, the further a shell is from the nucleus, the higher its energy is. This means that an electron in an atomic shell close to the nucleus will have a lower value of energy than an electron in a shell farther away from the nucleus. We will see in the next subsection that uppercase letters can be used to signify the different atomic shells (as illustrated in this example).

The atomic shells (also known as "atomic orbitals" or "electron shells") of any atom are actually solutions of the Schrödinger equation for that atom. The Schrödinger equation gives the probability of finding a particle—in this case an electron—at a particular location at a specific time. So for an atom it reveals that the electrons are most likely to be found in a series of discrete orbitals (the atomic shells) surrounding the nucleus, and conversely that there is little or no likelihood of them residing in the space between these atomic shells.

We have already seen that an atomic shell close to the nucleus will have a lower value of energy than a shell further away from the nucleus. But there is also a range of energies between any two shells that electrons cannot have because they never occupy that region of space. This means electrons in atoms can have certain values of energy (corresponding to the values of the different atomic shells) and not others.

When a property can only have certain discrete values like this, it is said to be "quantised". Electrons in atoms have four quantised properties, each of which can be labelled by a quantum number. Three of these quantum numbers—that characterise the atomic shells and hence the electrons in them—arise out of the solution to the Schrödinger equation. The principal quantum number, n, represents the total energy of an electron, while the orbital quantum number, l, represents the magnitude of an electron's orbital angular momentum. Finally, the magnetic quantum number, m_l, represents the direction in which the vector representing an electron's angular momentum points when the atom is placed in a magnetic field.

The fourth quantised property that electrons possess is called "spin", and the fourth quantum number—the spin quantum number, m_s—can, for the purposes of this simplified discussion, be considered to be representing the angular momentum that an electron has as a result of spinning about its own axis. We will look in more

TABLE C1
Atomic Shell Notation

Principal quantum number, n	1	2	3	4	5	...
Atomic shells		K	L	M	N	O	...

detail at each of these quantum numbers and what they represent in the next few subsections.

PRINCIPAL QUANTUM NUMBER

The principal quantum number, n, is effectively a label for the atomic shell itself, and so is the same for every electron in a particular shell. The farther a shell is from the nucleus, the larger n is. This means that n represents the size of the shell (as the radius increases with increasing distance from the nucleus), as well as indicating the energy of the shell and therefore the total energy of any electron in that shell.

The principal quantum number n can only have integral values (that is, it can only be an integer and not a fraction). Some of these are shown in Table C1, which lists the quantum numbers of the first few shells, starting with the shell closest to the nucleus and gradually moving outwards in consecutive order. As the table reveals, the different atomic shells are assigned the uppercase letters K, L, M, etc.—as illustrated in Figure C1.

ORBITAL QUANTUM NUMBER

Every atomic shell apart from the K shell is split into two or more "subshells" that have slightly different energies. (The K shell has just one subshell.) The subshells represent the minutely different energies of electrons in the same shell but with slightly different values of orbital angular momentum. These subshells are labelled by the orbital quantum number l which can have 0, 1, 2, ..., $(n - 1)$ values for each different value of the principal quantum number (n). The orbital quantum number, l, therefore represents the magnitude of an electron's orbital angular momentum.

Since the principal quantum number for the K shell is 1, it is clear that it can only have one subshell with $l = 0$. By contrast, the L shell can have $l = 0$ or $l = 1$, while the M shell can have $l = 0$, 1, or 2. Table C2 lists the different subshells corresponding to the different l values. Any electrons in the same subshell will have the same values of both n and l, and as Table C2 shows, lowercase letters are used

TABLE C2
Subshell Notation

Orbital quantum number, l	0	1	2	3	4	5	6	...
Subshell	s	p	d	f	g	h	i	...

to denote the different subshells. Using this notation, the K shell is said to have one *s* subshell, while the L shell consists of an *s* subshell and a *p* subshell, and so on for the other shells.

MAGNETIC QUANTUM NUMBER

If an external magnetic field is applied to an atom, every subshell except the *s* subshell splits into slightly different energy levels. These energy levels relate to the different directions the angular momentum vectors of the electrons in each subshell are able to point when the atom is in a magnetic field. The magnetic quantum number, m_l, characterises these energy levels.

For each value of *l*, there are $2l + 1$ different values that m_l can have, so $m_l = 0, \pm1, \pm2, \ldots, \pm l$. For example, as we have already seen, the *s* subshell ($l = 0$) can only have one energy level denoted by $m_l = 0$. Meanwhile, a *p* subshell ($l = 1$) has three possible energy levels—one with $m_l = -1$, another with $m_l = 0$, and a third with $m_l = 1$—which all have the same energy in the absence of a magnetic field.

SPIN QUANTUM NUMBER

As well as orbiting the nucleus like a planet orbiting a sun, electrons can also be imagined to be spinning about their own axes in the same way that the Earth rotates on its axis. (In fact, a classical model like this cannot be used to describe a quantum mechanical effect such as spin with any accuracy, but it does provide a mental picture that is easy to remember.) This intrinsic angular momentum or "spin" is represented by the spin quantum number m_s. The spin quantum number m_s can only have the values $+\frac{1}{2}$ or $-\frac{1}{2}$, which correspond to the electrons' spin axis pointing upwards or downwards. In fact, these two different energy states are often referred to as "spin up" and "spin down", respectively.

The Pauli exclusion principle (see Appendix D) states that no two identical fermions (elementary particles with half-integer spin) in any system can exist in the same quantum state. This means that no two electrons in an atom can have the same set of four quantum numbers. So a maximum of two electrons—one with $m_s = +\frac{1}{2}$ (spin up) and one with $m_s = -\frac{1}{2}$ (spin down)—can occupy any of the energy levels denoted by the magnetic quantum number m_l. In other words, for any value of m_l there are only two possible values of m_s.

As there are $2l + 1$ different values that m_l can have for any subshell, each subshell can only contain a maximum of $2(2l + 1)$ electrons. A full or "closed" *s* subshell ($l = 0$) can, for example, only hold two electrons. Meanwhile, a closed *p* subshell ($l = 1$) contains six electrons, and a closed *d* subshell ($l = 2$) ten electrons, and so on. Listing the quantum numbers for all the states that electrons in different shells can have is a useful way of illustrating all these rules. For example, the possible states for the K shell ($n = 1$) are

$$n = 1 \qquad l = 0 \qquad m_l = 0 \qquad m_s = -\tfrac{1}{2}$$
$$n = 1 \qquad l = 0 \qquad m_l = 0 \qquad m_s = +\tfrac{1}{2}$$

while for the L shell with $n = 2$ they are

$n = 2$	$l = 0$	$m_l = 0$	$m_s = -\frac{1}{2}$
$n = 2$	$l = 0$	$m_l = 0$	$m_s = +\frac{1}{2}$
$n = 2$	$l = 1$	$m_l = -1$	$m_s = -\frac{1}{2}$
$n = 2$	$l = 1$	$m_l = -1$	$m_s = +\frac{1}{2}$
$n = 2$	$l = 1$	$m_l = 0$	$m_s = -\frac{1}{2}$
$n = 2$	$l = 1$	$m_l = 0$	$m_s = +\frac{1}{2}$
$n = 2$	$l = 1$	$m_l = +1$	$m_s = -\frac{1}{2}$
$n = 2$	$l = 1$	$m_l = +1$	$m_s = +\frac{1}{2}$

ELECTRON CONFIGURATION OF AN ATOM

The notation introduced in the previous subsections allows the electron configuration of an atom to be described as in the following example:

$$F: 1s^2 2s^2 2p^5$$

This is the electron configuration of the element fluorine. Each shell is identified by the principal quantum number n, which is followed by the letter representing the orbital quantum number l and denoting the subshell. The superscripts after the letters reveal how many electrons there are in the corresponding subshell.

FILLING OF SUBSHELLS

Although the total energy of the shells increases as the distance from the nucleus increases, sometimes the energy of a subshell from a higher shell is less than the energy of a subshell from a lower shell, as Figure C2 shows. For example, the $4s$ subshell has a lower energy than the $3d$ subshell.

Since the states with the lowest energies are always filled with electrons first, this means that the $4s$ subshell will be filled before any electrons go into the $3d$ subshell. So the electrons in any atom fill up the subshells in the sequence illustrated by this diagram—taking into account the Pauli exclusion principle—rather than in straight numerical order with, for example, all of the N subshells being filled before any of the O subshells were occupied. As soon as putting another electron into a particular subshell would mean violating the Pauli exclusion principle—as it would have the same quantum numbers as an electron already in the shell—this electron has to move to the next subshell up.

The order of filling of the subshells is also indicated in Table C3, which lists out the electron configurations of the elements. If you look at krypton you will see that it has a full outermost subshell (the $4p$ subshell) containing six electrons. When the next element in the periodic table—rubidium—needs to house one more electron than krypton, instead of this electron going into the $4d$ subshell, it is accommodated in the $5s$ subshell. This is because the $5s$ subshell, as Figure C2 illustrates, has a lower energy than the $4d$ subshell.

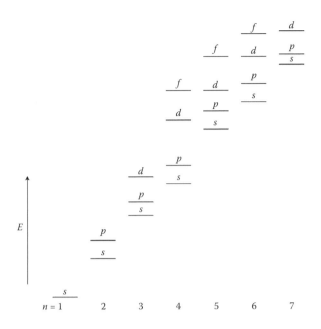

FIGURE C2   The energies of the atomic subshells. Electrons fill these subshells in order of increasing energy, starting from the lowest energy subshell—the $1s$ subshell.

Despite the overlapping of some subshells, it is always the case that when any two similar subshells—for example, any two s subshells or any two d subshells—are considered, the subshell with the lowest energy will be the one from the shell with the lowest total energy. For example, a $2s$ subshell will have a lower energy than a $3s$ subshell. In addition, within any given shell, the subshell energy increases as the value of the orbital quantum number l—which labels the different subshells—increases. For example, the $5f$ subshell is higher in energy than the $5d$ subshell.

C3   THE PERIODIC TABLE

Many of the properties—both physical and chemical—of the elements can be explained in terms of their electronic configurations. For example, the number of electrons in the outermost atomic subshells of any element determines how reactive that element is and also how well it can conduct electricity. The electronic configurations are essentially summarised in the periodic table (see Table C4), which lists the elements in order of increasing atomic number in seven horizontal rows known as "periods". Arranging the elements in this way produces vertical columns of elements known as "groups" that have similar electronic configurations and therefore similar properties. (The relationship between the electronic configurations of the various elements and their properties are explained where appropriate in the main text, and so will not be repeated here.)

Referring back to Table C3 will reveal that all of the elements in Group 1A (the alkali metals) have a single electron in their outermost subshell, while Group IIA

TABLE C3

Electron Configurations of the Elements (From *Perspectives of Modern Physics*, Beiser, A., 1981, © The McGraw-Hill Companies, Inc. Reproduced with permission.)

	K	L		M			N				O				P			Q
	1s	2s	2p	3s	3p	3d	4s	4p	4d	4f	5s	5p	5d	5f	6s	6p	6d	7s
1 H	1																	
2 He	2																	
3 Li	2	1																
4 Be	2	2																
5 B	2	2	1															
6 C	2	2	2															
7 N	2	2	3															
8 O	2	2	4															
9 F	2	2	5															
10 Ne	2	2	6															
11 Na	2	2	6	1														
12 Mg	2	2	6	2														
13 Al	2	2	6	2	1													
14 Si	2	2	6	2	2													
15 P	2	2	6	2	3													
16 S	2	2	6	2	4													
17 Cl	2	2	6	2	5													
18 A	2	2	6	2	6													
19 K	2	2	6	2	6		1											
20 Ca	2	2	6	2	6		2											
21 Sc	2	2	6	2	6	1	2											
22 Ti	2	2	6	2	6	2	2											
23 V	2	2	6	2	6	3	2											
24 Cr	2	2	6	2	6	5	1											
25 Mn	2	2	6	2	6	5	2											
26 Fe	2	2	6	2	6	6	2											
27 Co	2	2	6	2	6	7	2											
28 Ni	2	2	6	2	6	8	2											
29 Cu	2	2	6	2	6	10	1											
30 Zn	2	2	6	2	6	10	2											
31 Ga	2	2	6	2	6	10	2	1										
32 Ge	2	2	6	2	6	10	2	2										
33 As	2	2	6	2	6	10	2	3										
34 Se	2	2	6	2	6	10	2	4										
35 Br	2	2	6	2	6	10	2	5										
36 Kr	2	2	6	2	6	10	2	6										
37 Rb	2	2	6	2	6	10	2	6			1							
38 Sr	2	2	6	2	6	10	2	6			2							
39 Y	2	2	6	2	6	10	2	6	1		2							

TABLE C3 (continued)
Electron Configurations of the Elements

	K	L		M			N				O				P			Q
	1s	2s	2p	3s	3p	3d	4s	4p	4d	4f	5s	5p	5d	5f	6s	6p	6d	7s
40 Zr	2	2	6	2	6	10	2	6	2		2							
41 Nb	2	2	6	2	6	10	2	6	4		1							
42 Mo	2	2	6	2	6	10	2	6	5		1							
43 Te	2	2	6	2	6	10	2	6	5		2							
44 Ru	2	2	6	2	6	10	2	6	7		1							
45 Rh	2	2	6	2	6	10	2	6	8		1							
46 Pd	2	2	6	2	6	10	2	6	9		1							
47 Ag	2	2	6	2	6	10	2	6	10		1							
48 Cd	2	2	6	2	6	10	2	6	10		2							
49 In	2	2	6	2	6	10	2	6	10		2	1						
50 Sn	2	2	6	2	6	10	2	6	10		2	2						
51 Sb	2	2	6	2	6	10	2	6	10		2	3						
52 Te	2	2	6	2	6	10	2	6	10		2	4						
53 I	2	2	6	2	6	10	2	6	10		2	5						
54 Xe	2	2	6	2	6	10	2	6	10		2	6						
55 Cs	2	2	6	2	6	10	2	6	10		2	6			1			
56 Ba	2	2	6	2	6	10	2	6	10		2	6			2			
57 La	2	2	6	2	6	10	2	6	10		2	6	1		2			
58 Ce	2	2	6	2	6	10	2	6	10	2	2	6			2			
59 Pr	2	2	6	2	6	10	2	6	10	3	2	6			2			
60 Nd	2	2	6	2	6	10	2	6	10	4	2	6			2			
61 Pm	2	2	6	2	6	10	2	6	10	5	2	6			2			
62 Sm	2	2	6	2	6	10	2	6	10	6	2	6			2			
63 Eu	2	2	6	2	6	10	2	6	10	7	2	6			2			
64 Gd	2	2	6	2	6	10	2	6	10	7	2	6	1		2			
65 Tb	2	2	6	2	6	10	2	6	10	9	2	6			2			
66 Dy	2	2	6	2	6	10	2	6	10	10	2	6			2			
67 Ho	2	2	6	2	6	10	2	6	10	11	2	6			2			
68 Er	2	2	6	2	6	10	2	6	10	12	2	6			2			
69 Tm	2	2	6	2	6	10	2	6	10	13	2	6			2			
70 Yb	2	2	6	2	6	10	2	6	10	14	2	6			2			
71 Lu	2	2	6	2	6	10	2	6	10	14	2	6	1		2			
72 Hf	2	2	6	2	6	10	2	6	10	14	2	6	2		2			
73 Ta	2	2	6	2	6	10	2	6	10	14	2	6	3		2			
74 W	2	2	6	2	6	10	2	6	10	14	2	6	4		2			
75 Re	2	2	6	2	6	10	2	6	10	14	2	6	5		2			
76 Os	2	2	6	2	6	10	2	6	10	14	2	6	6		2			
77 Ir	2	2	6	2	6	10	2	6	10	14	2	6	7		2			
78 Pt	2	2	6	2	6	10	2	6	10	14	2	6	9		1			
79 Au	2	2	6	2	6	10	2	6	10	14	2	6	10		1			

continued

TABLE C3 (continued)
Electron Configurations of the Elements

	K	L		M			N				O				P			Q
	1s	2s	2p	3s	3p	3d	4s	4p	4d	4f	5s	5p	5d	5f	6s	6p	6d	7s
80 Hg	2	2	6	2	6	10	2	6	10	14	2	6	10		2			
81 Tl	2	2	6	2	6	10	2	6	10	14	2	6	10		2	1		
82 Pb	2	2	6	2	6	10	2	6	10	14	2	6	10		2	2		
83 Bi	2	2	6	2	6	10	2	6	10	14	2	6	10		2	3		
84 Po	2	2	6	2	6	10	2	6	10	14	2	6	10		2	4		
85 At	2	2	6	2	6	10	2	6	10	14	2	6	10		2	5		
86 Rn	2	2	6	2	6	10	2	6	10	14	2	6	10		2	6		
87 Fr	2	2	6	2	6	10	2	6	10	14	2	6	10		2	6		1
88 Ra	2	2	6	2	6	10	2	6	10	14	2	6	10		2	6		2
89 Ac	2	2	6	2	6	10	2	6	10	14	2	6	10		2	6	1	2
90 Th	2	2	6	2	6	10	2	6	10	14	2	6	10		2	6	2	2
91 Pa	2	2	6	2	6	10	2	6	10	14	2	6	10	2	2	6	1	2
92 U	2	2	6	2	6	10	2	6	10	14	2	6	10	3	2	6	1	2
93 Np	2	2	6	2	6	10	2	6	10	14	2	6	10	4	2	6	1	2
94 Pu	2	2	6	2	6	10	2	6	10	14	2	6	10	5	2	6	1	2
95 Am	2	2	6	2	6	10	2	6	10	14	2	6	10	6	2	6	1	2
96 Cm	2	2	6	2	6	10	2	6	10	14	2	6	10	7	2	6	1	2
97 Bk	2	2	6	2	6	10	2	6	10	14	2	6	10	8	2	6	1	2
98 Cf	2	2	6	2	6	10	2	6	10	14	2	6	10	10	2	6		2
99 E	2	2	6	2	6	10	2	6	10	14	2	6	10	11	2	6		2
100 Fm	2	2	6	2	6	10	2	6	10	14	2	6	10	12	2	6		2
101 Md	2	2	6	2	6	10	2	6	10	14	2	6	10	13	2	6		2
102 No	2	2	6	2	6	10	2	6	10	14	2	6	10	14	2	6		2
103 Lw	2	2	6	2	6	10	2	6	10	14	2	6	10	14	2	6	1	2

elements (the alkaline earth metals) have the maximum quota of two electrons in their outermost subshell. Group VIII elements (the noble gases) effectively have full outer shells (see Section 2.1.1 for further details), and the halogens of Group VIIB are one electron short of a full p subshell. Other groups of elements with similar properties include the transition metals—which are in fact several groups, as they encompass all the groups from IIIA to IIB—and the lanthanides (rare earths), which are the upper of the two periods shown separately below the main table.

TABLE C4
The Periodic Table

Period	Group I	Group II											Group III	Group IV	Group V	Group VI	Group VII	Group VIII
1	1 H 1.00																	2 He 4.00
2	3 Li 6.94	4 Be 9.01											5 B 10.81	6 C 12.01	7 N 14.01	8 O 16.00	9 F 19.00	10 Ne 20.18
3	11 Na 22.99	12 Mg 24.31											13 Al 26.98	14 Si 28.09	15 P 30.98	16 S 32.07	17 Cl 35.46	18 Ar 39.94
4	19 K 39.10	20 Ca 40.08	21 Sc 44.96	22 Ti 47.90	23 V 50.94	24 Cr 52.00	25 Mn 54.94	26 Fe 55.85	27 Co 58.93	28 Ni 58.71	29 Cu 63.54	30 Zn 65.37	31 Ga 69.72	32 Ge 72.59	33 As 74.92	34 Se 78.96	35 Br 79.91	36 Kr 83.8
5	37 Rb 85.47	38 Sr 87.66	39 Y 88.91	40 Zr 91.22	41 Nb 92.91	42 Mo 95.94	43 Tc (99)	44 Ru 101.1	45 Rh 102.91	46 Pd 106.4	47 Ag 107.87	48 Cd 112.40	49 In 114.82	50 Sn 118.69	51 Sb 121.75	52 Te 127.60	53 I 126.9	54 Xe 131.30
6	55 Cs 132.91	56 Ba 137.34	57–71 *	72 Hf 178.49	73 Ta 180.95	74 W 183.85	75 Re 186.2	76 Os 190.2	77 Ir 192.2	78 Pt 195.09	79 Au 197.0	80 Hg 200.59	81 Tl 204.37	82 Pb 207.19	83 Bi 208.98	84 Po (210)	85 At (210)	86 Rn 222
7	87 Fr (223)	88 Ra 226.05	89–103 **															

*Rare earths

57 La 138.91	58 Ce 140.12	59 Pr 140.91	60 Nd 144.24	61 Pm (145)	62 Sm 150.35	63 Eu 152	64 Gd 157.25	65 Tb 158.92	66 Dy 162.50	67 Ho 164.92	68 Er 167.26	69 Tm 168.93	70 Yb 173.04	71 Lu 174.97

**Actinides

89 Ac 227	90 Th 232.04	91 Pa 231	92 U 238.03	93 Np (237)	94 Pu (242)	95 Am (243)	96 Cm (247)	97 Bk (249)	98 Cf (251)	99 Es (254)	100 Fm (253)	101 Md (256)	102 No (254)	103 Lw (257)

Note: The atomic number is shown above the symbol for each element, while its atomic weight in atomic mass units (amu) is shown below. Elements with atomic weights in brackets are artificially made, so never occur in nature.

Appendix D: Revision of Quantum Mechanics

Since the behaviour of all atoms, molecules, and subatomic particles is governed by the principles of quantum mechanics, it is impossible to avoid the topic when studying solids. However, the quantum mechanics that is used to describe solid state physics can get very complicated, and as this book is an introductory-level text, the details of those techniques will not be discussed here. Instead, each chapter will focus, where appropriate, on applying the basic concepts of quantum mechanics to the behaviour of solids on the atomic scale. The following sections should therefore provide a brief overview of the quantum theory needed to understand the contents of this book.

D1   FUNDAMENTAL IDEAS OF QUANTUM THEORY

WAVE-PARTICLE DUALITY

In 1905, the German physicist Albert Einstein (1879–1955) suggested that electromagnetic radiation could not only be regarded as a wave, but also as a stream of individual particles—later termed photons. This concept allowed the photoelectric effect (in which electrons are ejected from a solid when electromagnetic radiation above a certain threshold frequency falls on its surface) to be explained. Following on from this idea, the French physicist Louis de Broglie (1892–1987) proposed in his doctoral thesis in 1924 that the dual wave-particle nature of radiation was also a feature of matter. In other words, that in the same way waves could sometimes behave like particles, particles could behave like waves. This concept is now known as "wave-particle duality".

The idea that any moving object—from an orbiting electron to a car moving down a road—has a matter wave associated with it was thought at first to have no foundation in the real world. In fact it is still impossible to detect any experimental evidence of wavelike motion for large objects such as cars. This is because the wavelengths associated with objects of large mass are so tiny, we do not have anything small enough to observe diffraction effects from the matter waves with.

By contrast, de Broglie's equations had predicted that a reasonably energetic electron would have a wavelength around the same size as the interatomic spacings in crystals. This was helpful because whenever the lattice parameter of a crystal is greater than or approximately equal to the wavelength of whatever is being shone at it, diffraction effects can be observed.

It was American physicists Clinton Davisson (1881–1958) and Lester Germer (1896–1971) along with British physicist George P. Thomson (1892–1975), who, in 1927, independently and simultaneously carried out the first electron diffraction experiments. Because electrons scatter in air, Davisson and Germer's experiment took place inside a glass vacuum tube. In this tube, electrons emitted from a hot filament were scattered off a single crystal of nickel. If electrons behaved entirely as particles, they would scatter randomly in all directions. But if they also behaved like waves, they would exhibit interference. When the intensity of the scattered electron beam was measured at different angles to the nickel, it became clear that in some directions there were large numbers of electrons while in others there were none. The pattern of these emitted electrons could only mean one thing—interference effects were being observed from the electron waves.

The electrons used in Davisson and Germer's experiment were relatively low energy, so only penetrated a small number of atomic planes into the crystal. In fact, nowadays their experimental setup is known as low-energy electron diffraction (LEED), and is used to investigate the atomic structure of surfaces.

While these experiments were taking place in New York, G. P. Thomson independently confirmed de Broglie's prediction in Scotland. Thomson projected higher energy electrons than Davisson and Germer used through thin crystal films and observed an interference pattern of light and dark bands the other side of the crystal films.

The bizarre nature of wave-particle duality is illustrated nicely when looking at the honours bestowed upon these pioneers. G. P. Thomson was awarded the Nobel Prize for physics along with Davisson in 1937 for discovering electron diffraction, which showed the electron behaved as a wave. However, 31 years earlier, Thomson's father, J. J. Thomson, had won the Nobel Prize for discovering the electron, which he had characterised as a particle!

Wave Functions

The mathematical function used to describe a matter wave is known as a "wave function" and is denoted by ψ. Although ψ itself has no physical meaning, the square of its absolute magnitude, $|\psi|^2$, has. If this quantity is calculated for a point x, y, z at a time t, then the answer will be proportional to the probability of finding the particle that the wave function describes at point x, y, z at that particular time. The larger the value of $|\psi|^2$, the more likely the particle is to be found there and then. Conversely, if the result is zero, the particle cannot be there at all.

This means that the wave function of a particle is effectively a "probability wave". So if the wave function of, say, an electron is spread out over a certain volume of space, this does not mean the electron itself is somehow spread over this region. Instead, it means that the electron can be found somewhere within this region and that when $|\psi|^2$ is evaluated for a particular point within this volume at a particular time, it will reveal how likely it is that the electron is there.

Although (as we saw in the previous subsection) particles have both wavelike properties and particlelike properties, for any experiment on a particle in which a single measurement is made, only one of these types of behaviour will apply. For example, if a particle is detected by a collision inside a particle accelerator, this collision will be governed by its particlelike properties and it would not be appropriate, or indeed helpful, to involve wave equations if any calculations on this collision needed to be made.

QUANTISATION

In classical physics (physics that does not rely on the ideas of quantum mechanics to explain it), physical quantities—in other words, physical properties that can be measured, such as time, mass, and length—can have any value. However, quantum physics says that some physical quantities, including energy and momentum, can only have particular discrete values.

Any properties whose values are restricted in this way are said to be "quantised", and the values they are allowed to have are known as "eigenvalues". For example, any particle oscillating with simple harmonic motion—including an atom vibrating in a hot solid—can only have eigenvalues of energy, E_n, given by the following equation:

$$E_n = \left(n + \frac{1}{2} \right) \hbar \omega \tag{D1.1}$$

where $n = 0, 1, 2, 3, \ldots$, etc., ω is the angular frequency of the vibration, and $\hbar = h/2\pi$, where $h = 6.626 \times 10^{-34}$ Js is known as Planck's constant, after the German physicist Max Planck (1858–1947). Planck had introduced both the constant h and the idea that energy could be quantised in 1900—and in so doing unknowingly sowed the seeds of quantum theory—while explaining the experimental observations of black-body radiation. (A "black body" is an opaque object that emits radiation over a large range of wavelengths when hot.)

Wave functions mathematically describe the "quantum state" of a quantum mechanical object. Any number that labels a quantum state is known as a "quantum number", and so in equation D1.1 the integer n is the quantum number. This is because it labels each different value of E, and each of these eigenvalues is associated with a different quantum state. Electrons in atoms have four quantised properties, so each electron has four quantum numbers describing its overall quantum state (see Appendix C).

HEISENBERG'S UNCERTAINTY PRINCIPLE

In 1927 the German physicist Werner Heisenberg (1901–1976) postulated that it is impossible to know the exact position of a particle while simultaneously knowing its momentum accurately and vice versa.

One very simple way of thinking about this idea is to say that the act of measuring one quantity—either the momentum or position—disturbs the particle, producing a change in the other quantity. For example, imagine the position of an electron could be measured in a hypothetical microscope that has to use gamma rays to image particles this small. The gamma rays will have to hit the electron and then be reflected by it in order to produce an image—in the same way that light reflected from an object allows us to see an image with an optical microscope. However, the wavelength of gamma rays is so tiny that their energy (which is inversely proportional to the wavelength) is large enough to change the momentum of the electron.

In reality, however, quantum objects are not at all like classical objects, and an act of measurement cannot prevent both position and momentum being known simultaneously because the particle simply does not simultaneously have a definite position and momentum in the first place. As it is an inherent property of the particle that it is impossible to know both its position and momentum simultaneously, Heisenberg's uncertainty principle will hold whether anyone looks at the particle or not.

D2   QUANTUM BEHAVIOUR OF PARTICLES

FERMIONS AND BOSONS

All subatomic particles can be loosely considered to be spinning about an imaginary axis. (Although this model helps us to picture spin easily, it is important to remember that spin is a purely quantum mechanical concept, and does not really have a classical analogue like this.) Any particles that have half-integer values of this spin are known as fermions. For example, electrons, protons, and neutrons are fermions as they all have a spin of ½. By contrast, bosons—which include photons and phonons—have integer values for their spins. (Sometimes particles are said to have "integral" or "half-integral" spins. In this context, the word "integral" has nothing to do with integration—it is an adjective describing a quantity with integer values.)

Every type of particle is either a fermion or a boson; for example, quarks are fermions, and while bosons obey Bose–Einstein statistics, Fermi–Dirac statistics need to be used to describe the behaviour of fermions (see Appendix E). Another distinction between bosons and fermions, which is in fact another way of defining them, is that fermions are particles that obey the Pauli exclusion principle, while bosons are particles that do not obey the principle.

THE PAULI EXCLUSION PRINCIPLE

In 1925, the Austrian physicist Wolfgang Pauli (1900–1958) proposed that no two electrons that are near enough to interact could have the same set of quantum numbers. This became known as the Pauli exclusion principle. Electrons in atoms have four quantised properties, each of which is labelled by a quantum number. One of these quantum numbers is the spin quantum number, and since there are only two possible values for the spin quantum number—+½ and –½—only two electrons can be in any atomic energy level that has the three other quantum numbers the same. This restricts the number of electrons that each atomic shell can hold. (See Appendix C for a thorough discussion of electrons in atomic shells.)

This idea was subsequently extended to cover all fermions, not just electrons. So the exclusion principle also prevents identical fermions from occupying the same quantum state. By contrast, any number of bosons can occupy the same quantum state in any quantum mechanical system.

PARTICLE IN A BOX

In solid state physics, as in quantum mechanics, it is often necessary to look at what happens when a particle can only move in a restricted area of space. A "particle in a

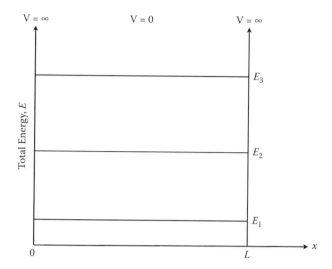

FIGURE D1 An infinitely deep one-dimensional square-well potential like this can be used to represent a "particle in a box" model, where the "box" is the area inside the well. The y-axis indicates the total energy of the confined particle, while the x-axis gives the position of the particle. As the box is 1-D, the particle can only move along the x-axis between 0 and L. It cannot "move" up or down the walls of the box by changing its position, as the y-axis is an energy axis and not a position axis. So the analogy of a box is only accurate in the sense that the box confines the movement of a particle. As indicated, the potential energy outside the box is infinite, while the potential energy inside the box is usually taken to be zero.

box" model (see Figure D1) can be used to represent this situation. It is assumed that the potential energy inside the hypothetical one-dimensional (1-D) box is a constant value (often taken to be zero to make the mathematics easier), while the potential energy outside the box is infinite. It is also assumed that the particle bounces off the sides of the walls whenever it hits them without losing any energy. This means its total energy (potential energy + kinetic energy) will remain constant.

According to classical mechanics, a particle confined within such a box can have any value of total energy so long as that energy falls within the walls of the box. (It cannot exist outside the box because no particle can have an infinite value of energy.)

In quantum mechanics, however, the particle cannot have any value of total energy. Instead its energy is quantised, so there are only certain discrete values of energy it can have. These values are given by the following equation:

$$E_n = \frac{n^2\pi^2\hbar^2}{2mL^2} \tag{D2.1}$$

where the quantum number $n = 1, 2, 3$, etc., m is the mass of the particle, L is the length of the box, and $\hbar = h/2\pi$. The first few allowed energies are illustrated in Figure D2.

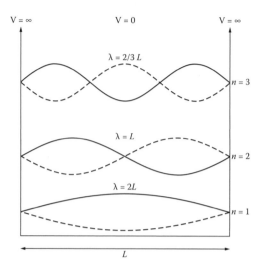

FIGURE D2 The movement of a particle in a 1-D box sets up standing waves like the waves on a guitar string. The standing waves shown have wavelengths of (from top to bottom) 2/3L, L, and 2L.

As equation D2.1 reveals, quantum mechanics dictates that a particle in a box cannot have a zero value of total energy. The lowest value of energy it can have is the value E_1, which is known as the "zero-point energy". This is because for the energy to be zero, the particle would have to be completely still and so would have a momentum of zero. This would violate Heisenberg's uncertainty principle because the momentum would be known exactly, and since the box can be extremely small, the position of the particle would be known accurately as well.

As we saw in the previous section, particles also have wavelike properties, and the Schrödinger equation—which is the most important equation in quantum mechanics—describes particles in terms of their wavelike characteristics. Equation D2.1 is actually obtained by solving the Schrödinger equation, but as this is beyond the scope of this book, the following simplified route will be used instead to give an idea of how equation D2.1 comes about.

In terms of waves, the allowed energies for a particle in a 1-D box are those for which the wavelength of the particle is such that a standing wave is formed when the particle bounces off the sides of the walls. In other words, if a particle is thought of as a wave, when it is reflected at the boundary of the box the reflected wave has to be in phase with the incident wave. (Phase is discussed in detail in Appendix B.) If this is the case, a standing wave is produced, as shown in Figure D2, which looks just like a standing wave for a stringed instrument such as a guitar.

The allowed wavelengths for a particle in a 1-D box are therefore

$$\lambda = \frac{2L}{n}$$

where $n = 1, 2, 3, \ldots$, etc.

The corresponding wave numbers, $k = \dfrac{2\pi}{\lambda}$, (see Appendix B for a revision of the terminology used when discussing vibrations and waves) are therefore

$$k = \frac{n\pi}{L}$$

and so the allowed values of momentum of the particle are

$$p = \frac{h}{\lambda} = \hbar k = \frac{n\pi\hbar}{L} \tag{D2.2}$$

As the potential energy is zero inside the box, the total energy will be equal to the kinetic energy. Therefore the energies corresponding to these values of momentum are given by

$$E = \frac{p^2}{2m}$$

and if we substitute for p from equation D2.2, we obtain equation D2.1 as follows:

$$E = \frac{\left(\dfrac{n\pi\hbar}{L} \right)^2}{2m} = \frac{n^2\pi^2\hbar^2}{2mL^2}$$

Appendix E: Revision of Statistical Mechanics

E1 USE OF STATISTICS IN SOLID STATE PHYSICS

As the main text of this book reveals, some of the properties of solids are caused by the movements of huge numbers—often around 10^{26}—of very small objects within them. For example, the specific heat is due to the motion of the atoms the solid is made from, while electrical conduction in metals is produced when electrons flow through a solid. It would clearly be impossible to work out the specific heat of a solid by looking at the movement of (and so solving the Schrödinger equation for) every single atom that solid is made from, then adding up the contributions. Instead statistical mechanics—or "statistical physics" as it is often called—is used to calculate quantities like this.

Statistical mechanics relates the movements on a microscopic scale of any "system" or collection of atoms, electrons, or other particles that make up a solid or gas to the macroscopic (large-scale) properties of the solid or gas in question by predicting the most likely behaviour of that system. The most probable solution—which is the solution given by the equations of statistical mechanics—has so much higher probability than any other solution that it is usually the only one that needs to be considered.

For example, if the behaviour of every single atom in a china teacup was taken into account, there might be a minuscule chance that on dropping it every atom would fly off in a different direction so the whole thing would disintegrate when it hit the floor. However, we know from experience that this never happens. Dropping a china teacup usually results in it smashing into two or more large pieces, and so statistical mechanics would reveal much larger probabilities for these scenarios than for the "disintegration".

The mathematics of statistical mechanics was first introduced by the Austrian theoretical physicist Ludwig Boltzmann (1844–1906) and is based on the concepts of chance and probability.

E2 PROBABILITY

THE CONCEPT OF PROBABILITY

If any task or experiment can have more than one possible outcome, it is possible to state a "probability" that a particular result will occur. For example, there are six different outcomes to rolling a die ("die" is the singular of the word "dice"), as there are six different faces that can land pointing upwards and therefore be counted as a score. So the probability of scoring a particular number—say, a 2—is 1/6, because the probability of any given outcome is defined as the fraction of times that outcome is expected to happen over a large number of times of carrying out the task or experiment. In this case there is an equal chance of any of the faces pointing upwards and there are six possible results.

Tossing a coin provides an even simpler example of how probability works. In this case the probability of obtaining heads is ½ and that of tossing tails is also ½. If an outcome to any task or experiment has a probability of 0, then it can never happen. By contrast, if any result has a probability of 1 it must always happen. In between 0 and 1, the outcome is uncertain, but the nearer any probability is to 1, the more likely it is that that particular outcome will occur. Since the maximum probability is 1, the sum of all the probabilities of all the alternative outcomes for any given task is equal to 1.

SIMPLE PROBABILITY CALCULATIONS

It is not actually possible to measure a probability because you would have to carry out a task an infinite number of times to obtain a result exactly equal to the theoretical probability. However, you can measure a "fractional frequency", which is the fraction of times a particular outcome occurs in a finite number of attempts at the task. If the outcome is measured for a large number of attempts, its value should approach that of the theoretical probability.

There are two basic rules for combining probabilities. If a number of outcomes occur independently, the probability of them happening together is worked out by multiplying their individual probabilities. For example, if you were to have a set of six dice and rolled them all, the probability of scoring a 6 on every single one of these dice is

$$\frac{1}{6} \times \frac{1}{6} \times \frac{1}{6} \times \frac{1}{6} \times \frac{1}{6} \times \frac{1}{6} = 2.14 \times 10^{-5}$$

As you might have expected, the likelihood of this occurring is not very large!

By contrast, when the probability of getting one or another of a number of mutually exclusive outcomes (in other words, outcomes that do not depend on each other in any way) is required, the probabilities of the individual outcomes are added together. For example, the probability of tails being scored if a coin is tossed or 5 being scored when a die is rolled is

$$\frac{1}{2} + \frac{1}{6} = \frac{2}{3}$$

E3   CLASSICAL STATISTICAL MECHANICS

GAS MODEL

The concept of a gas of molecules is used in physics not only to explain the behaviour of real-life gases, but also to explain the behaviour of other systems of moving particles such as the electrons in solids. In the simplest model of a gas, the molecules constantly move around randomly and the gas exerts a pressure on the walls of its container. This comes from the force exerted by the molecules when they hit the walls, as pressure is equal to force per unit area. In fact, the gas molecules give their kinetic energy to the walls, which in turn—because of Newton's third law—give an equal amount of energy back to the molecules so that the molecules effectively collide elastically with the wall. At higher temperatures, the molecules go faster and so hit the wall with greater force, so the pressure on the walls—and therefore the pressure of the gas itself—is greater at higher temperatures.

CONFIGURATIONS

In order to work out the probability of something happening after a particular event occurs, we need to identify all the possible outcomes that that event can lead to. When the probability of a tossed coin landing heads is worked out, it is not difficult to identify the two different outcomes—namely, heads or tails. However, trying to work out all the different states a gas can be in is a somewhat trickier matter.

This is because there are an enormous number of different microscopic states that a gas can have—each one corresponding to the molecules being in different positions and moving with different speeds. In order to make the problem manageable, we can make use of the fact that any measurement is only as accurate as the apparatus used to take the measurement. Since both the position and momentum of a molecule in a gas can only be known to a certain accuracy, it is possible to define very small three-dimensional (3-D) cubic cells known as "phase cells", each of which represents an experimentally measured position and speed. Any molecules within a phase cell will therefore have a position and speed within the tiny range specified by that phase cell. In fact, since each molecule has a mass, each phase cell is also defining a particular molecular energy. This means that any molecule falling within a given phase cell will have the same value of energy that this cell represents.

According to classical theory, each of the molecules of a gas can, in principle, be labelled, so Figure E1 shows two different microscopic states (often shortened to

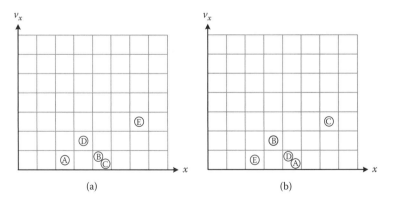

FIGURE E1   Two different configurations for a gas made up from just five particles. (This diagram originally appeared as Figure 2.23, p. 79, of *Classical Physics of Matter*, edited by John Bolton, © 2000, The Open University. Reproduced with permission.)

"microstates") for a gas made up from just five particles. In order to specify a microscopic state—which can also be known as a configuration—and therefore one possible "outcome", we simply need to know which molecules are in which phase cells.

BOLTZMANN'S DISTRIBUTION LAW

Having worked out how to specify different configurations, a probability then needs to be assigned to each configuration. Boltzmann discovered that for a gas in equilibrium at a temperature T, the probability of finding a given molecule in a given phase cell of energy E was

$$p = Ae^{-E/k_B T} \tag{E3.1}$$

where k_B is known as Boltzmann's constant and has the numerical value of 1.381×10^{-23} JK^{-1}, and A is another constant (so has the same value no matter which of the phase cells is being considered) and is used to make sure that the probabilities for all the phase cells add up to 1. This is vital because, as we saw in Section E2, for any event with more than one possible outcome, the sum of all the probabilities for each of these outcomes is equal to 1. (When a set of numbers or quantities is multiplied by a constant so that the sum of the set becomes equal to 1, this process is known as "normalization". So A in equation E3.1 is sometimes known as the "normalization factor".)

It turns out that, for a given molecule, it is more likely for it to be found in a low-energy phase cell than a high-energy one.

BOLTZMANN OCCUPATION FACTOR

With N particles in a classical system of distinguishable particles, the average number of particles within each phase cell will be given by Np. This quantity is known as the "Boltzmann occupation factor" and is denoted by $F(E)$. So from equation E3.1 we have

$$F(E) = NAe^{-E/k_B T} \tag{E3.2}$$

EQUIPARTITION OF ENERGY

Both Boltzmann's distribution law and the Boltzmann occupation factor are functions of the energy of the particles they are describing. Sometimes it is useful to know the average energy of a particle in a system described by these statistics, and this will depend on the number of "degrees of freedom" that the particles in this system have.

The degrees of freedom of a system are the number of different ways in which the particles in that system can take up energy. For example, the atoms in an ideal monatomic gas (in other words, a gas made up from atoms of just one type) do not have rotational energy like molecules, or vibrational energy relative to one another, but just translational energy as they move around. As the gas is 3-D, the atoms can move in three different directions, and so are said to have "three degrees of freedom".

When a system is in thermal equilibrium, its total energy is divided up equally among its degrees of freedom, and the average energy for each degree of freedom is $k_BT/2$. This is known as the "equipartition of energy" theorem. According to this theorem, if we consider an ideal monatomic gas, the mean energy of each atom per degree of freedom is also $k_BT/2$, so the overall average energy for each atom $\langle E \rangle$ is

$$\langle E \rangle = \frac{3k_BT}{2} \tag{E3.3}$$

as every atom has 3 degrees of freedom. Equation E3.3 also describes the average energy of a free electron in a metal when it is considered to be a classical particle in an ideal gas, although this model of free electrons is outdated and has been replaced by models that take quantum mechanics into account. Classical statistics are, however still used to describe the thermal motion of charge carriers in semiconductors and the distribution among energy levels of atomic magnetic moments.

E4 QUANTUM STATISTICS

In the previous section, it was assumed that it was possible—in theory—to label each of the molecules of a gas. It was also assumed that both the velocity and position of a molecule could be known to a fairly high degree of accuracy. However, with the advent of quantum mechanics, it became clear that neither assumption is true. Not only is it impossible to tell the molecules of a gas or the electrons in a solid apart because they are in fact identical, the Heisenberg uncertainty principle (see Appendix D) prevents both the position and velocity of a particle being known simultaneously.

Although the classical statistical mechanics of Boltzmann is incorrect, it works well enough to approximate the macroscopic properties of a genuine gas, and certain collections of atoms and molecules. For example, in magnetism the Boltzmann distribution can be used to describe the orientation of magnetic moments. However, for calculations on "gases" of electrons in solids and other systems of indistinguishable quantum particles, quantum statistics must be used instead.

CONFIGURATIONS OF INDISTINGUISHABLE PARTICLES

Instead of quantum particles being found in different phase cells like classical particles, they are found in different quantum states. (See Appendix D for a brief reminder of quantum mechanics.) So a configuration of a system of identical particles in quantum mechanics is defined by giving the numbers of particles in each quantum state.

As every type of quantum particle is either a fermion (particle with a half integer value of spin) or a boson (particle with an integer spin), two different types of quantum statistics—one for bosons and one for fermions—are required.

FERMI–DIRAC STATISTICS

Fermi–Dirac statistics are used to describe the behaviour of fermions (see Appendix D). Fermions obey the Pauli exclusion principle, so this means two identical

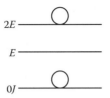

FIGURE E2 The only possible configuration for two fermions, which are identical and subject to the Pauli exclusion principle, in a hypothetical three-state system with total energy 2E.

fermions can never be in the same quantum mechanical state. Figure E2 shows the ways in which a hypothetical system consisting of just three energy levels and two identical fermions can have a total energy of 2E.

As you can see, there is only one arrangement for the fermions which gives a total energy of 2E because no two identical fermions can be in the same quantum state.

Of course this is just a hypothetical system. In a real system, although no two *identical* fermions can be in the same quantum state, it is possible to have more than one fermion in a given quantum state if its quantum numbers are different to any other fermion in that state—and so it is not identical. For example, two electrons can be in the same quantum state even if they have the same quantum numbers n, l, and m as long as their spin quantum numbers are not the same. So in this case, one electron must be "spin up" and the other "spin down" for them to occupy the same state. The average number of fermions in an energy state E is given by

$$F_F(E) = \frac{1}{e^{(E-\mu)/k_BT} + 1} \tag{E4.1}$$

where $F_F(E)$ is the Fermi occupation factor, k_B is Boltzmann's constant, T is the temperature in degrees kelvin, and μ is a characteristic energy for the system. The value of μ depends on the number of fermions in the system as well as the temperature, and if the system being considered is the free electrons in either a metal or semiconductor, μ is replaced in equation E4.1 by the "Fermi energy" E_F.

Figure E3(b) shows a graph of the Fermi occupation factor—also known as the Fermi distribution function—at temperatures above absolute zero, while (a) shows $F(E)$ at absolute zero. From the shape of the graph in (a) it is clear that for any value of energy above E_F in a hypothetical semiconductor or metal at absolute zero, $F(E) = 0$, which means that none of these energy states are occupied by an electron. By contrast, $F(E) = 1$ for all the energy states below E_F, so each of these states must be occupied by a single electron.

This means that at 0K, the Fermi energy is the dividing line between the filled and unfilled energy states in a solid. In the same way that electrons in atoms fill up the available energy states starting from the lowest level (see Appendix C), any free electrons in a solid will occupy energy levels in the order of lowest to highest. The Fermi energy will therefore represent the highest level filled by an electron. One way of visualising this is to imagine pouring some sand into a glass. The sand will fill the

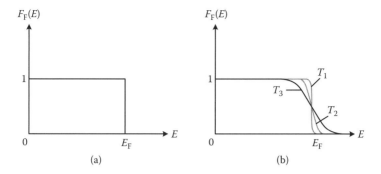

FIGURE E3 The Fermi distribution function for a hypothetical system of free electrons in a semiconductor plotted as a function of energy (a) at absolute zero and (b) at a temperature above 0K.

glass from the bottom upwards, and the upper surface of the sand is equivalent to E_F, as it marks the boundary between the filled and unfilled part of the glass.

Since absolute zero can never actually be reached, this graph of the Fermi distribution function actually looks more like the one shown in Figure E3(b). If you compare the two diagrams, you can see that the boundary at E_F is less abrupt at temperatures above 0K. This is because above 0K, electrons can be thermally excited into energy states above E_F. The higher the temperature, the greater the amount of blurring, as the more electrons can gain enough energy to be excited to levels above E_F.

For semiconductors at temperatures above absolute zero, the Fermi energy no longer represents the highest filled level. Instead it is the value of energy at which the Fermi–Dirac distribution function is equal to 0.5. In other words, the probability of an electron having the energy E_F is exactly a half.

BOSE–EINSTEIN STATISTICS

For indistinguishable systems in which the particles are bosons—and so have zero or integral spin and do not obey the Pauli exclusion principle—Bose–Einstein statistics are used. Figure E4 shows what happens for two bosons in the hypothetical system of three energy levels that we saw in the last subsection. As you can see, there are two configurations the bosons can have and still have a total energy of $2E$.

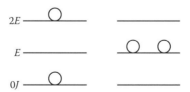

FIGURE E4 The only possible configurations for two bosons, which are identical but not subject to the Pauli exclusion principle, in a hypothetical three-state system with total energy $2E$.

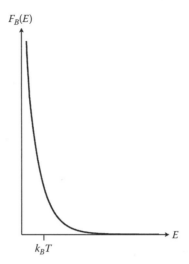

FIGURE E5   The Bose occupation factor for a gas of photons.

The average number of bosons in an energy state E is given by the Bose occupation factor $F_B(E)$ shown in the following equation:

$$F_B(E) = \frac{1}{e^{(E-\mu)/k_B T} - 1}$$
(E4.2)

where μ is an energy characteristic of the system under consideration, which in the case of a gas of photons is zero. Figure E5 shows the Bose occupation factor for a photon gas. The steep curve reveals that bosons—which can happily occupy the same states—are most likely to cram together in the lowest possible energy states.

Referring back to equations E4.1 and E4.2, it can be seen that the Boltzmann occupation factor (equation E3.2) is a limiting case of both the Fermi–Dirac and Bose–Einstein distributions when $E > k_B T$. This is because when $(E - \mu) \gg k_B T$, the exponential $(E - \mu)/k_B T \gg 1$, and so the 1 in the denominators can be ignored. Both distributions are then equal to

$$\frac{1}{e^{(E-\mu)/k_B T}}$$

which, since μ is a constant, can be rewritten as $Ae^{-E/k_B T}$ (with the constant $A = e^{-\mu/k_B T}$), which—as you will see if you look back to equation E3.2—has the same form as the Boltzmann distribution factor.

Appendix F: Glossary of Terms

acceptor: A dopant atom that has fewer valence electrons than needed to bond with the atoms of the semiconductor it is placed into. See also **p-type**.

allotrope: One of the possible crystal structures of an allotropic element. See also **allotropic**.

allotropic: A description of an element that has more than one possible crystal structure. Compare with **polymorphic**.

alloy: A mixture of two or more metals that can include nonmetallic elements.

amorphous: A description of a noncrystalline material in which the arrangement of atoms lacks long-range order.

annealing: A heat treatment that involves heating a solid to a high temperature (below its melting point) and holding the solid at that temperature for a period of time before slowly cooling it.

atomic coordinates: A set of three numbers used to indicate the position of atoms with respect to a unit cell.

atomic density: The number of atoms per unit volume in a solid.

bandgap: In general, a range of energy states between the allowed energy bands of a solid, but often used to mean the difference in energy between the top of the valence band and the bottom of the conduction band in an insulator or semiconductor.

basis: A single atom (or ion) or group of atoms (or ions) with a fixed spacing and orientation to each other and of a specific composition. An identical basis must be added to every point of a **lattice** in order to represent a crystal structure.

brittle: A material which tends to show little or no plastic deformation before it breaks, so is likely to shatter if hit with a hammer. Compare with **ductile**.

bulk modulus: The ratio of the pressure applied to a material (solid or fluid) as it is compressed to the fractional change in volume the material experiences as a result of this pressure.

ceramic: A compound such as zinc sulphide, silicon carbide, or calcium fluoride that contains either a metallic or semiconducting element as well as a nonmetallic element. Many ceramics have a mixture of ionic and covalent bonding.

chip: A thin layer of high-purity single crystal. When an integrated circuit is mounted on a single silicon crystal like this, the whole structure is commonly referred to as a "silicon chip".

cohesive energy: The energy per atom that would be needed to separate out a solid into the individual atoms (or ions) that it is made from.

colour centre: A type of point defect that can form in ionic crystals and which changes their colour.

composite: A combination of two or more different materials that have different properties.

compressive stress: The force per unit cross-sectional area squeezing a solid object along its longitudinal axis. Compare with **tensile stress**.

conduction band: The highest energy band in a metal that contains electrons, and the first unoccupied or partially occupied band directly above the **valence band** in semiconductors and insulators.

conductivity (electrical): A measure of the ability of a material to conduct electricity. The greater the value of the conductivity, the better a conductor the material is.

continuous random network (CRN) model: A model that can be used to describe the structure of amorphous materials in which the atoms are held together by **covalent bond**s.

coordination number: The number of **nearest neighbours** that an atom in a solid has.

covalent bond: A type of chemical bond in which valence electrons are shared between atoms. It is found in both elements and compounds, including the important semiconducting materials silicon and gallium arsenide.

critical field: The value of magnetic field above which a superconducting material loses its superconductivity.

crystalline: A description of a solid composed of a regularly repeating pattern of atoms.

crystal structure: The spatial arrangement of atoms inside a crystal. Compare with **lattice**.

deformation: A change in the shape or size of a solid material produced by the application of a force.

delocalized: A description of an electron in a solid that does not remain in the local environment of either a single bond or of its parent atom.

dense random packing (DRP) model: A model that can be used to describe the structure of amorphous materials in which the atoms are held together by metallic, ionic, or molecular bonds.

diamagnetism: A weak form of magnetism that only exists in the presence of an applied magnetic field, and opposes the applied field.

dielectric: An insulator that becomes polarised in the presence of an electric field.

diffusion: In solid state physics, this usually refers to the movement of charge carriers away from a region where their concentration is high.

diode: An electronic device with two electrodes.

direct-gap semiconductor: A semiconductor in which the maximum energy of the valence band occurs at the same value of wave vector, k, as the minimum energy level of the conduction band. Compare with **indirect-gap semiconductor**.

dislocation: A defect involving a whole line of atoms which can allow planes of atoms to slip over one another with a much lower value of applied stress than would be required to produce slip in a perfect crystal. Compare with **point defect**.

dislocation density: The number of dislocations intersecting a randomly chosen unit area of a crystal.

dislocation line: The line which for edge dislocations runs along parallel to the end of the extra part-plane of atoms, and in screw dislocations is the line which the planes of atoms spiral around.

domain: In magnetism, it is an area in a **ferromagnetic** material within which all the magnetic moments point in the same direction. In ferroelectricity, it is a region containing unit cells polarised in the same direction.

donor: A dopant atom that has more valence electrons than needed to bond with the atoms of the semiconductor it is placed into. See **n-type**.

drift current: The movement of charge carriers in a solid under the influence of an electric field.

drift velocity: The average velocity of the carriers in a **drift current**.

ductile: A material which tends to show a large amount of plastic deformation before it breaks, so it is likely to be flattened if hit with a hammer and can be drawn out into a wire. Compare with **brittle**.

effective mass: The ratio of force to acceleration that an electron has when it is moving under the influence of an electric field in a crystal. (It is not a true mass, but enables equations that describe electrons moving in a vacuum to still be used for electrons in crystals.)

elastic deformation: A deformation of a solid in which the solid returns to its original shape and size when the deforming stress on it is removed. Compare with **plastic deformation**.

elastic limit: The point that divides elastic and plastic behaviour. Below the elastic limit, only elastic deformation takes place, and materials that are under stress return to their original shape and size when the stress is removed. Above the elastic limit, stress will cause a permanent change in the shape of a material.

elastomer: An elastic polymer.

electric dipole moment: The product of the distance between a slightly separated pair of equal and opposite electric charges and either one of the charges.

electromagnet: A soft ferromagnetic material surrounded by a current-carrying coil.

electronegative (atom): An atom able to accept more electrons relatively easily or to share valence electrons, becoming a negative ion in the process.

electronegativity: A measure of an element's ability to accept an additional electron into its outermost atomic shell.

electropositive (atom): An atom which can lose its valence electron(s) relatively easily, becoming both a positive ion and more stable in the process.

energy band: A range of allowed energies for electrons in a solid.

epitaxy: A method of manufacturing semiconductor devices in which layers are grown with the same crystal orientation as that of the substrate.

extrinsic semiconductor: A semiconductor that has had the atoms of another semiconductor added to it (doping) to alter its properties. Compare with **intrinsic semiconductor**.

Fermi energy: At $T = 0K$, the energy of the highest filled energy level in a solid. More generally, the energy for which the probability of occupation is one half.

ferroelectrics: Materials that acquire a spontaneous electric dipole moment below a critical temperature.

ferromagnetic Curie temperature: The temperature at which the spontaneous magnetisation in a ferromagnet becomes zero. Above this temperature, the material behaves like a paramagnetic material.

ferromagnetic materials: Materials that have spontaneous atomic magnetic moments in the absence of any external field. They can be permanently magnetised by weak magnetic fields, and are attracted to magnets and magnetic fields.

filler: A material that can be added to a polymer to enhance its properties.

Frenkel defect: A point defect consisting of an interstitial atom and the vacancy it has left behind after moving from its normal position in the lattice.

glass transition temperature: A range of temperatures, rather than a single temperature, over which a supercooled liquid gradually changes into a glass. Despite not being a single temperature, a rapidly cooling amorphous material is considered to be a glass below its glass transition temperature but a supercooled liquid above it.

grain boundaries: Boundaries separating the individual grains (little crystals) of a polycrystalline solid.

hardness: The resistance of a material to being dented or scratched.

heat treatment: Any of a number of processes involving heating and cooling a solid to change its properties.

hole: An unoccupied state in the valence band that has a charge of exactly $+1e$ and that behaves like a positive charge carrier in a solid.

hydrogen bond: A type of chemical bond which occurs when a hydrogen atom covalently bonds to an atom of either oxygen, fluorine, or nitrogen. The single valence electron from the hydrogen atom spends more time nearer the atom of the other element than near its own nucleus. As a consequence, the hydrogen atom is left with a positive charge and can then attract a negatively charged part of either another molecule or its own molecule and form a hydrogen bond.

indirect-gap semiconductor: A semiconductor in which the maximum energy of the valence band occurs at a different value of wave vector, k, to the minimum energy level of the conduction band. Compare with **direct-gap semiconductor**.

integrated circuit: A complete circuit of electronic devices and their connections which is made on a single substrate.

interstitial: An atom (which can be either an impurity or a host atom) that sits in between the normal lattice sites of a crystal.

intrinsic semiconductor: A pure semiconductor that has not had impurities added. Compare with **extrinsic semiconductor**.

ionic bond: A type of chemical bond in which an electron is transferred from an electropositive atom to an electronegative atom. The resulting ions have opposite electrostatic charges, and their attraction to one another forms the bond. Ionic bonding is found in solids like NaCl which are composed of two elements, one of which is metallic.

lattice: An infinite array of points arranged in such a way that if you were able to stand on one of these points, no matter which of the points you chose to step onto, your surroundings would look exactly the same. Crystal structures can be represented by a lattice together with a **basis**.

lattice parameters: The lengths of the sides of any given unit cell, and the angles between them.

liquid crystal: A state of matter with a structure between that of a solid and a liquid in which there is long-range order of the molecules.

mean free path: The average distance a phonon or free electron travels before being scattered.

mean free time: The average time for which a charge carrier travels through a metal or semiconductor before having its path altered by a collision.

melt quenching: A method of forming amorphous materials by cooling them rapidly from their liquid state.

metallic bond: A type of chemical bond found in metals in which the atoms exist as positive ions surrounded by a "sea" of electrons. The attraction between the ions and the electrons forms the bond.

Miller indices: A set of three (or—for hexagonal crystals—four) numbers used to describe planes within a crystal.

monomer: A small molecule that can join with many other monomers to form a **polymer**.

nearest neighbours: The atoms closest to any given atom within a solid.

nonprimitive unit cell: Any unit cell containing more than one lattice point. Compare with **primitive unit cell**.

n-type: An **extrinsic semiconductor** that has had donor impurities added and so has electrons as its majority charge carriers. Compare with **p-type**.

packing fraction: The ratio of the volume of the atoms within a unit cell to the total volume of the cell (which is taken to be 1).

paramagnetism: A type of magnetism that causes a weak attraction to an applied magnetic field.

Pauli exclusion principle: The principle proposed by Austrian physicist Wolfgang Pauli in 1925 which states that no two identical fermions (elementary particles with half-integer spin) in any system can exist in the same quantum state. This means that, within the same atom, no electron can have a set of quantum numbers exactly the same as that of another electron.

phonons: Particles of vibratory energy in solids, in the same way that photons are particles of electromagnetic energy.

piezoelectrics: Materials in which the surfaces acquire an electric dipole moment when they are under stress.

plastic deformation: The permanent deformation of a solid subjected to a deforming stress. Compare with **elastic deformation**.

plasticiser: A substance that can be added to polymers to increase their flexibility.

p-n junction: A layer of **p-type** semiconductor back to back with a layer of **n-type** semiconductor.

point defect: A defect—such as an impurity atom—involving one or, in some cases, a small number of atomic sites in a crystal.

polycrystalline: A crystalline solid consisting of lots of little crystals known as grains rather than being one **single crystal**.

polymer: A huge molecule made up from lots of small molecules (called **monomers**) joined together.

polymerization: Any of a number of processes in which **monomers** are joined together to form artificial polymers.

polymorph: One of the possible crystal structures of a **polymorphic** compound.

polymorphic: A description of a compound that has more than one possible crystal structure. Compare with **allotropic**.

population inversion: A nonequilibrium state essential for laser operation in which more electrons are in a higher energy level than in a lower level.

primitive unit cell: Any unit cell containing only one single lattice point.

p-type: An **extrinsic semiconductor** that has had acceptor impurities added and so has holes as its majority charge carriers. Compare with **n-type**.

quasicrystal: A solid which produces a sharp X-ray diffraction pattern like a crystal but does not have the symmetry characteristics of a crystal.

quenching: A heat treatment in which a solid is heated and then cooled very rapidly.

recombination: An electron and hole "joining together" and cancelling each other out. (One way of thinking about recombination is to imagine an electron disappearing into a hole.)

resistance: A measure of how much a piece of material or electrical component resists the flow of an electric current.

resistivity: The reciprocal of the electrical **conductivity**, which is therefore a measure of the ability of a material to resist the flow of electricity through it.

saturation magnetisation: The maximum value of the magnetisation in a ferromagnetic material, above which it cannot be magnetised any further by an increase in applied magnetic field.

Schottky defect: A point defect, commonly known as a vacancy, which consists of a vacant lattice site.

single crystal: A crystalline solid made from a perfect continuous pattern of atoms.

slip: The sliding of one plane of atoms in a solid over another plane.

solenoid: A coil of conducting wire wound into a cylindrical shape so that the length of the wire is much greater than the diameter of the coil.

specific heat capacity: The amount of heat required to raise the temperature of 1 kg of a substance by 1K.

stabiliser: A substance added to a polymer to reduce the damage caused by its working environment.

stiffness: A measure of how much a material resists elastic deformation when it is under stress.

strain: The change in volume or shape that the stress acting on an object produces, defined as the change in length or volume divided by the original length or volume, respectively.

strength: The maximum amount of stress a material can withstand without fracturing.

stress: A force acting on an object, defined as the force per unit cross-sectional area.

substitutional (impurity): An impurity atom that replaces one of the host atoms at a normal lattice site.

substrate: A solid layer that acts as a base for the growth of other materials.

superconducting transition temperature (T_c)**:** The temperature separating the superconducting state from the normal state of a solid. Below T_c the solid becomes superconducting.

superconductor: A material that loses all its resistance to an electric current below a temperature called the **superconducting transition temperature**.

supercooled liquid: A liquid which has been cooled so quickly that it could not crystallise into a solid at its normal freezing point and is therefore still liquid below it.

tensile strength: The maximum **tensile stress** a solid object can withstand without breaking.

tensile stress: The force per unit cross-sectional area pulling on a solid object and stretching it along its longitudinal axis. Compare with **compressive stress**.

tetrahedral angle: The angle $(109°28')$ between covalent bonds in a covalent solid in which every atom can be considered to be at the centre of symmetry of a tetrahedron, and is bonded to four other atoms positioned at the corners of this tetrahedron.

thermal conduction: The transfer of heat from hot regions of a substance to cooler regions.

thermal conductivity: A measure of the ability of a material to conduct heat. The higher the value, the better a material is at conducting heat.

thermoplastic: A polymer that becomes soft when heated and hardens when cooled. It can be softened and reshaped several times and so can be recycled.

thermoset: A polymer that chars and begins to decompose when heated and so cannot be recycled like a thermoplastic.

unit cell: A building block for a crystal lattice, containing enough lattice points that repeating it over and over again with the relevant **basis** of atoms placed at each lattice point allows the entire structure of a given crystal to be represented.

valence band: The energy band containing the valence electrons of a solid. Compare with **conduction band**.

valence electrons: Electrons in the outermost occupied shell of an atom that take part in bonding processes between atoms.

van der Waals forces: Weak intermolecular or interatomic electrostatic forces (due to the formation of dipoles) that cause molecules and atoms to become attracted to one another when they are very close together.

vulcanisation: The process of adding sulphur to natural rubber. Small quantities of sulphur make it more elastic and less sticky, while larger quantities harden the rubber.

work hardening: Hardening a crystalline material by applying a stress greater than its elastic limit.

X-ray diffraction: A widely used technique for determining the positions of atoms inside crystals.

Young's modulus: The ratio of stress over strain for materials under **tensile stress** and beneath their elastic limit.

zero-point motion: The movement of atoms in a crystal at absolute zero.

Index